Cave and Karst Systems of the World

Series editor

James W. LaMoreaux, Tuscaloosa, USA

More information about this series at http://www.springer.com/series/11987

Horton H. Hobbs III · Rickard A. Olson
Elizabeth G. Winkler · David C. Culver
Editors

Mammoth Cave

A Human and Natural History

Springer

Editors
Horton H. Hobbs III (deceased)
Department of Biology
Wittenberg University
Springfield, OH
USA

Rickard A. Olson
Cave Research Foundation
Cave City, KY
USA

Elizabeth G. Winkler
Western Kentucky University
Bowling Green, KY
USA

David C. Culver
American University
Washington, DC
USA

ISSN 2364-4591 ISSN 2364-4605 (electronic)
Cave and Karst Systems of the World
ISBN 978-3-319-53717-7 ISBN 978-3-319-53718-4 (eBook)
DOI 10.1007/978-3-319-53718-4

Library of Congress Control Number: 2017936030

Printed on acid-free paper

This Springer imprint is published by Springer Nature
The registered company is Springer International Publishing AG
The registered company address is: Gewerbestrasse 11, 6330 Cham, Switzerland

Horton H. Hobbs III conceived the idea of this book and was the driving force behind it, up until his death in late August of 2016. Horton was a great man. He touched a lot of lives in powerfully positive ways as a husband, father, and academic father to his many students at Wittenberg University in Springfield, Ohio. There being no organized group of cavers, he founded the Wittenberg University Speleological Society with the delightful acronym WUSS! Over the decades, he led students on many exploration caving and cave research field trips. Some of these former students are now doing cutting edge research in the field of cave biology, microbiology, and geology, and at Carter Caves State Park, KY, they were so impressed with the contributions by Horton and his students that they named a cave for him.

Horton was beloved by his many friends and colleagues. There just did not seem to be a bad side to this man, and his academic productivity, much of it collaborative, was prodigious. One such collaboration was when he taught cave ecology at Mammoth Cave National Park as part of Western Kentucky University's Karst Field Studies Program. As park ecologist, I assisted with these classes. Horton knew just how far he could push these students without breaking them just as Chuck Yeager knew how to fly a fighter jet right to the edge of its performance capability. We all worked into the night, and the students came away with a much enhanced understanding of cave ecosystem structure and function via mini studies and presentations. Some of the students are doing cutting edge work in biospeleology. See a pattern here?

I was never technically a student of Horton's, but, in a significant way, I was. Horton and I first met at the Indiana University, Bloomington, where he was working on his Ph.D. He needed help with his cave crayfish research, and I was only too happy to help. I had dropped out of school after military draft was no longer a threat, but he encouraged me to return and apply myself. Nobody in my family had ever gone to college, and so I was not enthusiastic. Still, he kept up the pressure while teaching me to safely rappel into and ascend out of shafts using single rope technique (instead of going hand over hand!). After a quite lively discussion, I finally agreed to return to school. If not for Horton, I would never have earned a master's degree from University of Illinois at Urbana. Without Horton, I would never have been able to gain entrance to the Cave Research Foundation in 1973, and I certainly would not have become the ecologist at Mammoth Cave National Park twenty years later. This is just one personal example of the difference Horton Hobbs could and did make in people's lives.

In 2012, a navigable branch off Roaring River in Mammoth Cave was named the "Hobbs Fork of Roaring River" in his honor. This unusual distributary stream is habitat to both cave-adapted crayfish and the Kentucky cave shrimp. These are both decapod crustaceans, a taxonomic group Horton specialized in.

Rickard A. Olson

Preface

Mammoth Cave has figured large in the American consciousness, not only because of the rich cultural history of the cave but also due to its enormous size and attendant biological diversity. The history of its human use includes Native Americans, who explored the cave and mined its sulfate minerals. Later, African-American slaves, including Stephen Bishop and Mat Bransford, extensively explored and drew a map of the cave. Slaves also played a critical role in the mining of saltpeter for gunpowder. More recent stories include "wars" between commercial cave owners, the death of Floyd Collins in Sand Cave, epic exploration trips by modern cave explorers, and of course two centuries of tourism. The presence of a sandstone caprock over the cave has preserved the history of the geological development of the cave itself as well as providing a large variety of habitats. This has allowed earth scientists to unravel many of the basic, universal principles of cave formation (speleogenesis). The unprecedented length of the cave (now at more than 400 miles–640 km) led to the development of new exploration techniques and innovative methods of organizing expeditions. Mammoth Cave has also played an important role in the natural sciences. It was one of the first caves in North America visited by biologists; the first animals specialized for cave life in North America, including beetles, spiders, crayfishes, and fish, were described from Mammoth Cave in the 1840s.

As befits its importance, a number of books have been written about Mammoth Cave. Perhaps the earliest book is *Peter Pilgrim* by Robert Montgomery Bird in 1839 soon followed by *Rambles in the Mammoth Cave* in 1844 by Alexander Bullitt. The saga of exploration has been the subject of several books, including *The Caves Beyond* by Joseph Lawrence and Roger Brucker (1955), *The Longest Cave* by Roger Brucker and Richard Watson (1987), and *Beyond Mammoth Cave* by James Borden and Roger Brucker (2000). The death of Floyd Collins in Sand Cave was the focus of *Trapped* by Robert Murray and Roger Brucker (1979) and the fictional account *The Cave* by Robert Penn Warren (1959). A historical fiction treatment of Stephen Bishop, *Grand, Gloomy, and Peculiar*, by Roger Brucker, was published in 2009. On the scientific side, Mammoth Cave was the subject of *A Geological Guide to Mammoth Cave National Park* by Arthur N. Palmer (1981), *Karst Hydrology: Concepts from the Mammoth Cave Area edited* by William B. White and Elizabeth White (1989), and *Archeology of the Mammoth Cave Area* by Patty Jo Watson (1974). While no books have been devoted specifically to the biology of Mammoth Cave, cave life of Mammoth Cave has figured prominently in every English language textbook on cave biology. Additionally, thousands of newspaper and magazine articles, maps, and scientific publications concerning Mammoth Cave in south-central Kentucky have appeared since the early 1800s. It is certainly no wonder that this cave has been the recipient of such attention since, for no other reason, it is the longest known cave in the world.

What is missing in the 200-year history of study is any comprehensive treatment of Mammoth Cave that covers the full spectrum of disciplines that intersect with the world's longest cave. We assembled 16 of the leading experts in fields ranging from archeology to cultural history to life science to geosciences. Not only does this provide a convenient source of up-to-date information on Mammoth Cave, but also it allows for synergies among disciplines that intersect in Mammoth Cave. The audience for this book includes anyone interested in caves in general and Mammoth Cave in particular, experts in one discipline seeking information about other areas, and especially researchers and students interested in the many avenues of pursuit possible in Mammoth Cave. Technical jargon is kept to a minimum, and terms are explained carefully where they appear.

This book has four main sections divided into 18 chapters. The first section is an extensive description of the cave, its basic structural pattern, and how it relates to the surface landscape. This is designed to give the reader a word and visual (maps, photos) picture of the cave and provide orientation to this very complex, multi-level cave. The second section covers the human history of utilization and exploration of the cave. The time frame ranges back 5000 years, the date of the earliest charcoal fragments. Human uses include mining, tourism, and medical experiments in addition to exploration and cartography. Cave science is the topic of the third section, including geology, hydrology, mineralogy, meteorology, paleontology, ecology, biodiversity, and microbiology. The fourth section looks to the future, with an overview of environmental issues facing Mammoth Cave managers.

Springfield, OH, USA — Horton H. Hobbs III
Cave City, KY, USA — Rickard A. Olson
Bowling Green, KY, USA — Elizabeth G. Winkler
Washington, DC, USA — David C. Culver

Contents

1 An Orientation to Mammoth Cave and This Volume

Rickard S. Toomey III, Horton H. Hobbs III, and Rickard A. Olson

Abstract

Mammoth Cave is the longest known cave system in the world. The total surveyed length is currently 405 miles, and exploration will continue for many decades before the cave is fully mapped. This cave is famous also for prehistoric artifacts left by Indian cavers up to 5000 years ago and for its role in American history during the War of 1812. Mammoth Cave has diverse cave-adapted wildlife, so much so that it is considered a global biodiversity hot spot. It is part of the most studied karst landscape in the world; the cave and the water that formed it have been extensively researched. In both wet and dry passages, minerals have been deposited by percolating groundwater. Some of these are quite beautiful. The seemingly endless passages and interesting features have drawn people to visit Mammoth Cave for about 200 years. Tourism is still strong with roughly 500,000 people visiting Mammoth Cave National Park each year. This chapter introduces the reader to the Mammoth Cave System via an extensive set of maps and also provides an overview directing the reader to the following 17 chapters on important aspects of the cultural and natural history of the cave.

1.1 Introduction

Mammoth Cave is a very special place. So special that is was designated as the USA's twenty-sixth national park in 1941. Numerous parts of the cave are also listed on the National Register of Historic Places. Internationally its importance is recognized as well. In 1981, it was designated a World Heritage Site by the United Nations Educational, Scientific, and Cultural Organization. In 1990, its ecological significance was specifically recognized when it became the core of an International Biosphere Reserve of about 3200 square miles (830,000 ha). These recognitions derive from the unique combination of natural resources, cultural resources, natural process that formed the cave, and rich history of the cave. The goal of this book is to present some aspects of these resources and history through a series of chapters written by experts on various topics.

Mammoth Cave National Park is a park on two levels—52,830 acres (21,380 ha) of second-growth mixed hardwood and evergreen forest and 32 miles (51 km) of winding rivers, and below it, and the longest known cave system in the world. As of 2016, over 405 miles (650 km) of passages have been mapped in the Mammoth Cave System. However, length is not the only source of fame for Mammoth Cave. It is also known for its extremely rich biodiversity, its extraordinary prehistory, its historic uses, the intensity with which it has been studied over the years, and its role as a tourism destination. The park is located in the state of Kentucky, USA, within a day's drive of several major population centers (Fig. 1.1).

Although Mammoth Cave itself is the primary focus of this book, it will also, by necessity, include discussion of other aspects of its home—Mammoth Cave National Park, the karst landscape and component ecosystems of which it is a part, as well as other caves in the area. Including these

R.S. Toomey III (✉) · R.A. Olson
Cave Research Foundation, Cave City, Kentucky, United States
e-mail: myotis.toomey@gmail.com

R.A. Olson
e-mail: rickardolson16@gmail.com

H.H. Hobbs III
Wittenberg University, Springfield, USA

H.H. Hobbs III et al. (eds.), *Mammoth Cave*, Cave and Karst Systems of the World,
DOI 10.1007/978-3-319-53718-4_1

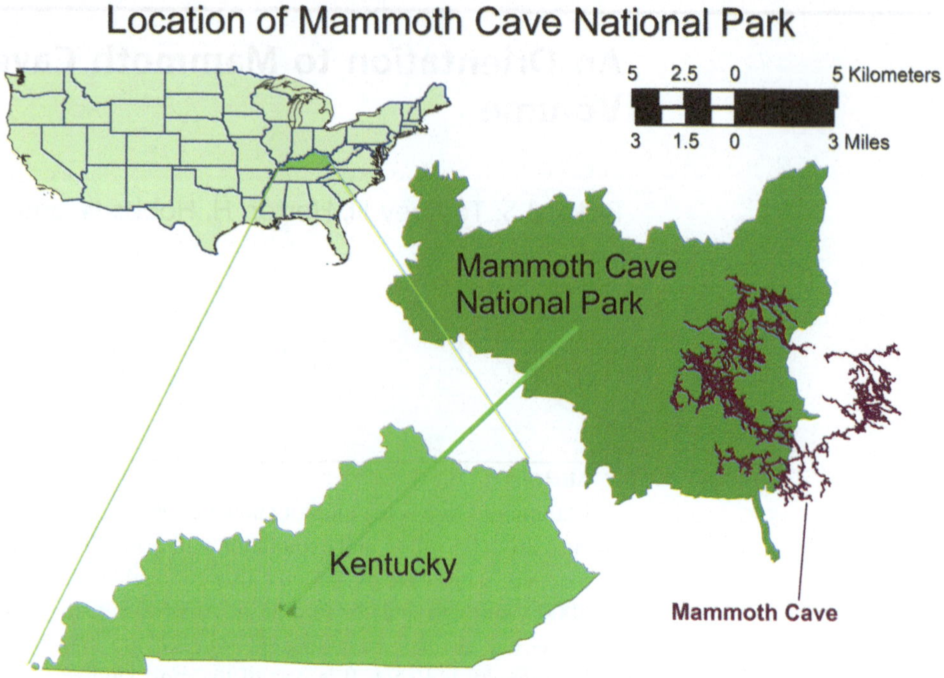

Fig. 1.1 Location of Mammoth Cave National Park within the USA and the Commonwealth of Kentucky. Also shows the location of the Mammoth Cave System with respect to the park

discussions provide the important context for understanding the cave, its resources, and its history. In this book, when authors are referring to Mammoth Cave National Park they generally use terms like, MCNP, the national park, or the park. In referring to Mammoth Cave, authors use a variety of designations, depending on to what they are specifically referring. Over time, Mammoth Cave System has been explored from almost all of its 30 entrances (Table 1.1; Fig. 1.2). Many of those entrances were once separate caves; others were man-made entrances constructed based on known cave passages or a variety of good guesses. Three entrances that were open in the past have collapsed, so the current number of entrances that can be used to access the cave is 27. Exploration of these various caves coalesced over time into four major areas of the cave: Mammoth Cave, the Flint Ridge System, the Joppa Ridge Section, and the Roppel System. These segments were connected at various times (discussed later in this chapter and in Chaps. 4 and 5). The result is the Flint Ridge–Mammoth–Joppa Ridge–Roppel Cave System, which will usually be shortened to the Mammoth Cave System. However, often in this book, the various component parts will be referred to separately (either individual caves or segments) because some resources or aspects of history will only apply to parts of the cave. In some cases, Mammoth Cave will be used to refer to the entire system; for others, it will refer to the portions under Mammoth Cave Ridge (context will provide this information) Figs. 1.15, 1.16, 1.17, 1.18, 1.19, 1.20, 1.21, 1.22, 1.23, and 1.24 provide maps showing the relationship of the various components. They also provide location of key places that are referred to in other chapters.

In addition to the Mammoth Cave System, the well-developed Central Kentucky Karst also contains numerous other large and small caves including the world's ninth (Fisher Ridge Cave System–125 miles [202 km]) and fifty-fifth (Whigpistle Cave System—35 miles [56 km]) longest caves. The park itself contains at least 400 caves in addition to the Mammoth Cave System. Including caves inside of and outside of the park, the Cave Research Foundation, Central Kentucky Karst Coalition, the James Cave Cavers, and other cave mapping groups have mapped over 600 miles (966 km) of caves in the area. This effort is far from over; exploration is ongoing, and new cave passages continue to be discovered. Quinlan and others (1983) suggested that the Mammoth Cave area potentially could have 1000 miles (1600 km) of human-enterable passages. The great length of caves in this area lead to management challenges as caves may extend into and out of the park. In the case of the Mammoth Cave System, approximately one-third of the 405 miles (650 km) of the system lies outside the park under lands with a variety of ownership. Portions of both the Fisher Ridge System and the Whigpistle System extend into the park; however, neither has a known entrance on the park, nor have they been connected to the system. A number of these caves (and others) will be mentioned in this book, because they contribute to the understanding and wonder of

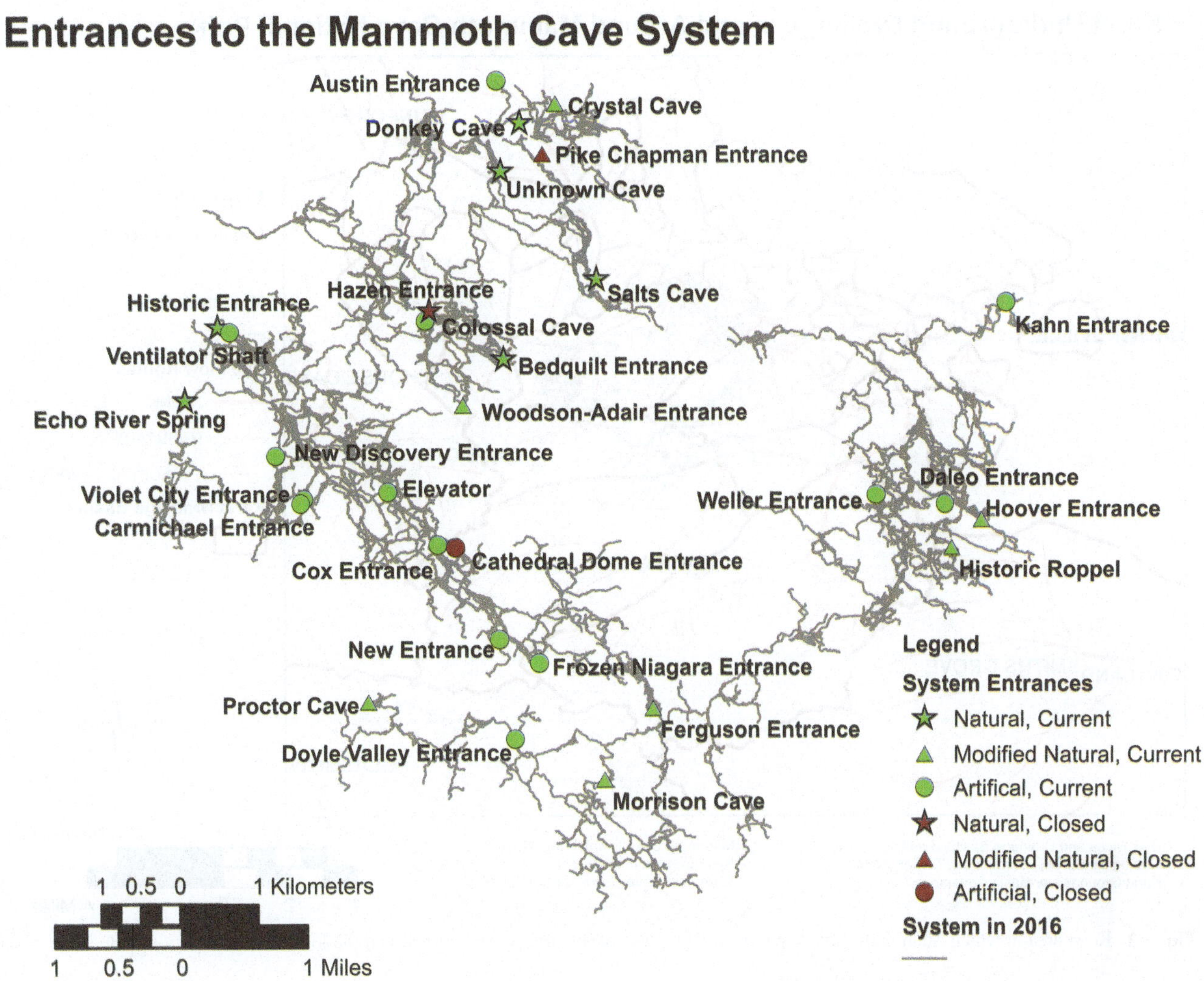

Fig. 1.2 Entrances to the Mammoth Cave System. Natural Entrances are ones that could be entered without significant modification. Modified Natural Entrances are ones that required significant modification to enter, but which probably had some natural aspects to them when discovered. Artificial Entrances are ones that were constructed in areas without a natural opening

Mammoth Cave. Chap. 4 (this volume) and Kambesis (2007) summarize modern cave exploration in the Mammoth Cave area; in addition; several books detail the exploration of portions of the Mammoth Cave System (Borden and Brucker 2000; Brucker and Watson 1976; Lawrence and Brucker 1955).

1.2 Mammoth Cave Geology

The karst area that houses Mammoth Cave is part of an extensive, well-developed, and well-studied karst area that extends from central Indiana through much of western Kentucky to northern Tennessee. It is developed in a series of Early Carboniferous (Mississippian) age rocks. These rocks are dominated by carbonates in the lower portion of the sequence. Higher in the sequence, sandstones and shales come to dominate. These beds gently dip toward the northwest. The main karst development is in an approximately 120-meter-thick carbonate sequence consisting of the St. Louis Limestone, the Ste. Genevieve Limestone, and the Girkin Limestone (and their equivalents). In addition, there is some karst development in several thinner carbonate layers, including the Haney and Glen Dean Limestones. Palmer discusses the geological setting of the area in Chap. 6. Colburn covers the Mississippian age fossils that are exposed in the cave in Chap. 11.

The groundwater basins developed in the central Kentucky karst are also extensive and complex. Over more than 50 years, researchers have made the Mammoth Cave area

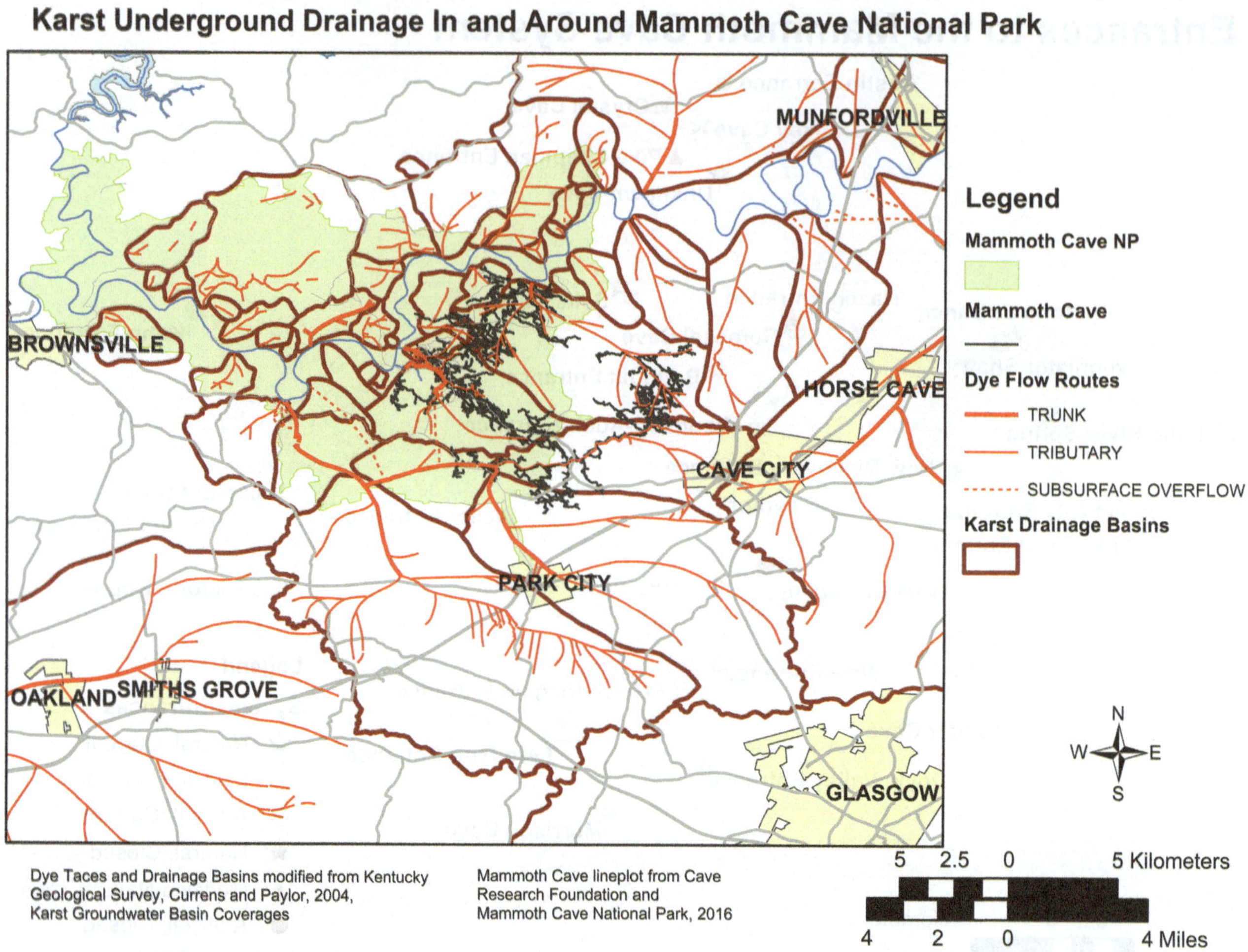

Fig. 1.3 Karst area in which Mammoth Cave has formed. Hundreds of dye traces have established the location of underground drainage basins

one of the best-understood large karst aquifers in the world. Over 500 dye traces have established the extent of groundwater basins in the Mammoth Cave area (White 2007). These studies have established that in most cases, drainage basins extend outside the park boundary (Fig. 1.3) and that much of the water flowing through it comes from sources far outside the park. White and White discuss the hydrology of the area in Chap. 8. Because the sources of water come from far outside the park, cave streams are vulnerable to pollution (see Chap. 18 for further discussion). The underground drainage in karst areas affects the surface and thus ecosystems via vegetation, fire ecology, and surface water availability (see Chap. 12).

Mammoth Cave developed over approximately the last 10–15 million years as a conduit for water to flow from the Pennyroyal Plateau (sinkhole plain) to the south and east of the park through the ridges of the Chester Upland to springs along the Green River. The cave is characterized by a series of horizontal levels of current and former phreatic (formed at or below the water table) passages that relate to levels at which the Green River stabilized during long-term down-cutting of its valley. By definition, a cave level is a large low-gradient passage that generally represents distinct stages of cave development and that demonstrates a clear relationship to former base levels (water table) in adjacent valleys. As surface streams erode, their valleys deepen, lowering the base level and the groundwater springs that appear in the valleys descend progressively to lower elevations over time. Concurrently, in response to the drop in the base level, cave passages that contribute flow to these springs also shift to lower levels and abandon the higher fossil passages. These nearly flat-ceilinged canyon-like (taller than wide) vadose (formed above the water table) passages continuously descend, are rarely completely filled by water, and erode downward where dissolution occurs primarily along their floors. When the downward erosion

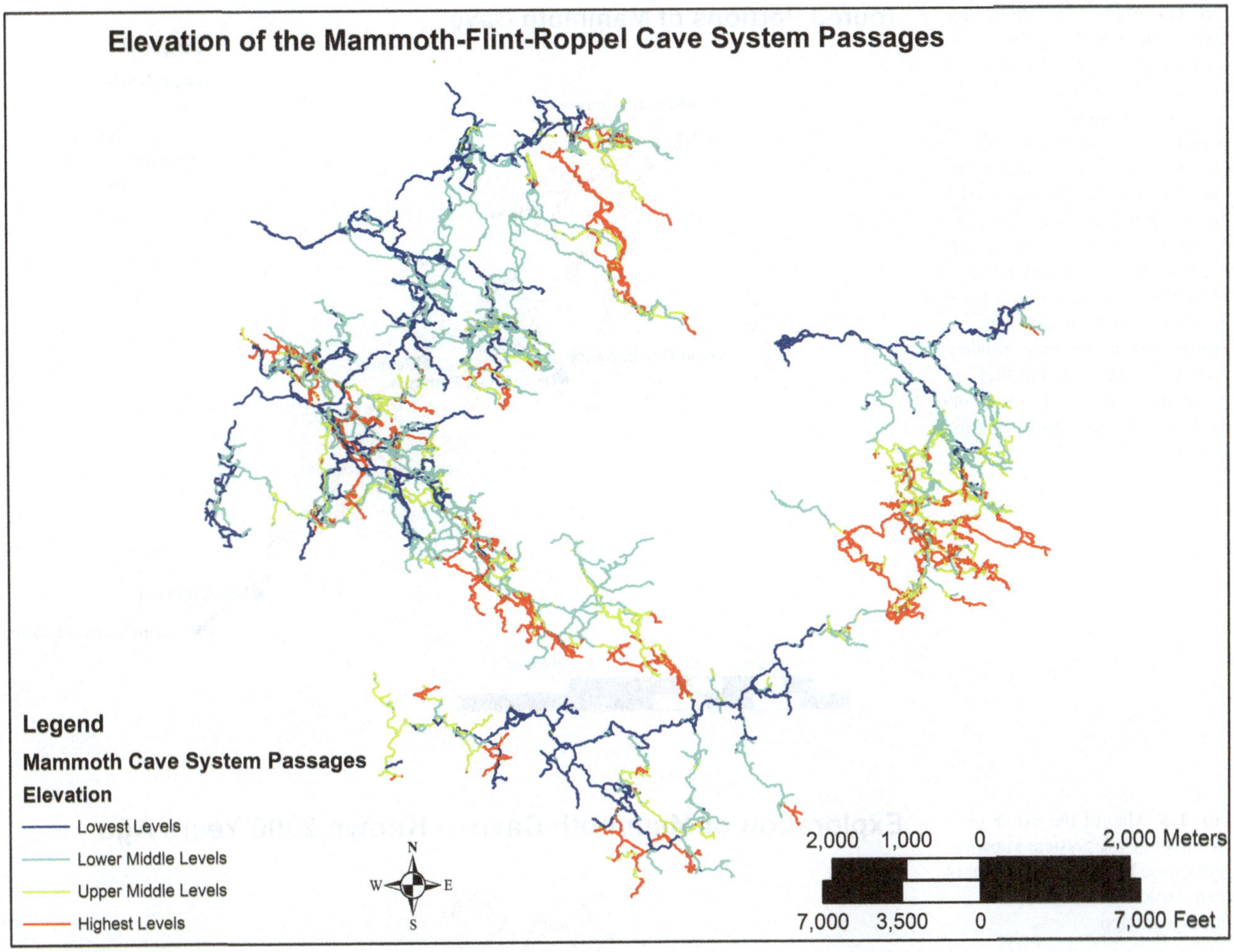

Fig. 1.4 Mammoth Cave passages colored by their elevations above the modern level of the Green River. Higher levels are older, and lower levels are younger. The levels on this figure do not correspond exactly with Palmer's cave levels due to changes in level elevations as one gets further from the Green River

slows or pauses (brought about by several complex processes such as glacial advances and retreats, changes in climate, sea level, and amounts of sediment), cave passages have sufficient time to enlarge. There is a decrease in gradient, and these phreatic passages take on an elliptical (oval tube, usually wider than tall) shape due to dissolution elongated along the bedding planes. It is here where phreatic passages form distinct cave levels. Thus, over time, cave passages eroded downward through the rock strata, occasionally experienced rather long periods of static base levels that resulted in distinct cave levels; the highest passages in Mammoth Cave are the oldest, the lowest ones are the most recent (Fig. 1.4). Four main paleo-levels are identified in Mammoth Cave. In addition, modern base-level cave streams feed the six primary springs (Pike, Great Onyx, Floating Mill, Styx, Echo, Turnhole) that drain the system. All of these levels are connected by a variety of vadose passages. Because the Green River not only cut down during the last five million years, but also periodically filled its valley with sediments raising its level, many of the passages in Mammoth Cave are partially (or entirely) filled with sediments. Granger and others (2001) discuss the dating of sediment fills in the various levels. The developmental history (speleogenesis) of the cave is covered in detail by Palmer in Chap. 6.

With the exception of a few areas like Frozen Niagara, Mammoth Cave is not particularly well known for its stalactites and stalagmites (speleothems). However, it actually has an impressive and well-studied array of secondary mineral deposits. Some of the most famous of these are gypsum flowers and snowballs. White discusses cave minerals in Chap. 9.

Many people assume that Mammoth Cave has a constant temperature and humidity throughout, but things would be

Fig. 1.5 Map of the Mammoth Cave section of the system showing the extent of toured passages. Three types of toured passages are noted. These are walking tours, wild cave tours (also known as caving or crawling tours), and areas formerly toured but which are no longer. Only the Mammoth Cave section is shown, because there are no regularly scheduled tours currently running in other sections of the cave. Former tour routes exist in other sections of the cave, but they are not shown due to scale issues and the lack of modern tours in those sections

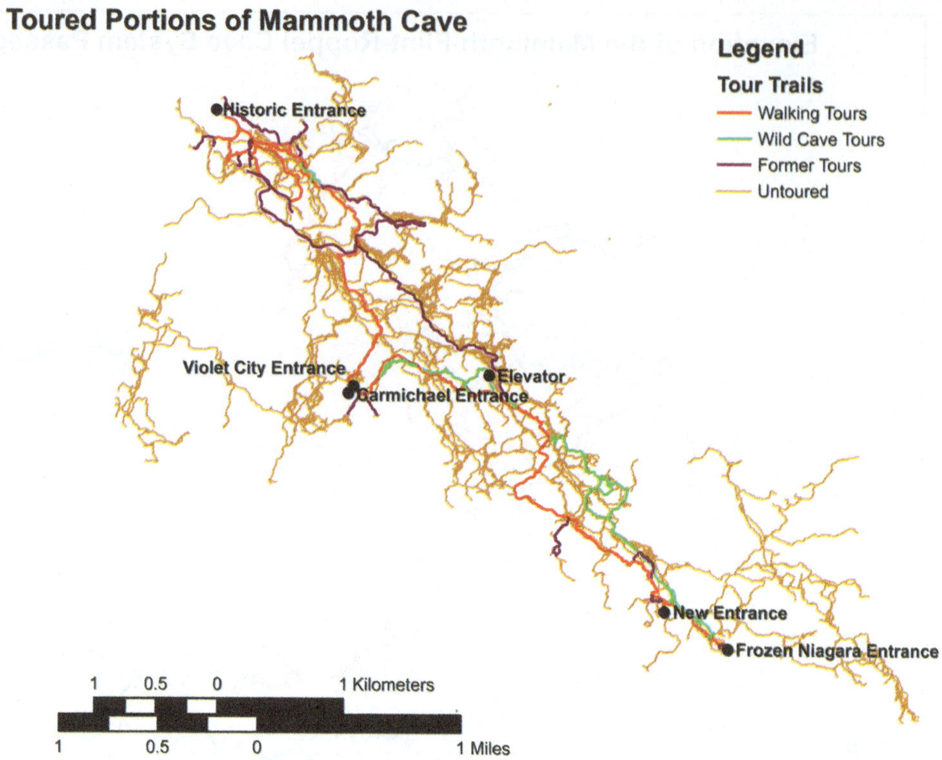

Fig. 1.6 Map of the extent of Mammoth Cave System known to prehistoric Indians at the height of their explorations around 2000 years ago

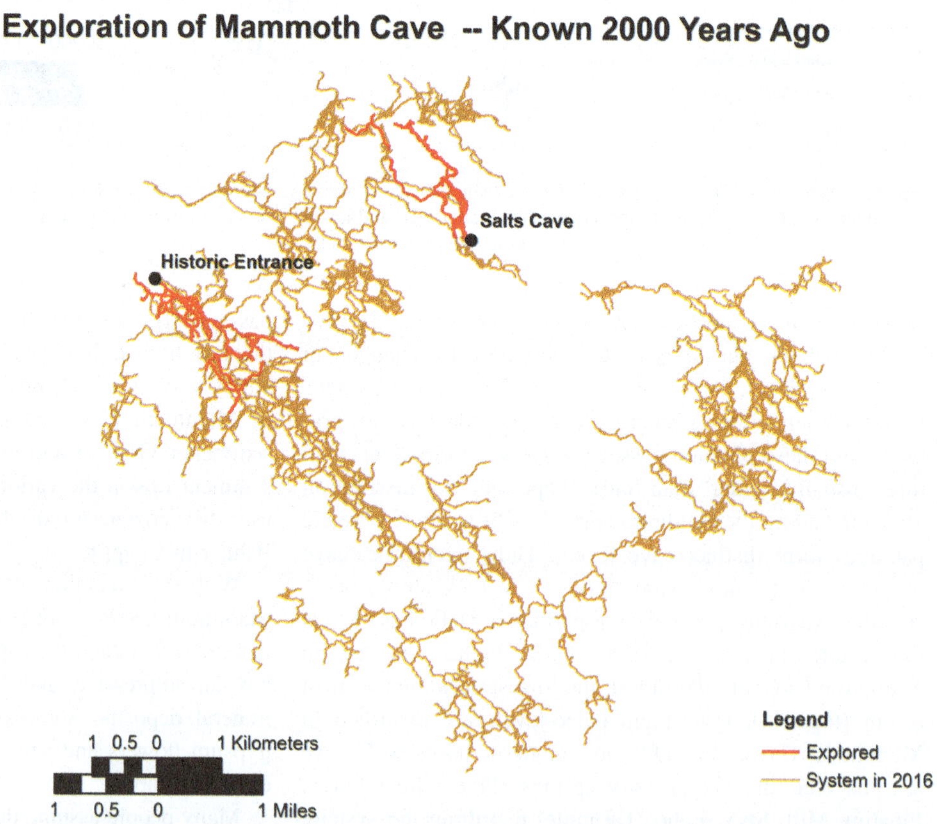

Fig. 1.7 Map of extent of Mammoth Cave System known in 1835

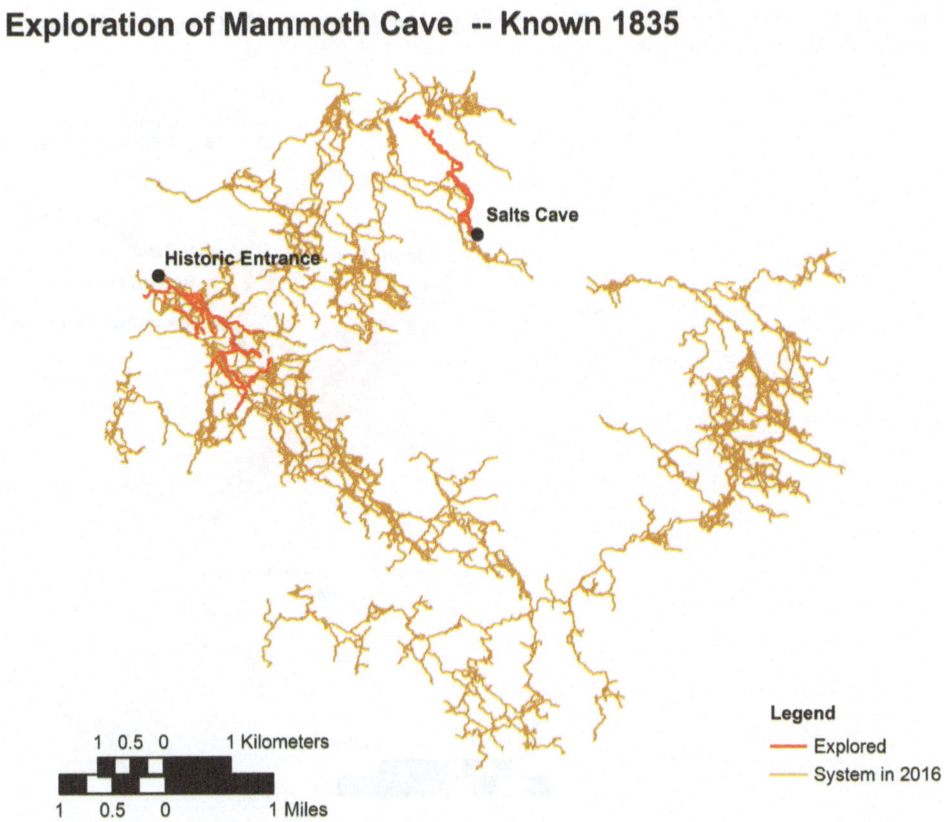

Fig. 1.8 Map of extent of Mammoth Cave System known in 1845

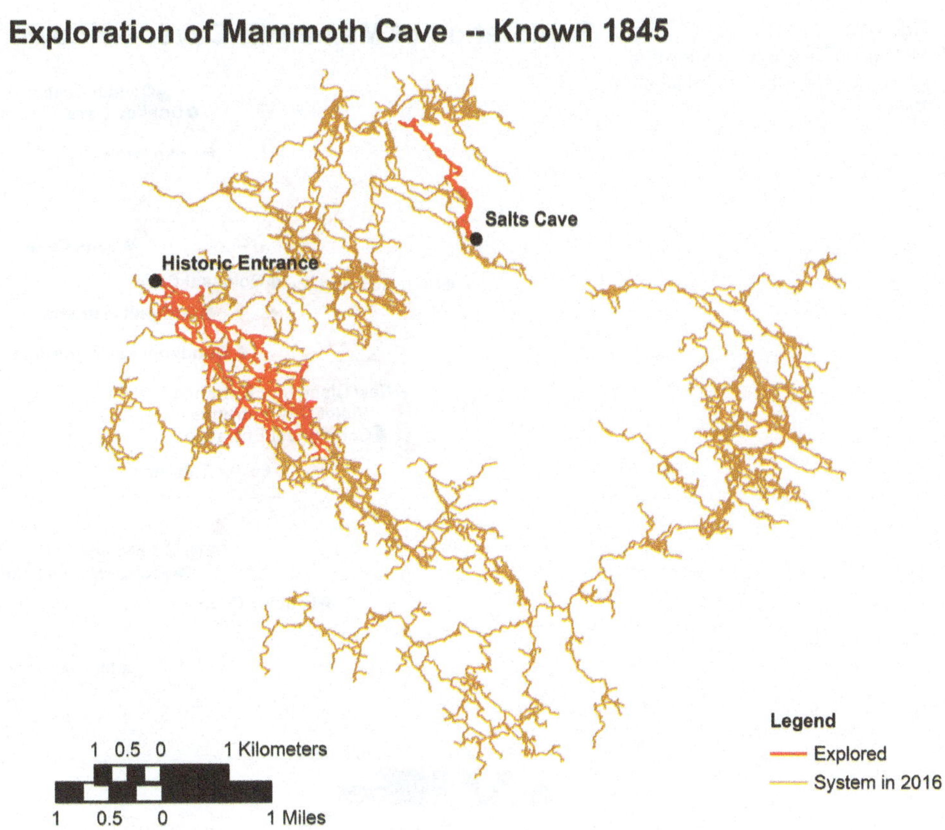

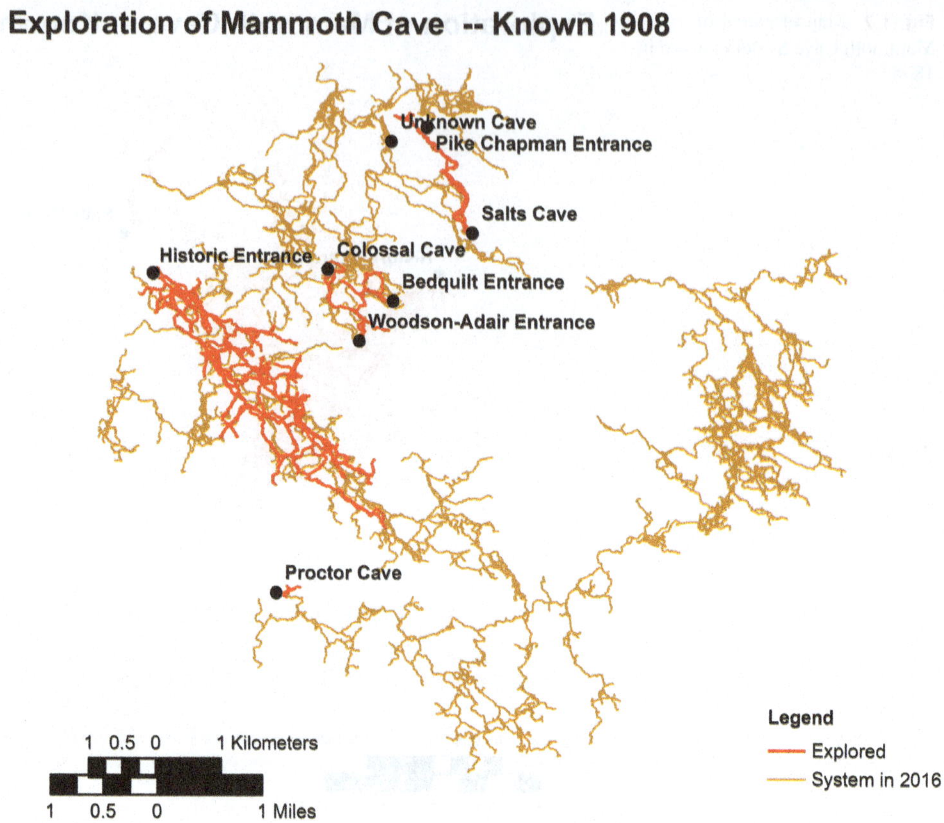

Fig. 1.9 Map of extent of Mammoth Cave System known in 1908

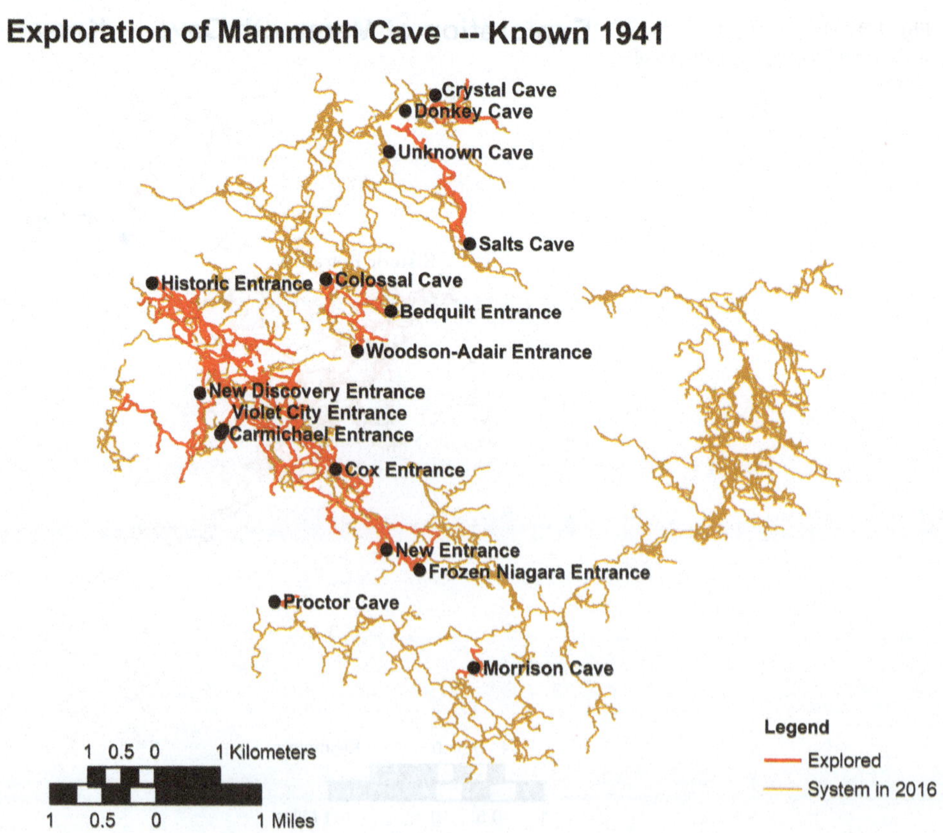

Fig. 1.10 Map of extent of Mammoth Cave System known when the park was established in 1941

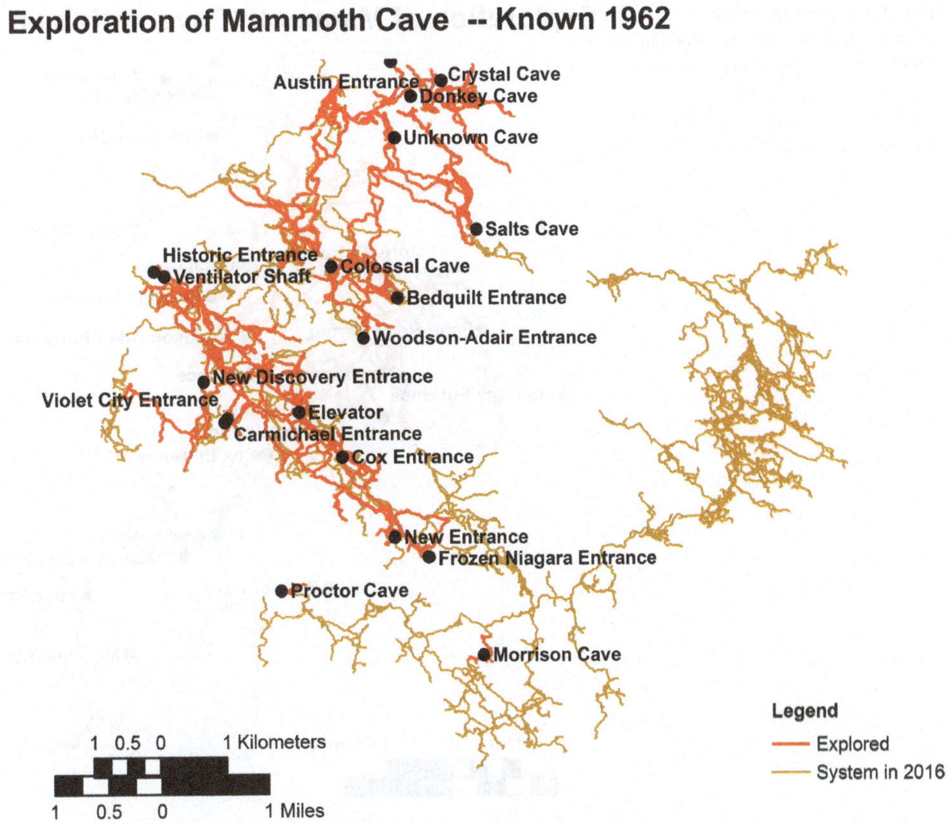

Fig. 1.11 Map of extent of Mammoth Cave System known in 1962 when connections integrated the Flint Ridge Cave System

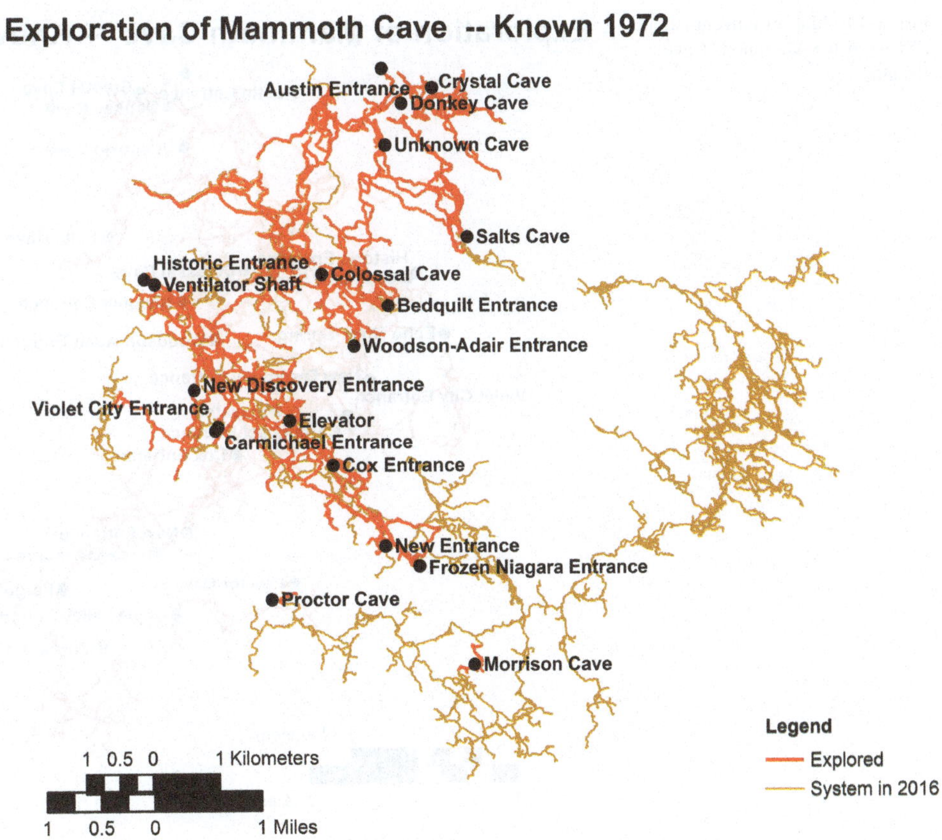

Fig. 1.12 Map of extent of Mammoth Cave System known in 1972 when the Mammoth and Flint Ridge Cave Systems were connected

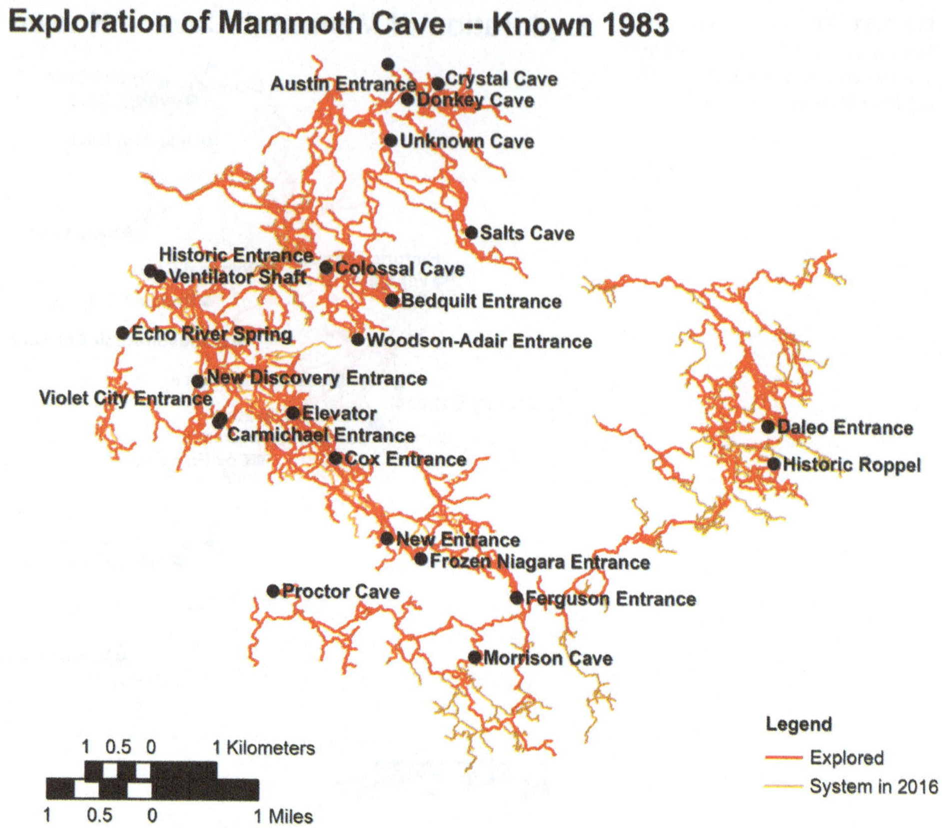

Fig. 1.13 Map of extent of Mammoth Cave System known in 1983 when the Mammoth–Flint Ridge–Joppa Ridge Cave Systems and the Roppel Cave System were connected

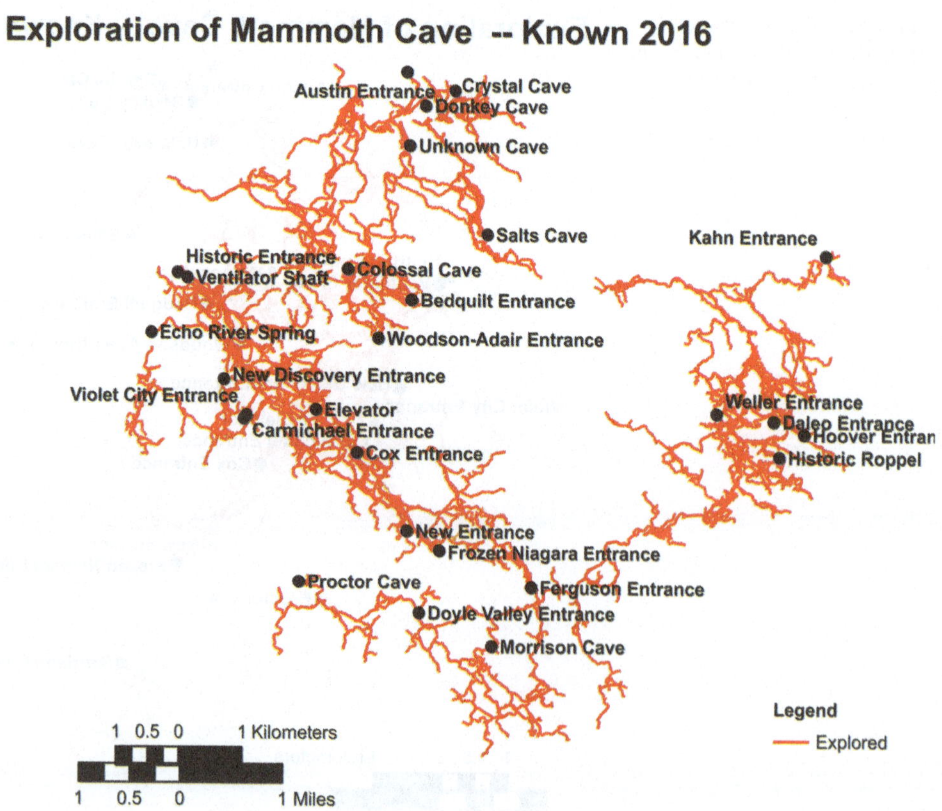

Fig. 1.14 Map of current extent (2016) of the Mammoth Cave System

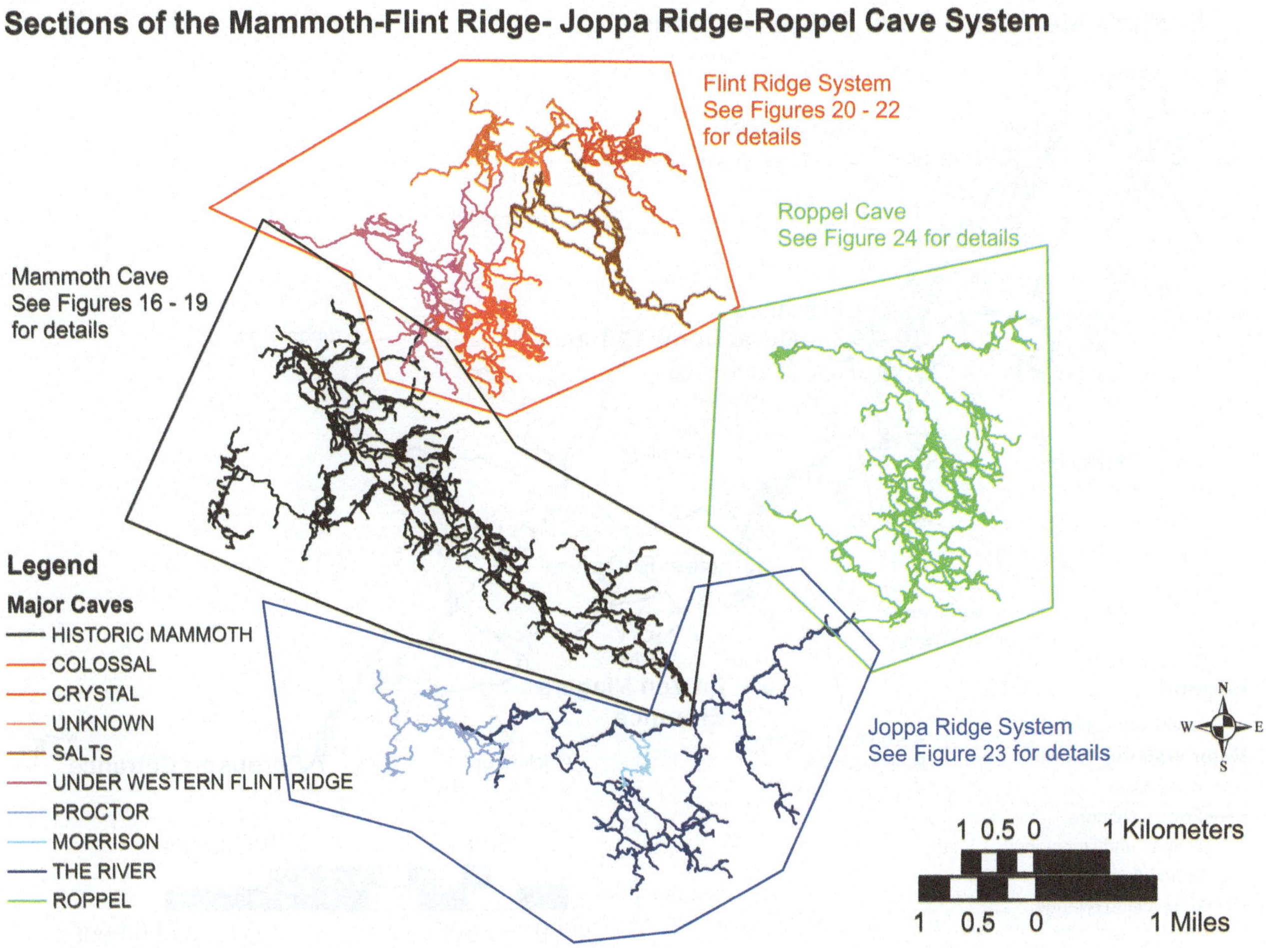

Fig. 1.15 Map of the Mammoth Cave System showing the relationship of the main sections

much more boring if it did. The variations in microclimate in the cave result in crucial differences in how well prehistoric and historic artifacts are preserved, where different animals live (and whether they can), and where various minerals will form. Although the basic deep-cave temperature averages 57 °F (14 °C), the actual temperature in many areas of the cave varies substantially and seasonally. These temperature variations relate to the temperature of water and air that come in from the surface. Because Mammoth Cave System is shallow and has thousands of places at different elevations where air can exchange with the surface, the main driver of airflow in the cave is the chimney effect. With this effect, during the winter, warm (cave temperature) air rises out of higher elevation entrances because it is less dense than the cooler surface air. Cold surface air is then drawn into lower elevation entrance. In the summer, the opposite occurs. Cool dense (cave temperature) air flows out of the lower elevation entrances, and warmer (surface) air is drawn into the upper entrances. Olson discusses Mammoth Cave's microclimate and its importance in Chap. 10.

1.3 Life in the Cave

Over time, animals have entered the cave through past and current entrances. These animals have left behind evidence of their use of the cave as bones, mummies, guano, or traces like tracks and scratches. As with the cave passages, some of the fossils may be over 1 million years old. Other remains, like some bat mummies and guano, are only a few hundred years old. Mammoth Cave is a rich source of fossils to aid in understanding past environments of the area. On the other hand, fossils from the cave provide important information on the history of the cave. Colburn summarizes the paleontology in Chap. 11. In addition, Thomas and Toomey discuss the recent bat remains from the cave in Chap. 17.

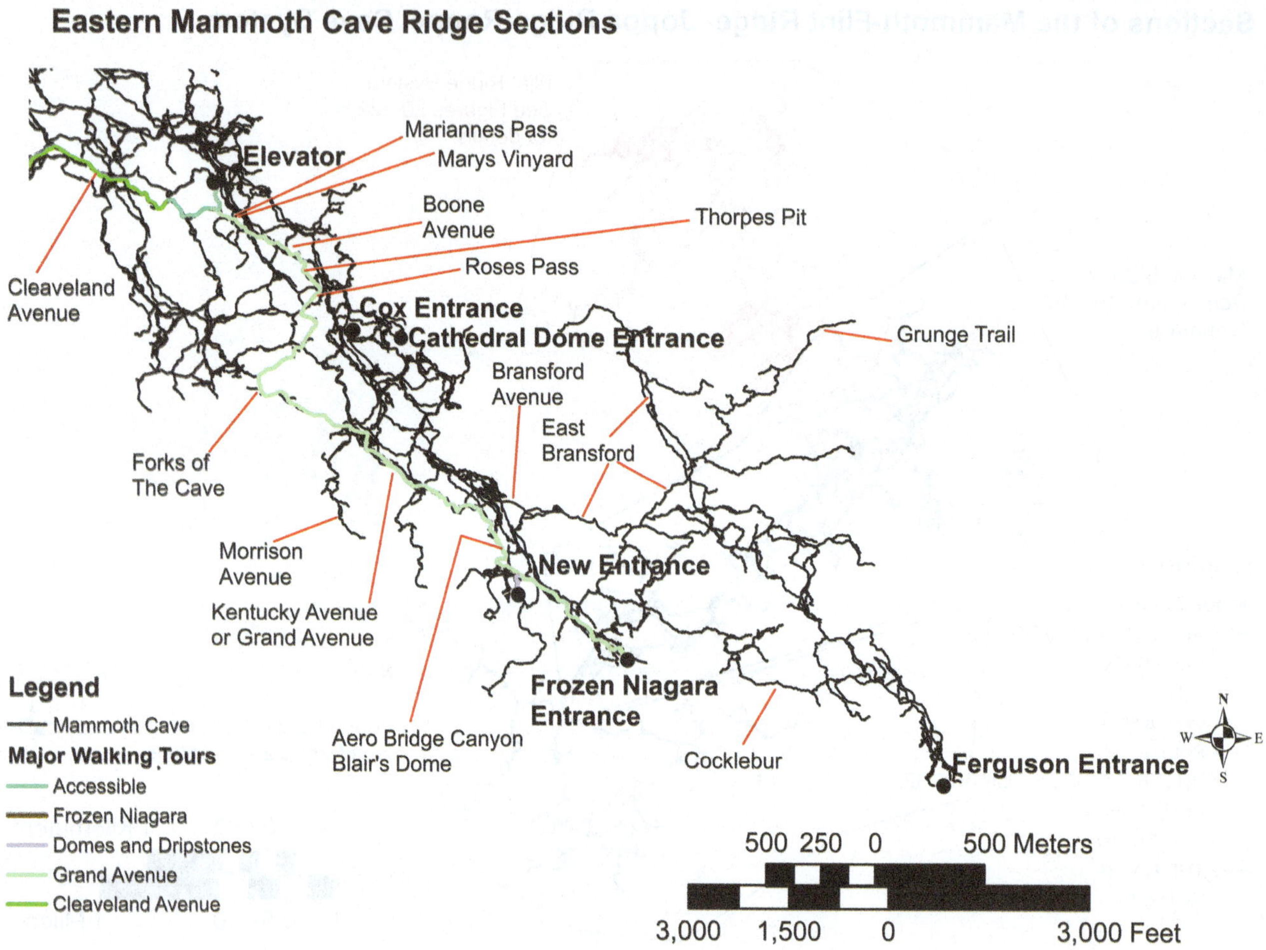

Fig. 1.16 Map of the eastern portion of the Mammoth Cave Ridge section of the system

The caves of the Mammoth Cave area have a very diverse and well-studied biology. Although at least 160 species of animals regularly occur in the cave, currently 52 species of troglobiotic and stygobiotic organisms are known from Mammoth Cave (Appendix 1.1), making it one of the highest subterranean diversities worldwide (Culver and Pipan 2013; Culver and Sket 2000). These are highly cave-adapted organisms that occur only in subterranean terrestrial (troglobiotic) and aquatic (stygobiotic) environments. Scientific study of the cave biology of the Mammoth Cave system began at the turn of the nineteenth century and actively continues today. These investigations have included taxonomic examination of specific organisms, ecological analyses of terrestrial and aquatic systems, and evolutionary studies of the adaptation of cave animals. Notable species include the endangered, endemic Mammoth Cave Shrimp (*Palaemonias ganteri*), two species of troglobiotic fish, and five co-occurring cave trechine beetle species. The caves of the park are also home to three threatened or endangered species of bats. The fact that over 50 species of organisms (Appendix 1.2) were first described from caves on the park is further evidence of the central role of Mammoth Cave in understanding cave biology. Mammoth Cave is among the best investigated and understood cave ecosystems in the world. Several chapters cover various aspects of the biology of the cave. In Chap. 13, Poulson discusses the terrestrial cave ecosystem. Helf and Olson cover aquatic cave ecology in Chap. 14. In Chap. 15, Culver and Hobbs summarize the cave biodiversity of the region and how it relates to subterranean biodiversity globally. Lavoie discusses cave microbiology in Chap. 16. Bats are presented in Chap. 17 by Thomas and Toomey.

1.4 Humans in the Cave

Humans have inhabited the park area for as long as 11,500 years. The park has many surface archaeological sites dating from the prehistoric (prior to written records, approximately mid 1500s) rock shelters to historic home sites. However, its most famous archaeological sites are in the caves of the area. Artifacts show that Mammoth Cave

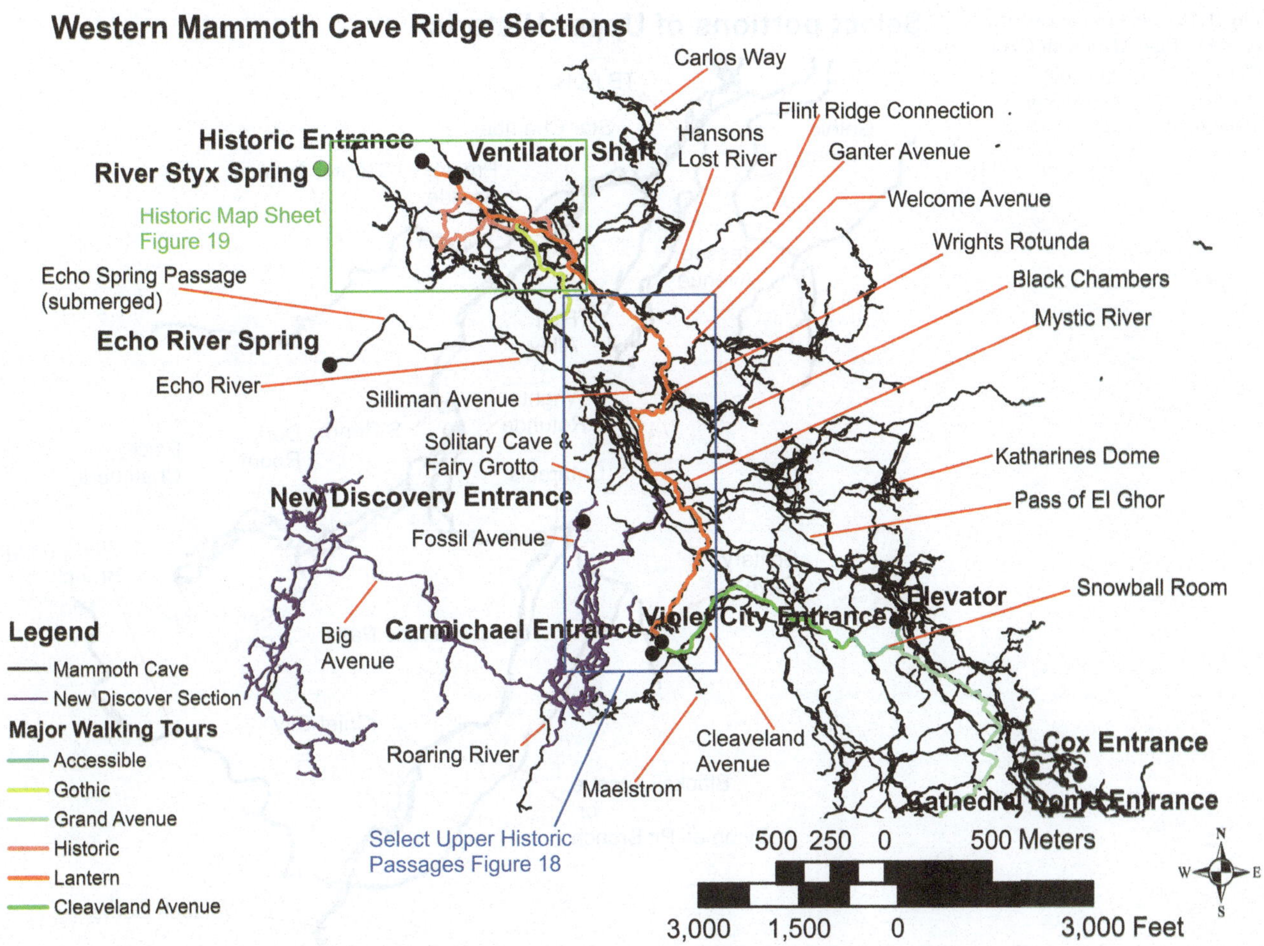

Fig. 1.17 Map of the western portion of the Mammoth Cave Ridge section of the system

was explored by Native Americans beginning approximately 5000 years ago. From about 3000–2200 years ago, several of the caves were utilized extensively by the local people who mined several types of evaporate minerals from the cave (including gypsum crusts and flowers, selenite crystals, epsomite, and mirabilite). They explored and utilized at least 19.5 miles (31.4 km) of passages in the Mammoth Cave System and smaller caves in the park. Modern cave archaeological studies at Mammoth Cave National Park, initiated in the late 1960 s, sought systematically to integrate cavers into projects and to bring dark-zone cave archaeology into the mainstream of scientific archaeological research (Crothers et al. 2007). The research in Mammoth and Salts Cave sections shed important light on the development of agriculture in the Midwest (Watson 1997); subsequent work in park caves has continued to provide many important findings. In Chap. 2, Crothers discusses the prehistoric archaeology of Mammoth Cave and several related caves.

During the latest 1700s and early 1800s, calcium nitrate was mined from Mammoth Cave (and other caves in the park) for the production of saltpeter. Some of the mining structures are still well-preserved, and some of the best conserved features are shown today on cave tours. Crothers (Chap. 2) and Olson (Chap. 3) discuss this aspect of Mammoth Cave history.

Mammoth Cave became a show cave in 1816 and has been open continuously as one since that time. The first tours were led by former saltpeter miners, and later slaves were brought in to guide tours. The slave guides were instrumental in exploring the cave, developing the tours that are the basis of modern tours, and in bringing Mammoth Cave international renown (Lyons 2006). Tour development of the caves of the Mammoth Cave area was continuous and dynamic throughout the last half of the nineteenth and first half of the twentieth centuries. Several caves that are now part of the Mammoth Cave System were originally developed as separate show caves. These include Crystal, Salts, Colossal, and Proctor. In addition, several entrances to Mammoth Cave were developed to allow tourism including New Entrance, Frozen Niagara, Violet City, Carmichael, and

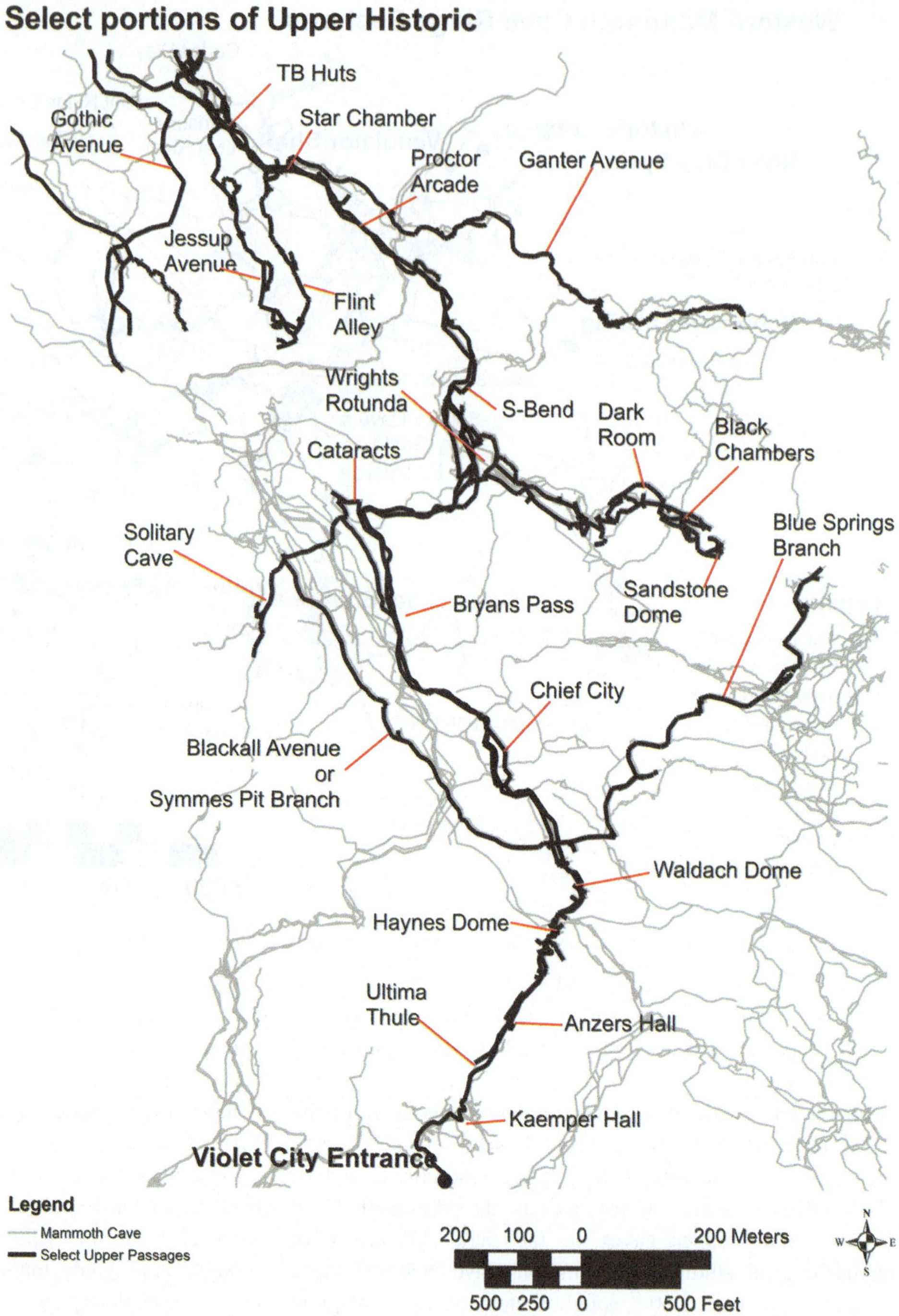

Fig. 1.18 Map of the eastern portion of the Mammoth Cave Ridge section highlighting selected *upper level* passages along the Lantern Tour route

the Elevator. The New Discovery Entrance was also developed, but it was never used for tours. Mammoth Cave is known more for its length and large passages than for its cave decorations. Today about 14 miles (22.5 km) of cave passages are visited in regularly scheduled cave tours (Fig. 1.5). Much more detailed information on the history of tourism in the Mammoth Cave area can be found in Olson (Chap. 3) and Sides (Chap. 4).

Over the past 200 years, many activities besides mining and tourism have occurred in Mammoth Cave. Some uses of Mammoth Cave have included an experimental tuberculosis sanitarium, a mushroom farm, a civil defense shelter, a sleep experiment, and a source of cooled air for buildings. In Chap. 10, Olson discusses the use of the cave for air conditioning. In Chap. 3, Olson discusses other aspects of the historic use of the cave.

Hundreds of people have explored the Mammoth Cave System over the past 5000 years looking for parts of the cave that nobody has seen before. As these people passed through the cave, they left evidence of their travels. In the case of

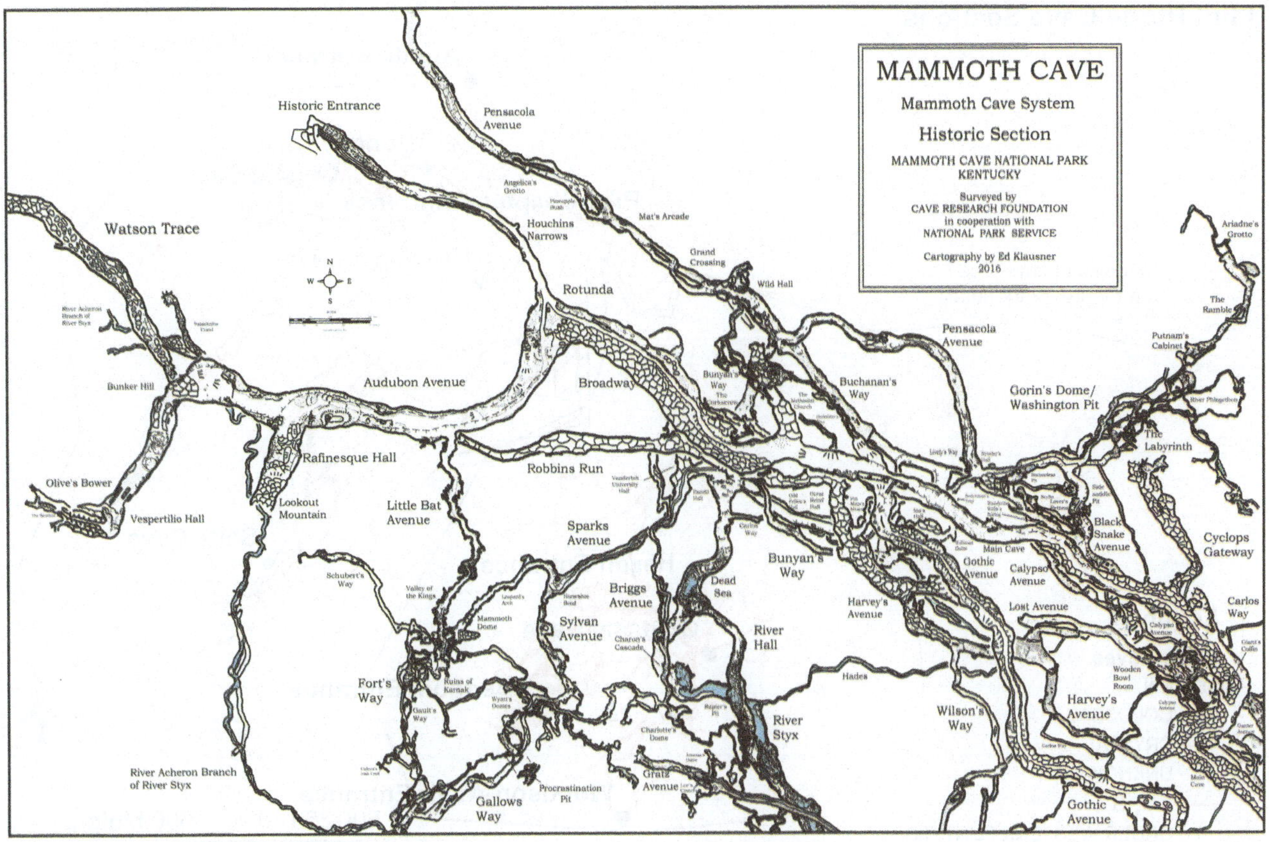

Fig. 1.19 Cave Research Foundation Historic Entrance sheet. Cartography by Ed Klausner

prehistoric people, this evidence includes cane torch fragments, other artifacts, paleofeces, stoke marks, and occasional pictographs. Historic people left artifacts, signatures, and survey marks. We do not know how prehistoric people recorded what they had found; however, starting in the early 1800s, people have also left maps of the cave and written reports of their exploration. We can use the things people have left behind and their written evidence to trace the exploration of the Mammoth Cave System. Sutton discusses the history of survey and cartography of the system in Chap. 5. Figures 1.6 through 1.14 are a series of maps showing the progression of discovery in the Mammoth Cave System. As the system has gotten longer, it has done so in two main ways. In some cases, more passages have been discovered off of areas already known to be part of Mammoth Cave. In other cases, new caves have been explored and those passages have been added to the cave system by connections. Stan Sides recounts the exploration of the cave in Chap. 4.

Beginning about 5000 years ago prehistoric people started exploring deep into caves in the area. Their explorations included several areas of the Mammoth Cave System. They explored from the Historic Entrance, the entrance to Salts Cave, and possibly from an entrance to Unknown Cave that has subsequently collapsed. By the time they stopped visiting the deep cave about 2200 years ago, they had explored about 16.9 miles (27.2 km) of passages in the system, as shown in Fig. 1.6. Note, several areas that were known by prehistoric people actually will not be on maps again until much later.

1.5 Historic and Modern Exploration and Mapping

Figure 1.7 shows the extent of cave exploration from the European discovery of the cave (late 1700s) through the early cave tour period. It includes the areas of Mammoth Cave that were mined for saltpeter and which occur on the Lee map (1835). Apparently only limited portions of Salts Cave were known during this time.

In 1845, a map of Mammoth Cave drawn by enslaved African-American guide Stephen Bishop was published in Bullitt (1845). Bishop built on the Lee map and drew passages that he and other guides had discovered. Although most of the map is not based on survey of the cave, it does

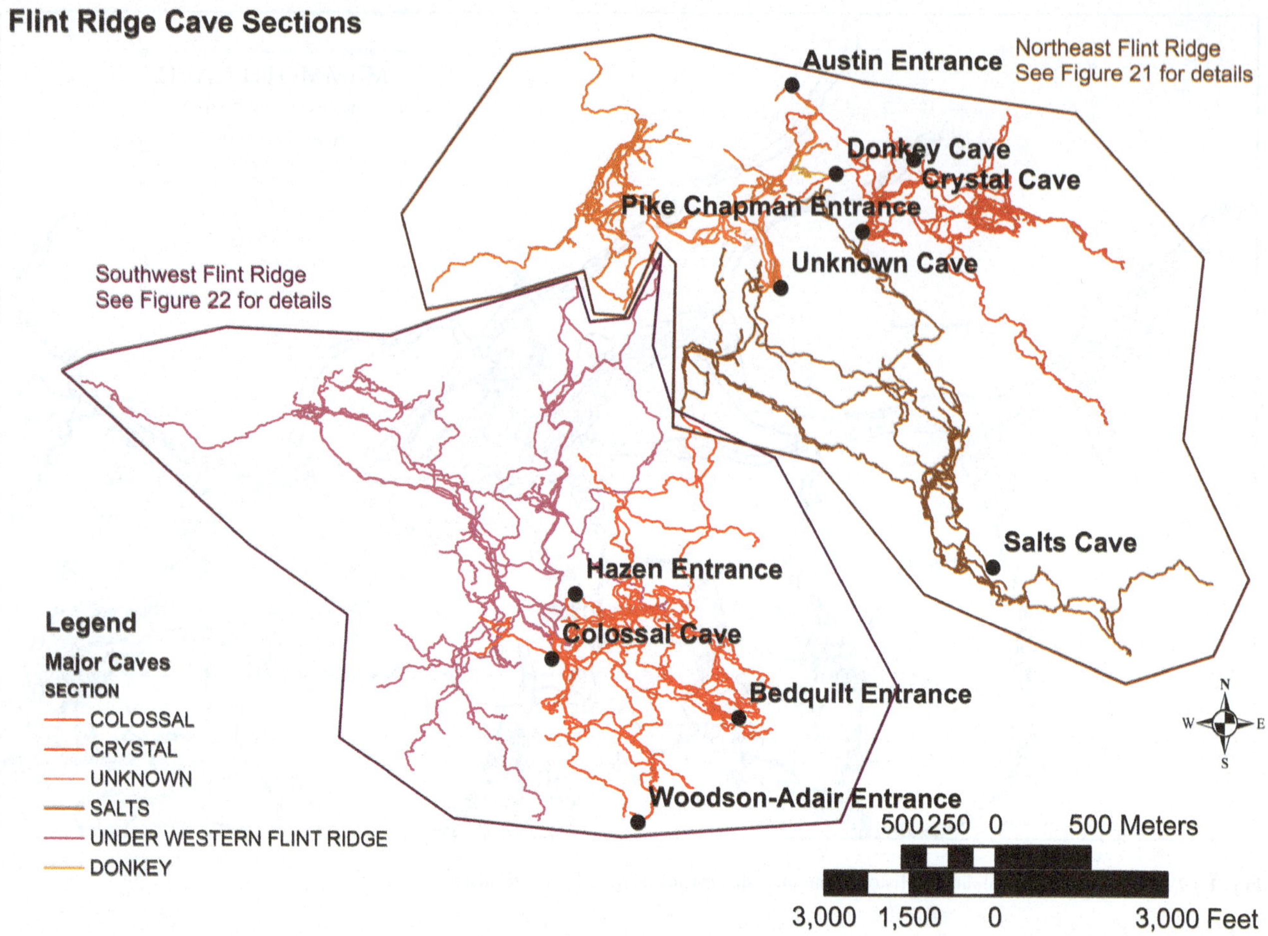

Fig. 1.20 Map of the Flint Ridge section of the system showing the extent of the various component caves

provide important information on what parts of the cave were known. Bishop shows an estimated 20 miles (32 km) of cave passages. Figure 1.8 shows this portion of Mammoth as well as known parts of Salts.

During the second half of the nineteenth century, several new caves, which later became part of the system, were found and explored: Proctor Cave (1863), Unknown Cave (1880s), and Colossal Cave (1895). Proctor and Colossal were developed for tours. Although Salts Cave continued to be toured during this time, no new passages were discovered in it. However, an entrance at the back of the Salts Cave trunk was developed (the Pike Chapman Entrance) in the 1870s. That entrance was closed and reopened several times. During eight months in 1908, Kämper and guide Ed Bishop surveyed approximately 35 miles (56 km) of Mammoth Cave and drew a map that is still used today. Most of the new passages in Mammoth Cave shown in Fig. 1.9 (representing the known cave in 1908) are a result of mapping by Max Kämper.

In 1941, when Mammoth Cave National Park was dedicated, approximately 44 miles (71 km) of Mammoth Cave had been explored and surveyed (Fig. 1.10). By this time, several caves and areas had been added to what was known in 1908. In 1917, Floyd Collins discovered Crystal Cave. He developed it as well as Donkey Cave for tours. About this time, George Morrison dug the Cox Entrance into Mammoth Cave and explored what later became known as Morrison Cave. He followed this with the New (1921) and the Frozen Niagara (1924) entrances. From these entrances, he explored and commercialized the portion of Mammoth Cave under the eastern edge of Mammoth Cave Ridge. In 1930, he also developed an entrance at Cathedral Dome (to help service tours); however, that entrance collapsed in about 1931 and was not redeveloped. During the early 1920s, the Pike Chapman entrance to Salts was reopened as part of a non-tourist commercial operation; it later closed and has not been reopened. On October 10, 1938, Carl and Pete Hanson and Leo and Claude Hunt discovered the New Discovery section of Mammoth Cave off the Roaring River. The Civilian Conservation Corp developed an entrance (completed in 1940), and trails in this highly decorated section, but tours never occurred there.

Between the founding of the park and 1962, almost all the new exploration took place in the Flint Ridge portion of the cave (Fig. 1.11). The flurry of exploration started with Bill

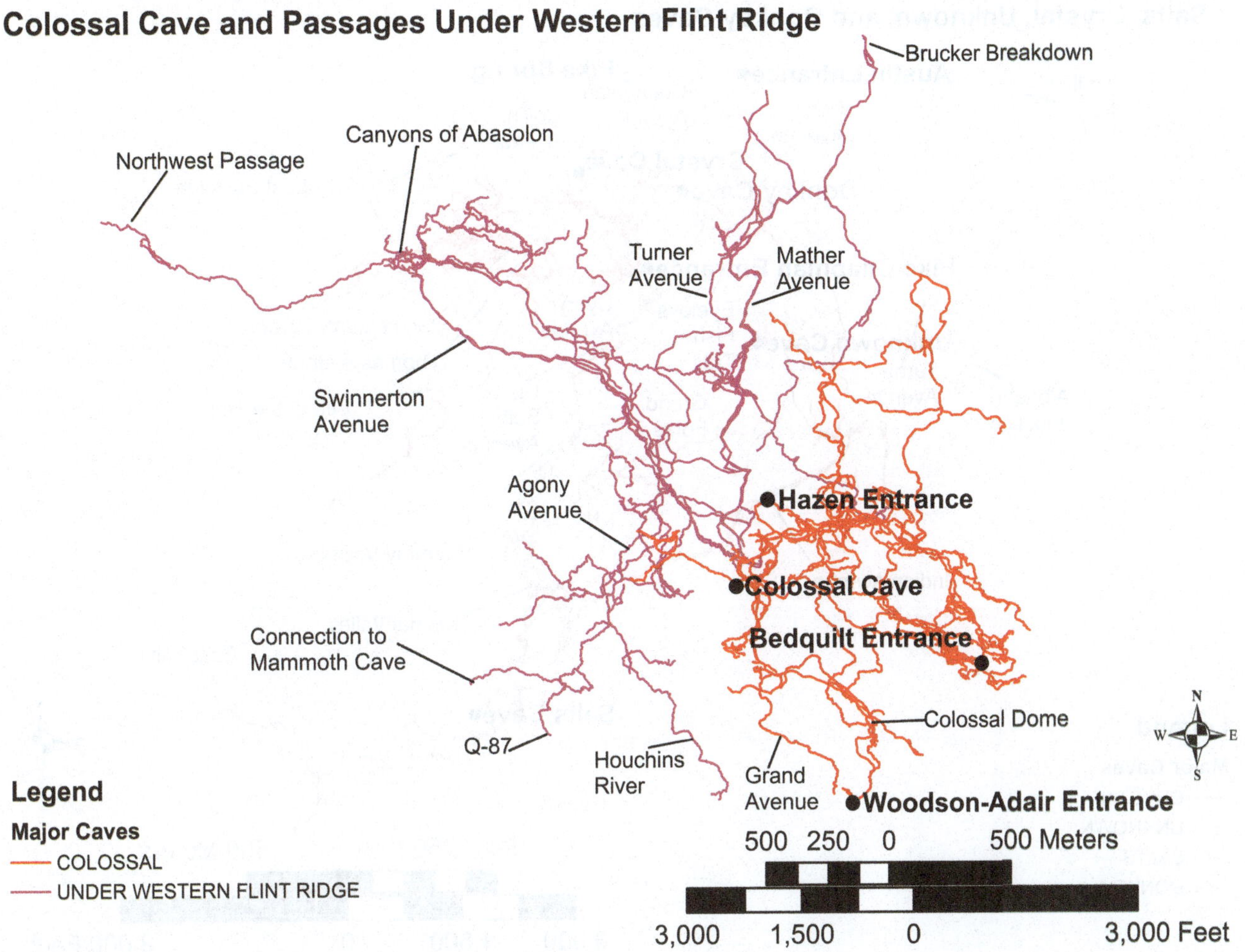

Fig. 1.21 Map showing Salts, Unknown, Crystal, and Donkey Caves within the Flint Ridge section of the system

Austin and Jack Lehrberger initiating survey through Crystal Cave and later via the Austin Entrance (constructed in 1956). In February 1954, the National Speleological Society's C-3 expedition began systematic mapping of Flint Ridge caves, tested techniques for exploring large caves, and led to the creation of the Cave Research Foundation (CRF) in 1957. In 1955, Austin and Lehrberger connected Unknown Cave with Crystal Cave. CRF cavers connected Colossal Cave with Salts Cave in 1960. In 1961, Unknown-Crystal and Colossal-Salts were connected by CRF cavers resulting in the integration of the Flint Ridge Cave System, which had a length of approximately 30 miles (48 km) (Watson and Brucker 1976). The Flint Ridge segment of Fig. 1.11 is largely based on Brucker and Burns (1964).

The Flint Ridge section continued to be the focus of survey between 1962 and 1972 (Fig. 1.12). Continued CRF exploration resulted in the Flint Ridge Cave System becoming the longest known cave in the world in March 1969 at 65 surveyed miles (105 km). At this time, historic Mammoth Cave was 44 miles (71 km) in length. In May 1969, CRF received permission to explore and survey in Mammoth Cave as well. On September 9, 1972, CRF cavers entered Unknown Cave via the Austin Entrance, travelled and surveyed for 14.5 hours, and left via the Elevator in Mammoth Cave thus connecting the Flint Ridge System and Mammoth Cave. The result was a system that was 144.4 miles (232 km).

Survey in the Mammoth–Flint Ridge section continued between 1972 and 1983. For example, on January 15, 1981, a team dove Echo River Spring and surveyed 3340 feet (1018 m) to connect it to the system. However, the main areas of discovery shifted south and east to the Joppa Ridge caves and Roppel Cave (Fig. 1.13). On April 2, 1976, Roppel Cave was discovered. Its survey was led by the Central Kentucky Karst Coalition (CKKC). On July 2, 1979, Morrison and Proctor caves were connected by a CRF team (forming the Joppa Ridge System). The "French Connection" linked the Mammoth–Flint Ridge System to the Joppa

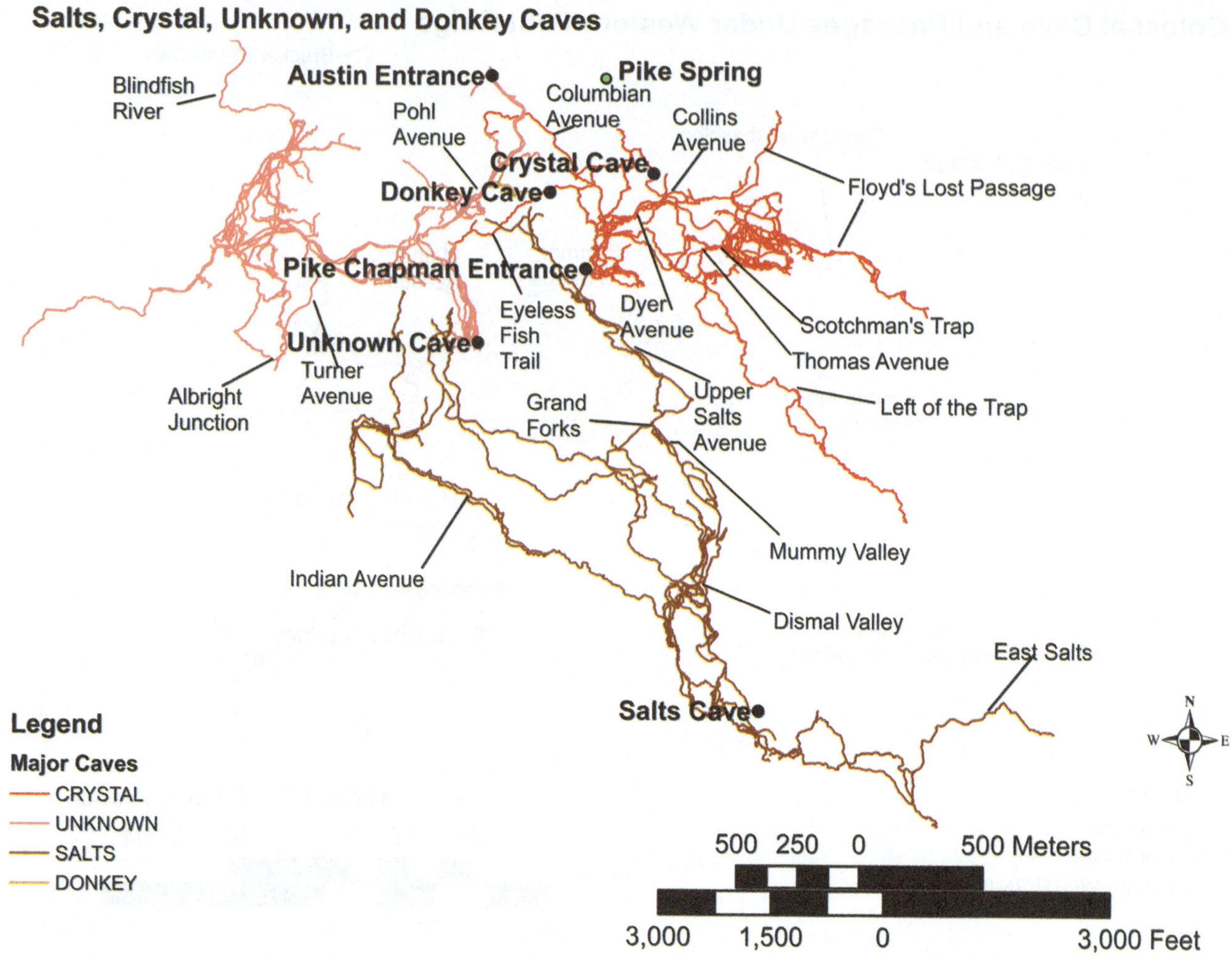

Fig. 1.22 Map showing the Colossal Cave section of the Flint Ridge section as well as passages under the western portion of Flint Ridge. Various of these passages are reached via the Austin Entrance or Colossal Cave; however, they cannot be easily assigned to any one cave

Ridge System on August 11, 1979. In 1981, the bottom of the Ferguson Entrance was located from inside the cave; it was found from the surface later that same year. Exploration of the Roppel section was made easier by the construction of the Daleo Entrance in 1980–81. On September 9, 1983, CRF and CKKC cavers jointly connected the Mammoth Cave–Flint Ridge–Joppa Ridge Cave System with Roppel Cave to form the Mammoth Cave System as we recognize it today. At that point, the system was 293 miles (472 km) long.

Since 1983, survey has continued in many areas of the system. Survey in the Joppa Ridge portion of the cave was greatly aided by the construction of the Doyle Valley Entrance in 1984, and construction of the Khan (1989) and Weller (1991–92) entrances assisted with access to areas of the Roppel section. Two additional separate caves were connected to the system in recent time. On March 19, 2005, Hoover Cave was connected to the Roppel Section by CKKC cavers. On May 28, 2011, Donkey Cave was connected to the Flint Ridge section of the Cave. The only person who managed to get through that hideously tight connection was Joyce Hoffmaster. As of 2016, the surveyed length of the Mammoth Cave System was 405 miles (652 km) (Fig. 1.14).

The history of exploration of the Mammoth Cave System is complex, as described above and in Chaps. 3 and 4. The cave has 27 current and three closed entrances. Many of these entrances represented caves with separate histories before they were connected to the system. In this book, the reader will find many different caves, cave areas, passages, and entrances referred to. The following primer will assist in understanding the complex geography of the cave. It will

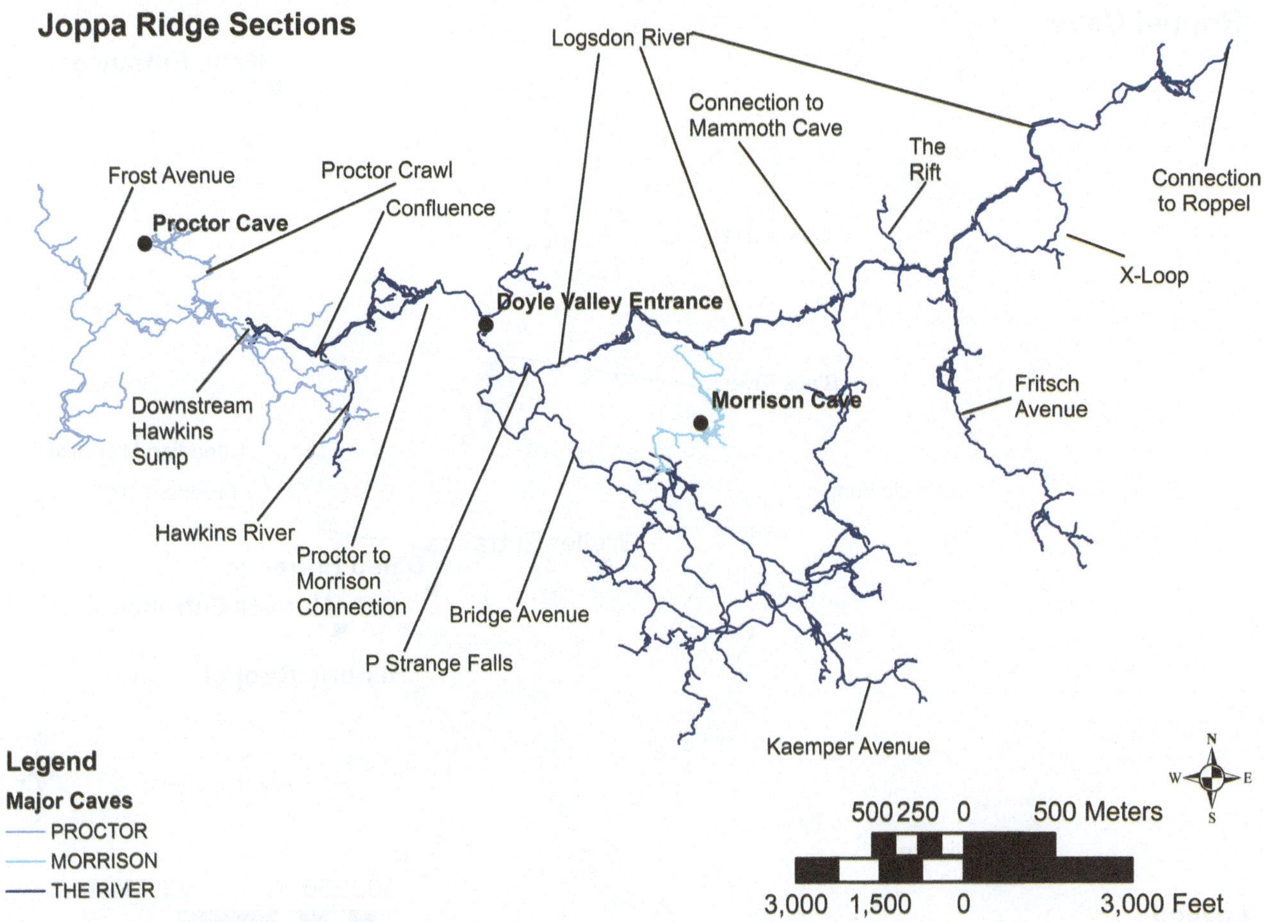

Fig. 1.23 Map showing the components of the Joppa Ridge section of the system

also provide maps (Figs. 1.15 through 1.24) and a general introduction for places you will find in other chapters. The maps accompanying this primer will provide a general idea of how different segments relate to each other and provide locations of some of the main Mammoth Cave System caves, entrances, and places referenced in other chapters. However, these maps are not comprehensive. Many places mentioned only once or twice are not shown on these maps due to spatial constraints. Many other caves in the area are critical to understanding Mammoth Cave. Figure 1.25 shows some of the main caves that are mentioned in various chapters of this volume. However, this figure does not include every cave mentioned in the book.

The main components of the Mammoth Cave System are the Mammoth Cave Ridge section, the Flint Ridge Section, the Joppa Ridge Section, and the Roppel (or Toohey Ridge) Section (Fig. 15). Each of these sections has a large amount of cave that was explored from entrances in those sections before they were connected to the system as a whole. Figures 1.16, 1.17, 1.18, and 1.19 show somewhat more detailed maps of the Mammoth Cave Ridge section. Figures 1.20, 1.21 and 1.22 show the relationships of the caves in the Flint Ridge section. Figure 1.20 shows how all the Flint Ridge caves are related. Figures 1.21 and 1.22 provide details and passage names in the northeastern and southwestern portions of Flint Ridge. Figure 1.23 provides information on the passages in the Joppa Ridge section of the system. Figure 1.24 shows the Roppel section of the system.

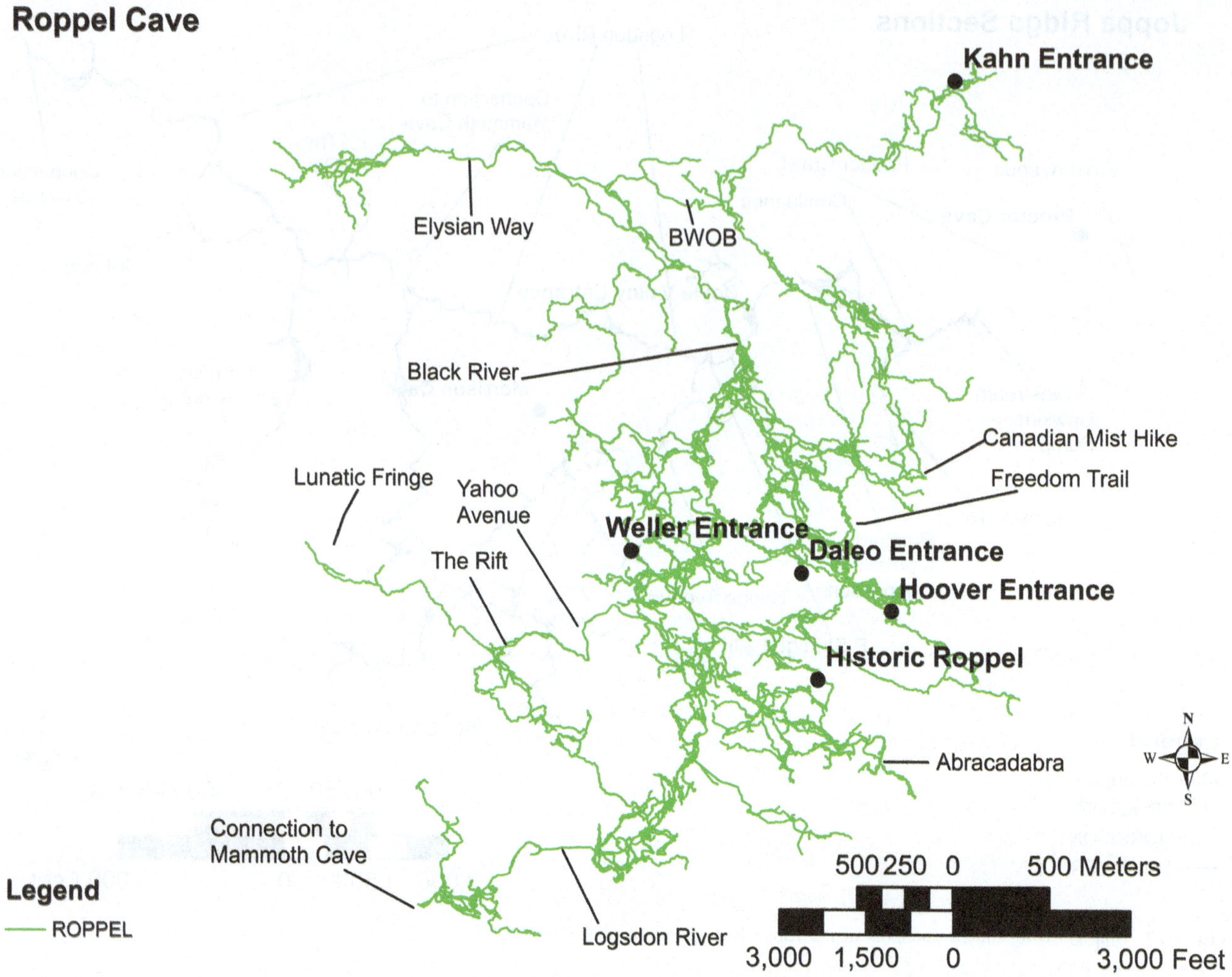

Fig. 1.24 Map of the Roppel section of the system

Table 1.1 Past and current entrances to the Mammoth Cave System

System segment	Cave	Entrance	Natural?	Comments
Mammoth Cave Ridge				
	Mammoth Cave			
		Historic	Natural	known >5000 years
		Cox	Artificial	Morrison 1916–17
		New	Artificial	Morrison 1921
		Frozen Niagara	Artificial	Morrison 1924
		Violet City	Artificial	Ganter 1915 or CCC 1931
		Cathedral Domes	Artificial	Morrison 1930; collapsed and closed
		Carmichael	Artificial	CCC 1931
		New Discovery	Artificial	CCC 1940
		Elevator	Artificial	NPS 1958
		Ventilation Shaft	Artificial	NPS 1959
		Echo River Spring	Natural	Connected 1981
		Ferguson	Modified Natural	Brucker 1981
Flint Ridge				
	Salts Cave			
		Historic Salts	Natural	Known >5000 years
		Pike Chapman	Modified Natural	Vials 1870s; Hazen 1893; closed by 1910 reopened 1902s subsequently closed
	Unknown Cave			
		Historic Unknown	Natural	Known from 1880s
		Austin	Artificial	Austin 1956
	Colossal Cave			
		Woodson-Adair	Modified Natural	~Lee 1890
		Bedquilt	Natural	Lee 1895
		Colossal	Artificial	L & N Railroad 1896
		Hazen	Natural	Hazen 1896, blasted shut shortly thereafter
	Crystal Cave			
		Crystal	Modified Natural	Collin 1917
	Donkey Cave			
		Donkey	Modified Natural	Collins 1920s
Joppa Ridge				
	Proctor Cave			
		Historic Proctor	Modified Natural	Doyel 1863
		Doyle Valley	Artificial	Quinlan 1984
	Morrison Cave			
		Morrison	Modified Natural	Morrison ~1920
Roppel/Toohey Ridge				
	Roppel Cave			
		Historic Roppel	Modified Natural	CKKC 1976
		Daleo	Artificial	Weller 1980–1981
		Khan	Artificial	Borden and Weller 1989
		Weller	Artificial	Weller 1991–1992
		Hoover	Modified Natural	Wells 2005

Some entrances were found, explored, and then connected to the system. Others were constructed into caves that were already known. Entrances designated as "Natural" could be entered by people without digging or blasting. "Modified Natural" entrances were located because they showed small openings that were expanded to allow people to enter. "Artificial" entrances were constructed at locations without surface clues using cave surveys to know where to approach known cave passages. The comment column provides information on when the entrance was found or constructed and by whom

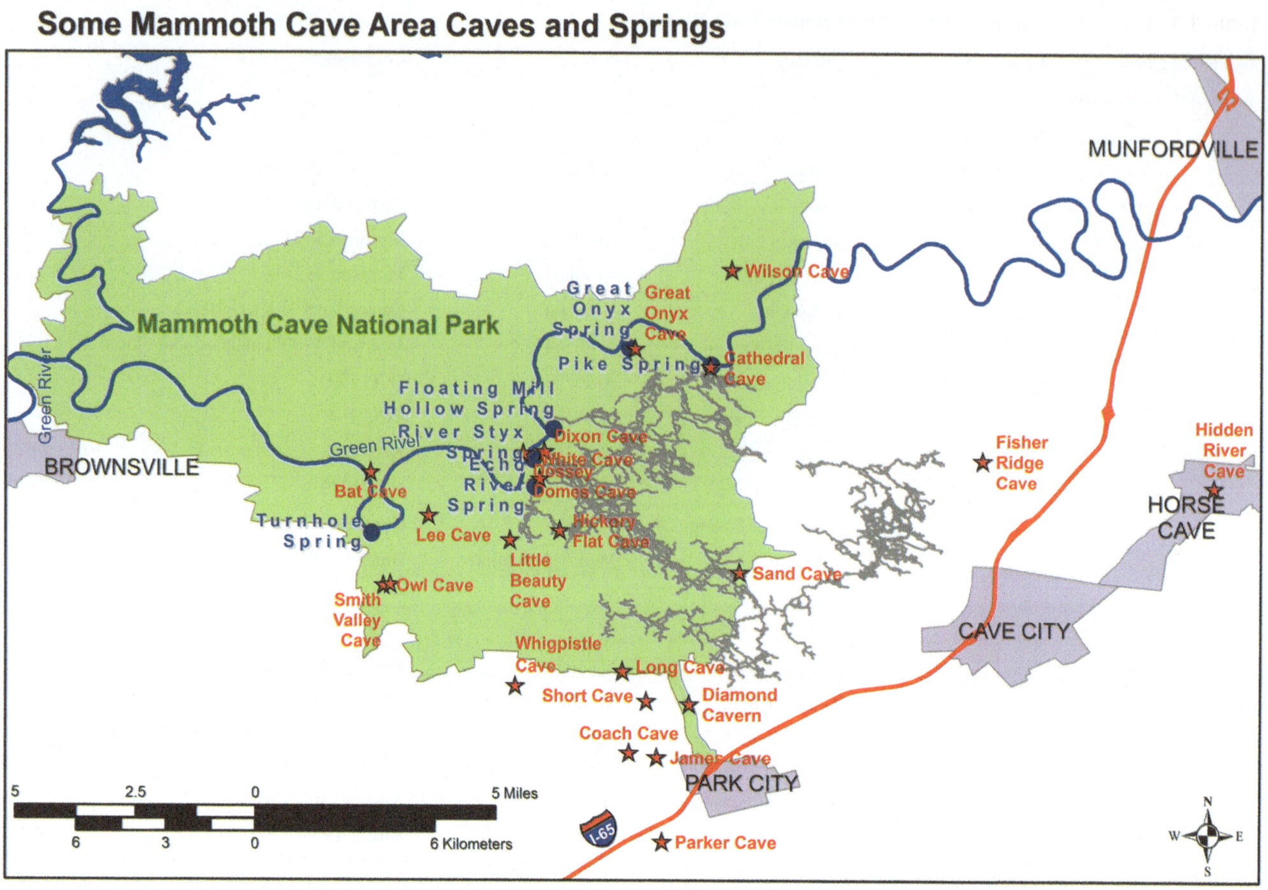

Fig. 1.25 Map showing the general location of many of the other caves mentioned in the book. It also shows the locations of the six springs through which the various rivers in Mammoth Cave drain to the Green River

Appendix

Table 1.2 Cave-adapted animals found in the caves of Mammoth Cave National Park

Species from Caves of Mammoth Cave National Park							
Troglobionts	**Stygobionts**	**Troglophiles**	**Stygophiles**	**Obligate Trogloxenes**	**Facultative Trogloxenes**	**Facultative Stygoxenes**	**Aquatic Accidentals**
Obligate, permanent resident of subterranean terrestrial habitat	**Obligate, permanent resident of subterranean aquatic habitat**	**Facultative permanent resident of terrestrial subterranean habitat**	**Facultative permanent resident of aquatic subterranean habitat**	**Obligate resident of terrestrial subterranean habitat but associated with surface habitats for some part of its life cycle**	**Facultative resident of terrestrial subterranean habitat but associated with surface habitats for some part of its life cycle**	**Facultative resident of aquatic subterranean habitat but associated with surface habitats for some part of its life cycle**	**Aquatic species appearing sporadically in subterranean habitat**
Snails	**Flatworms**	**Fungus**	**Diatoms**	**Mites**	**Nematodes**	**Nematodes**	**Sponges**
Carychium stygium	*Sphalloplana percaeca*	*Pseudogymnoascus destructans*	*Cymbella clausii*	*Trombiculid*	*Pterygodermatis coloradensis*	Numerous species	*Spongilla fragilis*
Helicodiscus punctatellus	*Sphalloplana buchanani*	*Beauveria amorpha*	*Cymbella gerloffii*	*Ichoronyssus* sp.	**Horsehair Worms**	**Amphipods**	**Flatworms**

(continued)

Table 1.2 (continued)

Species from Caves of Mammoth Cave National Park							
Troglobionts	**Stygobionts**	**Troglophiles**	**Stygophiles**	**Obligate Trogloxenes**	**Facultative Trogloxenes**	**Facultative Stygoxenes**	**Aquatic Accidentals**
Glyphyalinia specus	**Snails**	**Snails**	*Cymbella hohnii*	*Liponyssus* sp.	*Chordodes morgani*	*Crangonyx forbesi*	*Catenula* sp.
Mites	*Antroselates spiralis*	*Zonitoides arboreus*	*Gomphonema hotchkissii*	**Crickets**	**Snails**	**Terrestrial Accidentals**	**Rotifers**
Traegaardhia holsingeri	**Amphipods**	*Glyphyalinia specus*	**Oligochaete worms**	*Hadenoecus subterraneus*	*Mesodon appressus*	**Terrestrial species appearing sporadically in subterranean habitat**	*Keratella* sp.
Belba bulbipedata	*Crangonyx barri*	**Mites**	*Chaetogaster* sp.	**Flies and Gnats**	**Spiders**	**Oligochaete worms**	*Asplanchna* sp.
Linopodes mammouthia	*Stygobromus vitreus*	*Arctoseius* spp.	**Copepods**	*Amoebaleria defessa*	*Nesticus carteri*	*Allolobophora trapezoides*	**Oligochaete worms**
Laelaps cavernicola	*Stygobromus exilis*	*Ceratozetes* spp.	*Moraria* spp.	*Amoebaleria sackeni*	*Dolomedes* sp.	*Dendrobaena rubida*	*Aeolosoma hemprichi*
Macrocheles troglodytes	**Copepods**	**Spiders**	**Decapods**	*Heleomyza brachypterna*	**Harvestmen**	*Eisenia joetida*	**Copepods**
Galumna alata	*Bryocamptus morrisoni elegans*	*Meta ovalis*	*Cambarus tenebrosus*	*Oeothea specus*	*Leiobunum longipes*	**Spiders**	*Maraenobiotus* spp.
Spiders	*Megacyclops donnaldsoni*	*Cybaeus giganteus*	**Flies and Gnats**	**Salamanders**	**Pseudoscorpions**	*Coras juvenalis*	*Nitocra* spp.
Anthrobia monmouthi	*Attheyella pilosa*	*Liocranoides unicolor*	*Chironomus* sp.	*Eurycea lucifuga*	*Tyrannochthonius hypogeus*	**Beetles**	*Parastenocaris* spp.
Phanetta subterranea	*Cauloxenus stygius*	**Isopods**	*Parakiefferiella* sp.	**Mammals**	**Millipedes**	*Heterothops campbelli*	*Phyllognathus viguieri*
Bathyphantes weyeri	**Decapods**	*Miktoniscus mammothensis*	*Parametriocnemus*	*Corynorhinus rafinesqueii*	*Oxidus gracilis*	**Salamanders and Frogs**	*Orthocyclops modestus*
Porrhomma cavernicola	*Palaemonias ganteri*	*Haplophthalmus danicus*	*Polypedilum*	*Myotis grisecens*	**Crickets**	*Rana catesbeiana*	*Tropocyclops prasinus*
Harvestmen	*Orconectes pellucidus*	**Centipedes**	*Tanytarsus*	*Myotis lucifugus*	*Ceuthophilus stygius*	**Turtles**	*Acanthocyclops vernalis*
Phalangodes armata	**Isopods**	*Lithobia* spp.	**Fish**	*Myotis septentrionalis*	*Ceuthophilus latens*	*Terrapene carolina*	*Acanthocyclops robustus*
Pseudoscorpions	*Caecidotea stygia*	*Tidabius* sp.	*Forbesichtys agassizi*	*Myotis sodalis*	**Butterflies and Moths**	**Lizards and Snakes**	*Diacyclops thomasi*
Hesperochernes mirabilis	*Caecidotea bicrenata whitei*	**Springtails**	*Cottus carolinae*	*Perimyotis subflavus*	*Scoliopteryx libatrix*	*Scincus laticeps*	*Mesocyclops edax*
Kleptochthonius cerberus	**Ostracods**	*Neanura spp.*		*Neotoma magister*	**Fleas**	*Diadophis punctatus*	*Macrocyclops fuscus*
Kleptochthonius hageni	*Sagittocythere stygia*	*Arrhopalites pygmaeus*			*Epitedia wenmanni*	*Carphophis amoenus*	*Macrocyclops albidus*
Millipedes	*Sagittocythere barri*	*Sinella curviseta*			*Phallocropsylla* sp.	*Agkistrodon contortrix*	*Eucyclops agilis*
Chaetaspis fragilis	**Fish**	*Pseudosinella duodecimpunctata*			*Ceratopsyllus* sp.	*Crotalus horridus*	*Eucyclops elegans*
Scoterpes copei	*Amblyopsis spelaea*	*Pseudosinella argentea*			*Myoxopsylla sexdentatus*	**Birds**	*Paracyclops chiltoni*

(continued)

Table 1.2 (continued)

Species from Caves of Mammoth Cave National Park							
Troglobionts	**Stygobionts**	**Troglophiles**	**Stygophiles**	**Obligate Trogloxenes**	**Facultative Trogloxenes**	**Facultative Stygoxenes**	**Aquatic Accidentals**
Springtails	*Typhlichthys subterraneus*	*Ceratophysella denticulata*			*Myodopsylla insignis*	*Otus asio*	*Skistodiaptomus pallidus*
Pygmarrhopalites altus		*Folsomia candida*			*Nycteridopsylla chapini*	**Mammals**	*Osphranticum labronectum* **Water Fleas**
Pseudosinella espanita		*Lepidocyrtus atropurpureus*			**Salamanders and Frogs**	*Mephitis mephitis*	*Bosimina longirostris*
Willemia sp.		*Pogonognathellus bidentatus*			*Eurycea longicauda*	*Mustela longicauda*	**Isopods**
Onychiurus sp.		*Pogonognathellus flavescens*			*Eurycea cirrigerra*	*Sciurus carolinensis*	*Lirceus fontinalis*
Diplurans		**Diplurans**			*Plethodon glutinosus*	*Marmota monax*	**Ostracods**
Litocampa cookei		*Metajapyx subterraneus*			*Pseudotriton ruber*	*Rattus norvegicus*	*Candona sp*
Beetles		**Bristletails**			*Plethodon dorsalis*	*Odocoileus virginianus*	"*Limnocythere* sp."
Neaphaenops tellkampfi		*Triura cavernicola*			*Rana clamitans melanota*		**Tardigrades**
Pseudanophthalmus menetriesi		**Book Lice**			*Rana palustris*		*Macrobiotus* spp.
Pseudanophthalmus striatus		*Psillipsocus ramburii*			*Rana sphenocephala*		**Fish**
Pseudanophthalmus inexpectatus		**Beetles**			**Birds**		*Lampetra aepyptera*
Pseudanophthalmus pubescens		*Quedius erythrogaster*			*Sayornis phoebe*		*Moxostoma erythrurum*
Pseudanophthalmus audax		*Quedius fulgidus*			*Cathartes aura*		*Dorosoma cepedianum*
Ptomaphagus hirtus		*Atheta troglophilia*			*Coragyps atratus*		*Campostoma oligolepis*
Batrisodes henroti		*Atheta lucifuga*			**Mammals**		*Ictalurus punctatus*
Flies and Gnats		*Atheta annexa*			*Eptesicus fuscus*		*Semotilus atromaculatus*
Spelobia tenebrarum		*Lesteva pallipes*			*Lasionycteris noctivagans*		*Phoxinus erythrogaster*
Megaselia cavernicola		**Flies and Gnats**			*Lasiurus borealis*		*Hypetelium nigricans*
		Psychoda spp.			*Myotis leibii*		*Catostomus commersoni*
		Rymosia spp.			*Peromyscus leucopus*		*Lepisoteus osseus*
		Bradysia spp.			*Procyon lotor*		*Cyprinus carpio*
		Macrocera nobilis					*Notropis volucellus*
		Chironomus sp.					*Pimephales notatus*
		Parakiefferiella sp.					*Fundulus sp.*
		Parametriocnemus					*Gambusia affinis*

(continued)

Table 1.2 (continued)

Species from Caves of Mammoth Cave National Park							
Troglobionts	**Stygobionts**	**Troglophiles**	**Stygophiles**	**Obligate Trogloxenes**	**Facultative Trogloxenes**	**Facultative Stygoxenes**	**Aquatic Accidentals**
		Polypedilum					*Ambloplites rupestris*
		Tanytarsus					*Lepomis macrochirus*
							Lepomis megalotis
							Micropterus dolomieu
							Micropterus punctulatus
							Etheostoma caeruleum
							Etheostoma nigrum
							Etheostoma zonale
							Percina copelandi
							Percina evides
							Percina caprodes

In addition, this list contains animals which are not cave adapted, but occur in the caves regularly. It is not meant to be a comprehensive list of every animal that has ever been found in the cave. The animals are grouped into functional categories by where they occur (terrestrial or aquatic) and how cave adapted they are

Table 1.3 Animals that have been described from the Mammoth Cave karst (including the Mammoth Cave System, other caves on the park, and park springs)

Described as	Current valid taxon (if any)	Author	Date described	Type locality
Diatoms				
Cymbella clausii	*Cymbella clausii*	VanLandingham	1966	Mammoth Cave, bottom mud small pools
Cymbella gerloffii	*Cymbella gerloffii*	VanLandingham	1966	Mammoth Cave, bottom mud small pools
Cymbella gerloffii	*Cymbella hohnii*	VanLandingham	1966	Mammoth Cave, bottom mud small pools
Gomphonema hotchkissii	*Gomphonema hotchkissii*	VanLandingham	1967	Mammoth Cave
Flatworms				
Dendrocoelum percoecum	*Sphalloplana (Sphalloplana) percoeca*	Packard	1879	Mammoth Cave
Sphalloplana (Speophila) buchanani	*Sphalloplana (Speophila) buchanani*	Hyman	1937	Mammoth Cave
Snails				
Antroselates spiralis	*Antroselates spiralis*	Hubricht	1963	Echo River Spring
Carychium stygium	*Carychium stygius*	Call	1897	Mammoth Cave, Mammoth Dome
Melania latitans	*Lithasia obovata*	Anthony	1854	Mammoth Cave, River Styx?
Helicodiscus punctatellus	*Helicodiscus punctatellus*	Morrison	1942	White Cave
Mites				
Rhagidia cavernarum	*Rhagidia cavernarum*	Packard	1888	Mammoth Cave
Gamasus (or Hypoaspis?) troglodytes	unclear	Packard	1885	Mammoth Cave, Labyrinth
Laelaps (= Iphis?) cavernicola	unclear	Packard	1885	Mammoth Cave, Labyrinth
Damaeus bulbipedata	unclear	Packard	1885	Dixon Cave
Oribata alata	unclear	Packard	1885	Dixon Cave
Linopodes mammouthia	*Linopodes mammouthia*	Banks	1897	Mammoth Cave,
Spiders				
Anthrobia monmouthi	*Anthrobia monmouthi*	Tellkampf	1844	Mammoth Cave
Caelotes juvenalis	*Coras juvenilis*	Keyserling	1881	Mammoth Cave
Harvestmen				
Phrixis longipes	*Phalangodes armata*	Cope	1872	Mammoth Cave
Phalangodes armata	*Phalangodes armata*	Tellkampf	1844	Mammoth Cave
Pseudoscorpions				
Kleptochthonius cerberus	*Kleptochthonius cerberus*	Malcolm and Chamberlin	1961	White Cave
Kleptochthonius hageni	*Kleptochthonius hageni*	Muchmore	1963	Mammoth Cave
Tyrannochthonius hypogeus	*Tyrannochthonius hypogeus*	Muchmore	1996	MCNP, Bruce Hollow, stump litter
Amphipods				
Stygobromus vitreus	*Stygobromus vitreus*	Cope	1872	Mammoth Cave, Richardson Spring
Cragonyx barri	*Cragonyx barri*	Zhang and Holsinger	2003	Mammoth Cave, Cathedral Domes
Gammarus propinquus	*Gammarus minus*	Hay	1902	spring north of Mammoth Cave
Decapods				
Palaemonias ganteri	*Palaemonias ganteri*	Hay	1901	Mammoth Cave, Roaring River
Astacus pellucidus	*Orconectes pellucidus*	Tellkampf	1844	Mammoth Cave
Cambarus bartonii tenebrosus	*Cambarus tenebrosus*	Hay	1901	Mammoth Cave, Echo and Styx Rivers

(continued)

Table 1.3 (continued)

Described as	Current valid taxon (if any)	Author	Date described	Type locality
Copepods				
Attheyella pilosa	*Attheyella (Ryloviella) pilosa*	Chappuis	1929	Mammoth Cave
Canthocamptus cavernarum	*Canthocamptus cavernarum*	Packard	1879	Mammoth Cave, Wandering Willie Spring
Isopods				
Asellus stygus	*Caecidotea stygia*	Packard	1871	Mammoth Cave
Miktoniscus mammothensis	*Miktoniscus mammothensis*	Muchmore	1964	White Cave and Cedar Sink
Ostracods				
Sagittocythere stygia	*Sagittocythere stygia*	Hart and Hart	1966	Mammoth Cave, River Styx
Millipedes				
Spirostrephon copei	*Scoterpes copei*	Packard	1871	Mammoth Cave
Antriadesmus fragilis	*Chaetaspis fragilis*	Loomis	1943	White Cave
Springtails				
Arrhopalites altus	*Pygmarrhopalites altus*	Christiansen	1966	Mammoth Cave, rotting shirt Eyeless Fish Trail
Smynthurus mammouthia	*Pygmarrhopalites pygmaeus*	Banks	1897	Mammoth Cave, Labyrinth
Entomobrya cavicola	*Folsomia candida*	Banks	1897	Mammoth Cave
Diplurans				
Campodea cookei	*Litocampa cookei*	Packard	1871	Mammoth Cave
Japyx subterraneus	*Metajapyx subterraneus*	Packard	1874	White Cave Jr.
Bristletails				
Triura cavernicola	Not found since, may be valid	Tellkampf	1844	Mammoth Cave
Book Lice				
Dorypteryx (?) hageni	*Psyllipsocus ramburii*	Banks	1897	Mammoth Cave, Labyrinth
Stoneflies				
Leuctra schusteri	*Leuctra schusteri*	Grubbs	2015	Cooper Spring
Beetles				
Anophthalmus menetriesi	*Pseudanophthalmus menetriesi*	Motschulsky	1862	Mammoth Cave
Anophthalmus ventricosus	*Pseudanophthalmus menetriesi*	Motschulsky	1862	Mammoth Cave
Anophthalmus angulatus	*Pseudanophthalmus menetriesi*	LeConte	1863	Mammoth Cave
Anophthalmus striatus	*Pseudanophthalmus menetriesi or P. striatus*	Motschulsky	1862	Mammoth Cave
Anophthalmus interstitialis	*Pseudanophthalmus menetriesi (or P. striatus* if valid*)*	Hubbard	1880	Mammoth Cave, Washington Hall
Pseudanophthalmus inexpectatus	*Pseudanophthalmus inexpectatus*	Barr	1959	White Cave
Adelops hirtus	*Ptomaphagus (Adelops) hirtus*	Tellkampf	1844	Mammoth Cave
Fungus Gnats				
Limosina stygia	*Spelobia tenebrarum*	Coquillett	1897	Mammoth Cave, River Hall
Fish				
Amblyopsis spelaeus	*Amblyopsis spelaea*	DeKay	1842	Mammoth Cave, River Styx

References

Borden JD, Brucker RW (2000) Beyond Mammoth Cave. Southern Illinois University Press, Carbondale and Edwardsville

Brucker RW, Watson RA (1976) The longest cave. Alfred A Knopf, New York

Brucker RW, Burns DB (1964) The Flint Ridge Cave System. Cave Research Foundation, Washington, DC

Bullitt AC (1845) Rambles in the Mammoth Cave during the Year 1844. Morton and Griswold, Louisville

Crothers G, Willey P, Watson PJ (2007) Cave Archaeology and the NSS: 1941–2006. J Cave Karst Stud 69: 27–34

Culver DC, Pipan T (2013) Subterranean ecosystems. In: Levin SA (ed) Encyclopedia of biodiversity, 2nd edn. Elsevier, The Netherlands

Culver DC, Sket B (2000) Hotspots of subterranean biodiversity in caves and wells. J Cave Karst Stud 62:11–17

Granger DE, Fabel D, Palmer AN (2001) Pliocene-Pleistocene incision of the Green River, Kentucky, determined from radioactive decay of cosmogenic. ^{26}Al and^{10}Be in Mammoth Cave sediments. GSA Bulletin 113: 825–836

Kambesis P (2007) The importance of cave exploration to scientific research. J Cave Karst Stud 69: 46–58

Lawrence J, Brucker R (1955) The caves beyond. Funk and Wagnalls, Reprinted 1975 Zephyrus Press

Lee EF (1835) Notes on the Mammoth Cave, to accompany a map. James and Gazley, Cincinnati, OH

Lyons JM (2006) Making their mark. The signature of slavery at Mammoth Cave. Eastern National, Fort, Washington, PA

Quinlan JF, Ewers RO, Ray JA, Power RL, Krothe NC (1983) Hydrology and geomorphology of the Mammoth Cave region, Kentucky, and the Mitchell Plain, Indiana. In: Shaver RH and Sunderman JA (eds) Field trips in midwestern geology: Bloomington, Indiana, Geological society of America and Indiana Geological Survey, vol 2, pp 1–85

Watson PJ (ed) (1997) Archaeology of the Mammoth Cave Area. Cave Books, St. Louis, Originally published 1974 by Academic Press, NY

White WB (2007) A brief history of karst hydrology: contributions of the NSS. J Cave Karst Stud 69: 13–26

The Prehistoric Archeology of Mammoth Cave

2

George M. Crothers

Abstract

The prehistory of the Mammoth Cave area has been of interest to archeologists for more than a century because of the remarkable preservation in the dry cave environment. Beginning with the antiquarian search for mummified bodies in the early nineteenth century up to the most modern scientific research, Mammoth Cave has revealed important information about the Native Americans who lived near and explored these cave systems beginning more than 5000 years ago. The first interest in Mammoth Cave may have simply been human curiosity, but by 3000 years ago Native Americans began mining several exotic minerals that form in the large, dry passages. They left behind tools and other remains from this mining effort that has been important for understanding the beginnings of the first farming societies in eastern North America and provides insight into their ritual and ceremonial life. By 2000 years ago, that mining ceased and Native Americans do not seem to have returned to the cave in any large numbers, although other caves in the area continued to be visited for other purposes. This is the story of the archeology of Mammoth Cave and what we have learned about the first Native Americans who explored its passages.

In the bluffs along the Green River and in the coves and hollows that carve up the sandstone plateau of Mammoth Cave National Park, known as the Chester Upland, cave entrances and rock overhangs or rockshelters can be found in abundance. Where these natural shelters are suitable—dry and roomy with relatively level floors—prehistoric Native Americans, ancestors of American Indians, left numerous remains from their past activities. They built fires, made stone tools from locally occurring chert (a fine-grained, silica-rich rock commonly found in limestone formations), butchered and cooked game animals and fish that were abundant in the region, gathered numerous plants for food and to make tools or shelter, and occasionally buried deceased members of their group in these shelters. They also explored the deep passages of many caves. This is the story of what we know about these prehistoric cave explorers, how they used the deep caves, and about the archeology that has helped piece together this story.

Dr. Patty Jo Watson, who has spent a lifetime studying the prehistory of the Mammoth Cave area (Watson 1969, 1997), has called the aboriginal people who once traveled the labyrinthine routes of Mammoth and Salts Caves "the world's greatest cave explorers." This is not an exaggeration. Beginning some 4500–5000 years ago, prehistoric people began venturing into the remote depths of the large, dry passages of several caves that comprise the Mammoth Cave system. Bundling together dry river cane, weed stalks, or woody stems to make torches, they lit their way through large passages and tight crawlways leaving behind a trail of charred torch material and marks from their torches where they stoked them against the walls to keep them lit (Figs. 2.1 and 2.2). Archeologists have found evidence of this exploration upward of 6–8 km (4.7–5.0 miles) from any known natural cave entrance. Although some caves in other parts of the world were explored earlier, and perhaps used more intensively, no caves show such deep exploration and extensive use as Mammoth and Salts Caves in Mammoth

G.M. Crothers (✉)
William S. Webb Museum of Anthropology, University of Kentucky, Lexington, USA
e-mail: gmcrot2@uky.edu

H.H. Hobbs III et al. (eds.), *Mammoth Cave*, Cave and Karst Systems of the World,
DOI 10.1007/978-3-319-53718-4_2

Cave National Park. Indeed, Watson once wrote: "Virtually everywhere we have gone in Upper Salts, whether via the main passages or routes through the breakdown, we have found plentiful remains left by the aboriginal cave explorers" (1969: 3).

However, it was not just curiosity that enticed prehistoric people to explore Mammoth Cave. Beginning approximately 3000 years ago, the minerals that form in some dry passage ways, especially gypsum or calcium sulfate dihydrate, became of interest to these cave explorers and considerable effort was made to remove them. This mining ended about a millennium later for reasons unknown to us. From 2000 years ago until the first European-Americans and African-Americans began exploring Mammoth Cave in the late eighteenth century, there appears to have been little or no use of these large caves by Native Americans. Other caves in the region do show continued use, but not in the manner of Mammoth or Salts Caves. Before we go farther into the world of cave archeology, however, it is useful to set the stage above ground.

Fig. 2.1 Torch marks on a cave wall. *Photograph* by Ben Miller

Fig. 2.2 The unburned ends of torch remains, which can be found in dry caves

2.1 The Prehistory of Central Kentucky

The verdant Green River valley would have been a paradise for the first hunters and gatherers who ventured into this area during the Paleo-Indian Period. Very distinctive stone tools, identified as the Clovis artifact complex, first appeared in Kentucky at the end of the last Ice Age or the Late Pleistocene Epoch, possibly as early as 11,500 years ago. The climate was cooler and wetter than today and megafauna, such as wooly mammoths, mastodons, and giant ground sloths, still roamed North America. Some of these animals were most certainly hunted by Paleo-Indians, but the Green River valley was rich with numerous plant and animal food sources that we are more familiar with today: whitetail deer, wild turkey, fish, hickory nuts, and acorns. Several Clovis sites are known from the lower Green River valley outside of Mammoth Cave National Park, including sites in rockshelters and cave entrances. However, we have no evidence of Clovis people venturing into deep caves. What they thought of the yawning entrance to Mammoth Cave when it was first encountered, we will never know.

After the Ice Age, the Holocene or current climate period stabilized into the pattern we are familiar with today. Although megafauna became extinct, whitetail deer and other forest animals continued to be hunted. An increasingly important component of the diet during the early Holocene, or what archeologists call the Archaic Period, was fish and shellfish. In modern times, the Green River downstream of Mammoth Cave has been dredged and the river level raised by locks and dams for navigation, but formerly it teamed with a great diversity of freshwater mussels that could be easily collected from the shallow rapids. Fish such as catfish, drum, and buffalo fish also could be taken easily in the deeper pools by hand line or collected from shoals when they were spawning. Add to this diet, nutritious nuts from hickory and black walnut trees, which are common in central Kentucky, and you have the basis of an extremely healthy diet and plentiful food resources for the Archaic Period Indians who lived in central Kentucky. There is abundant, well-preserved evidence of this Archaic Period lifestyle from deeply stratified shell middens that are found along the middle and lower Green River dating between 7000 and 3000 years ago. These sites are rich with the shells of freshwater mussels, the bones of fish and mammals, carbonized nutshells and seeds, and many artifacts made of bone and stone (Marquardt and Watson 2005).

The Archaic Period Indians also were very familiar with the Mammoth Cave region, as indicated by the numerous sites dating from this time period found throughout the park. The rugged country in Mammoth Cave National

Park—narrow river valley, steep ridges, and intervening karst valleys—is not particularly conducive to preserving large open-air sites, but rockshelters and cave entrances provided commodious living spaces. In 1916, Nels C. Nelson of the American Museum of Natural History was one of the first professional archeologists to conduct work at Mammoth Cave (Nelson 1917). Although Nelson collected representative artifacts from the interior of Mammoth and Salts Caves, his primary interest was in conducting excavations in the vestibule or antechamber to Mammoth Cave. From this work, Nelson determined that considerable camp refuse was present, indicating a relatively long period of occupancy by prehistoric Indians. He concluded "that all traces of maize growing, pottery making, and the production of polished stone implements characteristic of the Mound-builder culture as a whole are entirely absent" (1917: 68). That is, the artifact assemblage was from a stone-age culture, pre-dating the invention of pottery and the construction of earth works and burial mounds. In 2003, a small archeological testing project in Mammoth Cave vestibule—part of trail construction improvements—was able to sample the camp refuse first identified by Nelson. Radiocarbon determinations on charcoal recovered from these deposits indicate that this camp refuse is at least 5380–5650 years old.[1]

It was during the Archaic Period occupation of Mammoth Cave vestibule that the first people began venturing into the deeper recesses of the cave. The two oldest radiocarbon dates from the interior of Mammoth Cave date between 4570 and 4910 years ago (one sample on torch charcoal from Audubon Ave. in Upper Mammoth and the other sample from a torch remnant found in Jessup Ave., a side branch of Ganter Ave. in Lower Mammoth). There are nearly identical early dates from Lee Cave in the Joppa Ridge portion of Mammoth Cave National Park, dating between 4470 and 4920 years ago. Lee Cave, although not connected to the Mammoth Cave system, was also explored very early, but apparently the prehistoric entrance collapsed sometime after this initial exploration and the cave was not rediscovered until modern times when a group of cavers found a much smaller entrance (Freeman et al. 1973). The earliest date for cave exploration anywhere in eastern North America is from a cave in north-central Tennessee. This cave, which has some 274 human footprints preserved in the mud floor of a remote passage, has torch charcoal samples associated with this exploration dating between 5010 and 5660 years ago.

All of these early cave dates are associated only with exploration: a few fragments of torch charcoal, perhaps some stoke marks, and, in the case of the Tennessee cave, the remarkably preserved footprints of nine individuals: a caving party composed of males, females, and possibly one adolescent (Watson et al. 2005). By 4500 years ago, prehistoric Native Americans living in the mid-continental region of eastern North America, which includes Mammoth Cave, were quite adept at traveling far into very complex caves, carrying enough torch material to last several hours, perhaps up to a full day.

2.2 The Antiquarian Interest in Caves

As American settlers began filing land patents in Kentucky, many of them as compensation for military service during the Revolutionary War, caves were potentially valuable features of the land because they were known to be sources of niter. Niter, in this case, calcium nitrate, could be derived from cave sediments and converted to saltpeter or potassium nitrate, an essential component of gunpowder. With the advent of the second war with Great Britain, or the War of 1812, the price of saltpeter skyrocketed because the shipment of foreign supplies was blockaded by the British navy. Caves in Kentucky, Tennessee, and Virginia became important sources of domestic niter. In Kentucky, leading up to and during the War of 1812, the mining of niter went from a cottage industry to an industrial-scale business. Unprecedented amounts of niter dirt were dug from many caves and large rockshelters. And with this digging, new discoveries of the prehistoric past also became a sensation.

The trade in mummies—desiccated bodies of prehistoric Indians found in the dry cave fill that was also rich in nitrate—became a secondary reason for interest in these caves, one that would help launch new business interests following the War of 1812: tourism. Because very little was actually recorded about these discoveries, the location of most of the so-called mummies is poorly known or was intentionally obfuscated to claim ownership for exhibit purposes. (The single, best source that describes the various discoveries in the Mammoth Cave area is by Meloy (1977). Much of the information related below is from Meloy's work.)

The first well-preserved body, that of a very young child, was apparently found in Short Cave in 1811, just a few miles

[1]Radiocarbon dates are usually reported as *radiocarbon years before present* with an associated counting error (for example, 4720 ± 60 rcybp). "Before present" is established as AD 1950, the year radiocarbon dating was first commonly available. To convert a radiocarbon determination to calendar years, it is necessary to calibrate the determination using a dendrochronology/radiocarbon calibration curve. Radiocarbon dates used here have been calibrated, increased by 64 years (the difference between AD 1950 and 2014, the current "present"), and rounded to the nearest 10 years to arrive at a date of "years ago." For example, the radiocarbon date of 4720 ± 60 rcybp, the sample from Mammoth Cave vestibule, calibrates to 5320–5585 years BP. Then add 64 years to the minimum and maximum range (= 5384–5649) to arrive at 5380–5650 years ago, rounded to the nearest 10 years. Because decay of the radiocarbon isotope (^{14}C) is a random process (hence the counting error), we can only say that the true date of a sample is estimated (with a 95% probability) to fall somewhere within the calibrated range.

from Mammoth Cave. However, these remains were destroyed by the miners. Upon learning of the discovery, Charles Wilkins, part owner in the mining operation at Mammoth Cave, offered a reward to the miners if they would preserve the next mummy they found. Indeed, a second mummy was found by the miners sometime between 1811 and 1813 in a small stone crypt. Wilkins collected this mummy and the artifacts that had been buried with it and took them to Mammoth Cave where they were placed in the cave and shown to visitors. The body, complete with short cut hair, long fingernails, and unblemished skin, was judged to be that of a tall female. More spectacularly, she had been buried with her wardrobe, which consisted of two deer skins and a woven textile of course fabric, a pair of woven shoes or slippers, a knapsack, and a small side bag, both also made of woven plant material. The two bags contained a variety of personal items, including a small woven cap, bird quills woven together, several hundred strings of small beads made from seeds, a string of fawn hoofs colored with red ocher, a large bird talon strung on a cord, a bear jaw strung on a cord, two rattlesnake skins, vegetable dyes, animal sinew, seven bone needles, a piece of deer skin with a hole for the thumb to be worn as protection for the hand, and two cane whistles. This mummy came to be popularly known as Fawn Hoof because of the necklace of deer hooves found with her.

After a short time on exhibit in Mammoth Cave, during which period several visitors wrote widely published descriptions of the find, a Mr. Nahum Ward of Marietta, Ohio, agreed to transport the mummy and its accoutrements to Peale's Museum in Philadelphia, in 1815. Being an entrepreneur, however, Mr. Ward took the mummy on a circuitous route, first to Lexington, Kentucky, then to Philadelphia (but by-passing Peale's Museum), and then on to Boston, stopping frequently and charging admission to see this amazing discovery. The mummy was finally deposited with the American Antiquarian Society in Worcester, Massachusetts, in 1817, but only after Wilkins and the Society threatened to press charges against Ward. Ward also published a very fanciful adventure story of his exploration into the many cave passages and *his* finding of the mummy. This description was picked up and published widely, including by international newspapers. Ward may have been a bit of a scoundrel, but he was responsible for generating more popular interest in Mammoth Cave than almost anyone else in his day.

At least two other mummies were found in Short Cave in 1814. One ended up at Scudder's Museum in New York, later purchased by P.T. Barnum, but was subsequently destroyed by fire in 1865. The second may have been sent to a museum in Cincinnati, but it was also reported to have been destroyed in a fire prior to 1844. One mummy was also purported to have been found by saltpeter miners in the Audubon Ave. area of Mammoth Cave in 1814. They supposedly concealed the mummy by covering it with rocks with the intent of recovering it at some later time. For some reason this was never effected, but in 1840 the hotel manager at Mammoth Cave, having learned of this earlier discovery, relocated the remains. However, the body was so badly crushed by the rocks that it was of little value for display. It is not clear what happened to these remains. If this story is true, it was the only mummy actually found in Mammoth Cave by saltpeter miners.

Not far from Mammoth Cave, in the Flint Ridge portion of Mammoth Cave National Park, Salts Cave was also discovered to contain the desiccated remains of a prehistoric Indian. Although Salts Cave is a large, dry cave system, it was not mined for saltpeter, and these mummified remains were not found until 1875. Rather than being buried in the dry sediments, these remains, of a boy approximately nine years old, were found curled up on a rock ledge at a considerable distance from the entrance. Analysis of the remains indicated that he died about 2000 years ago, possibly from internal hemorrhaging as indicated by excess blood in his chest cavity. Hemorrhaging of this type can be caused by a blow to the chest or back (Robbins 1997: 144). It is easy to imagine that this young boy died after a fall in the cave. He was found in a side passage a short distance from the bottom of a steep-sided valley in Upper Salts, an area now known as Mummy Valley. Like the bodies found before him, he was traded among various cave owners and displayed to the public with many fanciful stories of his death. Represented to the public for many years as a young Indian girl or an Indian Princess, later examination of the body clearly indicated that this individual was a boy.

While desiccated human bodies were rare discoveries in the cave, but made for spectacular press, the environmental qualities of Mammoth and Salts Caves preserved many other types of archeological remains. Torch remains could be found in abundance strewn over the floor, some still bound together with strips of bark to form a torch bundle. In places, the cave walls and breakdown are blackened with hundreds of charcoal marks where burning torches had been stubbed into the surface. Soot covers large portions of cave passage from the countless prehistoric fires. Fragments of twined cordage and braided strips of bark and, occasionally, fragments of woven textiles and woven bags can also be found (Fig. 2.3). In particular, interesting are the woven slippers, masterfully made of plant fiber, some decorated with fringes or tassels, and having a drawstring to pull them tight around the ankle. Numerous worn-out examples of this footwear were cast aside by aboriginal cavers (Fig. 2.4). It was clear even to the earliest European-American explorers, however, that the cave had been used for more than just exploration. Long poles that had been hauled into the cave were found still wedged into piles of rock enabling access to high ledges or intersecting side passages. Rock cairns, apparently built

Fig. 2.3 Examples of woven fabric made from twined plant fiber. Pending

Fig. 2.4 An example of a woven slipper. Slipper is *upside down* with the heel to the *left*, which is worn through. The fiber tied around the *middle* was a last ditch effort to keep it on someone's foot. Scale in centimeters. *Photograph* by Charles Swedlund

for the same purpose, could be found along many passages. Digging sticks, their ends worn smooth from use; mussel shells, their edges ground down from scraping minerals off the cave walls (Fig. 2.5); large gourd bowls (Fig. 2.6) and, occasionally, wooden bowls were found throughout the passages visited by Indians. Gypsum crust, which normally occurs as thick sheets that exfoliates from the cave wall and as spectacular flower-like crystals (Fig. 2.7), has been battered, crushed, and scraped from the walls (Fig. 2.8) where the cave contains prehistoric archeological materials.

Fig. 2.5 A mussel shell used to scrape minerals from the wall. *Note* edge damage from scraping. Scale in centimeters. *Photograph* by Charles Swedlund

Fig. 2.6 A complete gourd bowl. The bowl is made from the *bottom* part of a large bottle gourd. *Photograph* by Charles Swedlund

Although many people speculated about the possible uses for gypsum by the prehistoric Indians who mined it, it was at least quite obvious that they went to great lengths to obtain it.

Undoubtedly through the nineteenth and early twentieth centuries, many prehistoric finds were collected by cave visitors as curios and keepsakes. A few collections were made for scientific purposes and did end up in museum collections. Perhaps the two best-known collections are those made by Frederick W. Putnam, for the Peabody Museum of Natural History at Harvard University and Col. Bennett H. Young, a Civil War Confederate officer and prominent Louisville, Kentucky, lawyer. Young's private collection was eventually acquired by the Museum of the American Indian, Heye Foundation (now part of the National Museum of the American Indian, Smithsonian

Fig. 2.7 Gypsum crust in its natural state. *Photograph* by Charles Swedlund

Fig. 2.8 Gypsum crust that has been battered and scraped. The *dark* staining is from historic soot and dust. Scale in centimeters. *Photograph* by Charles Swedlund

Institution). More disturbing than the occasional collecting of artifacts over the years, however, are reports that early cave guides gathered prehistoric debris in the cave to build fires that would light up the large passages for the benefit of visitors.

2.3 Scientific Archeology

Not surprisingly, it was the discovery of another prehistoric mummified body that generated renewed interest in Mammoth Cave. In 1935, preparations were underway to acquire and donate the lands encompassing Mammoth Cave and other large caves in the region to the federal government as a National Park. In anticipation of eventual National Park status, work was already underway using Civilian Conservation Corps labor to enlarge and improve the trail system through portions of the cave. Two local men, Grover Campbell and Lyman Cutliff who had been hired to keep an eye on the young CCC workers in the cave, were exploring a high ledge near the trail when they unexpectedly discovered the remains of another unfortunate aboriginal caver. This body lay partially crushed under a massive rock.

Because the trail construction was under federal management, the National Park Service took charge of the discovery. A Park Service archeologist by the name of Alonzo W. Pond was sent to investigate. Over the next several months, a scaffold was constructed and a system of pulleys and tethers put into place to lift the six-ton rock and remove the remains from underneath it. Pond documented the remains very thoroughly, but the recovery was also done for maximum publicity. Pond wrote a richly illustrated popular article that appeared in *Natural History* magazine (Pond 1937), and for the rest of his career the story of the discovery and removal of the Mammoth Cave mummy—illustrated with lantern slides—became one of his stock lectures. The Park, also wishing to cash in on the publicity, placed the remains of the prehistoric miner and his artifacts in a glass case to be exhibited in the cave. The remains are no longer on display, however, and at the request of Native American groups, the body remains at a secret location in the cave.

Material associated with the Mammoth Cave mummy was radiocarbon dated to 2380–2790 years ago. He was estimated to be approximately 45 years old at the time of death and would have stood about five feet, three inches tall (160 cm). He wore a woven plant-fiber blanket around his hips, knotted at the waist. He also had a small mussel shell on a cord around his neck. This was described as an amulet by Pond, but was probably just a convenient means of carrying his scraping tool. His torch lay near his body, along with a large gourd fragment. Archeologists surmise that he was on the ledge digging in the sediment for gypsum crystals when he either undermined or otherwise dislodged the large rock he was kneeling under so that it rolled over on him. He may have been trying to scramble out from under the rock as it moved. His right arm snapped above the elbow where he tried to brace himself against the weight; his left rib cage was crushed, and his skull was fractured. The soft sediment and other smaller rocks kept his body from being further damaged, but he would have died quickly from his injuries (Robbins 1997: 138).

Unlike the human remains unearthed by saltpeter miners in Short Cave, which appear to have been intentional burials, the Mammoth Cave and Salts Cave mummies were accident victims. Clearly, prehistoric cave exploration and mining could be dangerous. By the mid-1950s, archeologists had collected a range of perishable material found in these caves, and thought that they were due to the mining of gypsum, but very little systematic documentation had been made of these cave sites nor had any specific analyses been performed on the range of ancient materials present within them.

In 1957, the Cave Research Foundation (CRF) was formed to promote the exploration and scientific study of caves and karst environments in the Mammoth Cave region. This multi-disciplinary organization had a difficult time finding an archeologist who was willing to take on the task of systematically documenting prehistoric human activity in Mammoth and other caves nearby. It seems that most archeologists are not too fond of working underground without sunshine or warm breezes. The work eventually fell to Patty Jo Watson, then a young Ph.D. in archeology whose primary fieldwork was in Southwest Asia (Iraq, Iran, and Turkey) but who was also a caver. Working primarily in Salts Cave, because the archeological situation there had not been so greatly affected by historic activities, as was the case in Mammoth Cave, she teamed up with other CRF scientists and archeological colleagues to build a long-term, interdisciplinary research program, known as the Cave Research Foundation Archeological Project. This project, while always relatively low key, has been continuously operating for more than 50 years (beginning in 1963, the first year of fieldwork). In addition to Watson's two major publications (1969, 1997), she and her colleagues and students have published dozens of scholarly articles and reports on many facets of Mammoth Cave archeology.

Watson conducted excavations in the vestibule of Salts Cave, where—as was the case in Mammoth Cave—there are significant ancient cultural deposits (Fig. 2.9). However, there are also important differences. Salts Cave entrance is small and descends steeply into the vestibule. Very little light enters the cave, and during heavy rains water pours through the entrance because it is at the bottom of a large sinkhole. Although debris in the midden consisted of the usual kinds of material found in domestic sites—animal bones, charred plant remains, and debris from making stone tools—it also contains a large amount of human bone, yet this was not a normal burial site. Approximately 2000 fragments of human bone were recovered intermixed in the deposits, many of them burned, some with cut marks, and even a few that had been fashioned into tools or artifacts. The 2000 or so bones represented at least 41 different individuals that ranged in age from fetal remains and infants to fully adult males and females (Robbins 1997). Although Salts Cave may have been used as a temporary camp from time to time, it was also an important mortuary facility where defleshing and cremation was part of the burial ritual. Salts Cave vestibule was occupied for a shorter period of time than was the entrance to Mammoth Cave. Radiocarbon dates from Salts Cave vestibule excavations span the time from 2520 to 4060 years ago, from Late Archaic to Early Woodland time periods.

Watson also systematically documented the range and intensity of prehistoric activity in Salts Cave. She sampled numerous kinds of material to understand the time depth

Fig. 2.9 Salts Cave vestibule excavations in progress. Cave Research Foundation Archaeological Project. *Photograph* courtesy of the William S. Webb Museum of Anthropology, University of Kentucky

represented by these artifacts. The upper levels in both Mammoth and Salts Cave interiors, unlike most archeological situations, are completely dry. Sediments that would normally be laid down in stratified layers as a gage to identifying the relative ages of cave activity, have not been deposited in these upper levels since the Pleistocene. The surface we walk on today in the cave is the same surface that prehistoric Indians walked on in the past, making it difficult to know the age of any material without submitting it to radiocarbon dating. Watson did just that, submitting more than 40 samples of various kinds to obtain radiocarbon determinations on the full range of activities in both Mammoth and Salts Caves. The surprising part, with the exception of the two early dates for exploration of the cave noted earlier, is that all of the dates range between 2220 and 3450 years ago. The period of aboriginal mineral mining appears to have begun at the end of the Late Archaic Period and occurred most intensively during the Early Woodland Period. The Early Woodland Period, which archeologists typically date from 2200 to 3000 years ago, was an important time of change among prehistoric societies in eastern North America.

2.4 The First Farmers

The origins of agriculture, when hunters and gatherers began to settle into more permanent communities and take up cultivation of several indigenous plants (and, in some parts of the world, animals as well) to supplement their diets, are considered the most fundamental reorganization of human society following the evolution of biologically modern humans. The creation of agricultural economies did not take place until the Holocene, first in western and eastern Asia and then slightly later in South and North America, but it was an independent process in only a few places in the world. Eastern North America was one such location, and Mammoth and Salts Caves contain the best evidence for this transition in the Americas.

The stable cave environment preserves organic material, as we have learned, in an exceptionally good state. Among the many prehistoric items left in the cave, one of the most numerous, other than torch material, is dried human excrement, or what archeologists call paleofeces. Literally thousands of these paleofeces can be found throughout dry portions of the cave. There is good reason why they are so numerous. Even though gypsum (calcium sulfate dihydrate) forms in many portions of the cave, other sulfate minerals also form under the right conditions. They are less common, but are abundant where found. Two of these minerals are mirabilite (sodium sulfate decahydrate) and epsomite (magnesium sulfate heptahydrate). Medicinally, both these minerals are very effective saline laxatives. Hence, archeologists think that another purpose for visiting these caves, in addition to mining gypsum, was to obtain these salts, which were consumed in the cave for their laxative effect.

Watson's colleague, Richard Yarnell, examined many of these paleofeces to determine dietary content. Fragments of nutshell indicated that hickory nuts were an important constituent of the diet, but more importantly, for the story of early agriculture, were the seed remains of several indigenous plants: sumpweed or marshelder (*Iva annua*), sunflower (*Helianthus annuus*), goosefoot or lambsquarters (*Chenopodium berlandieri*), and maygrass or Carolina canarygrass (*Phalaris caroliniana*). The hard seeds of these plants are abundant in the paleofeces, and at least three of them showed signs of early domestication. Domestication of plants bearing edible seeds is usually indicated by an increase in seed size beyond that present in naturally occurring stands. For example, as a wild annual, sunflower seed shells (achenes) are rarely longer than 7 mm. Achenes consistently larger than 7 mm indicate that the plant is undergoing cultivation resulting in morphological changes. Later genetic changes induced by modern breeders have created the monstrous single-headed flower we grow today, but modern wild varieties of sunflower have several much smaller seed heads, as would also have been the case for ancient ancestors of the first domestic sunflowers.

The cave data are an important contribution to documenting the origins of agriculture in eastern North America. They show that this process was well underway by 3000 years ago, and that the seeds from domestic plants made up a substantial portion of the diet, not just an occasional or isolated occurrence in a few paleofeces. In addition to plants with edible seeds, Native American farmers were also growing bottle gourds (*Lagenaria siceraria*) and gourd-like squashes (*Cucurbitae pepo*) for their hard-shelled fruits that were used as containers by the ancient cavers. The seeds of these plants could be and were eaten, but the fleshy varieties of *C. pepo,* like pumpkin, zucchini, and scallop squashes, with which we are more familiar today, were developed later. The first domesticated squashes were more similar to the ornamental gourds (*C. pepo* var. *ovifera* and *C. pepo* var. *verrucosa*) common in modern markets in September and October.

2.5 Ritual Use of Caves

What compelled Early Woodland cave explorers to spend so much time and effort in these large, dry caves collecting minerals? The mineral salts have medicinal uses that may explain the desire to obtain them, but gypsum has no medicinal value or practical uses that archeologists have been able to identify. If ground and mixed with water, gypsum makes a good white paint or plaster (essentially, plaster of Paris), but gypsum residue has never been identified on any archeological remains from this time period (for example, as wall plaster or as paint on pottery or other artifacts). So what use did it have? The paleofeces are a key to answering this question.

In an ingenious study by Watson and her colleagues (Sobolik et al. 1996), they were able to extract residual steroids from the paleofecal specimens. The relative amounts of testosterone (T) and estradiol (E_2) can be used to estimate the biological sex of the defecator. That is, males typically have much higher hormonal ratios of testosterone to estradiol than do females. Males and females may overlap in these ratios, but only males have extremely high quantities of T to E_2. Twelve fecal specimens (six each from Mammoth and Salts Caves) were subject to this analysis. All twelve samples were determined to have high T: E_2 ratios in the range we would expect for mature males. Recall that both cave accident victims were also male: a middle-aged man and a young boy verging on adolescence. Mineral mining appears to have been an exclusively male activity.

Caves have special significance in many cultures. They can be foreboding, a perpetually dark underworld, and cave formations (speleothems—e.g., stalactites) can be exotic, fantastic creations not found in the everyday aboveground world. When you enter a cave, you descend into the unknown. When you emerge, you return to a world of light. One cross-cultural use of caves and cave-like structures is for secretive or exclusionary practices, including male initiation ceremonies: the rite of passage for young boys into adolescence and adulthood. I have suggested that Mammoth and Salts Caves were used to seclude young boys from the rest of the society and to initiate them into the fraternity of adult men (Crothers 2012). Medicinal salts used as laxatives in the cave, physically purged the initiate, a literal and symbolic act of cleansing, preparing him for the next stage of life. Typically, in initiation ceremonies, after a period of seclusion during which initiates learn ritual knowledge they will need as adults, they are reintroduced to society wearing new clothing, body decoration, or exhibiting bodily mutilation signifying their new status. The reintroduction ceremony may include performance of ritual dances or singing songs learned while in seclusion. Gypsum could have played a role as body paint or as some other forms of personal adornment used in such ceremonies. Fantastic crystal forms of gypsum may have had significance as amulets, or powdered gypsum may have been imbued with mystical power, an important substance carried throughout life and occasionally replenished when necessary. The important point is that gypsum, found only in certain cave contexts, identifies the person in possession of it as having obtained that status, as being a full adult member of society.

Finally, we need to explain why this interest in gypsum and cave ritual became so important during the Early Woodland Period. Recall that prehistoric people had been exploring deep into caves at least 4500 and possibly as early as 5600 years ago, a full 1000–2100 years before there is any interest in mining gypsum or collecting other sulfate minerals in any archeologically visible way. Also note that mineral mining was not exclusive to Mammoth and Salts Caves. Archeologists have identified at least five other caves in Kentucky and Tennessee that have evidence of gypsum mining. All the mining activity in these caves has been dated to the Early Woodland Period. It is not a coincident that this mining—the ritual use of caves—occurs contemporaneously with the beginnings of intensive plant cultivation in eastern North America. Even though all cultures observe some rites of passage, male initiation ceremonies are more common, or more aggressively practiced, in agricultural societies. Agricultural subsistence requires a fundamental reorganization of society to be successful. It is a technological revolution (new tools and techniques to produce crops), an economic revolution (new forms of property rights that specify who owns the product of agricultural labor and perhaps more importantly, the surplus food produced), and a social revolution (there are those who labor and those who control that labor pool). Ceremonies, like initiation rites, strengthen social ties. They create communities of similarly aged cohorts—sodalities in anthropological terms—that are important for political and economic integration. Sodalities cross-cut kinship ties, the latter being the most fundamental form of social organization, to create non-kin groups spanning multiple villages or extended families. Sodalities often have specific purposes: defense, warfare, or simply to provide a larger labor pool for economic and social activities. Agriculture is an economic endeavor that requires specific communal property rights, defense of those rights, and sustained communal labor.

2.6 The End of Cave Mining

The intensive period of mineral mining appears to end rather abruptly around 2200 years ago, based on the available radiocarbon dates. Why the mining ended is not clear, but gypsum was not a scarce commodity in these caves. Perhaps one had to travel farther and work a little harder, but Mammoth and Salts Caves are enormous places and pristine gypsum can still be found in remote passages. Perhaps the mystery of the cave had run its course. The end of the mining, which interestingly enough corresponds with the end of the Early Woodland Period, is followed by an intensive period of mound building and earthwork construction throughout much of eastern North America and best known from the Hopewell Culture (1500–2200 years ago) in the middle Ohio River Valley. I suspect new forms of ritual, possibly centered at these impressive earthwork sites, began to eclipse caves as ritual places.

Caves continued to be used in other important ways, however, during and after the Early Woodland Period (Crothers 2009; Crothers et al. 2002). Perhaps the most widespread use was as mortuary sites. As discussed earlier, Salts Cave vestibule was used for special kinds of mortuary processing. In many areas of eastern North America, pit caves—caves with vertical openings to the surface—were also used as mortuary sites. Bodies were dropped into these shafts, often with personal artifacts like smoking pipes and shell beads (Crothers and Willey 2009). While a few of these pit cave sites are known from the Mammoth Cave area, it was an especially common use of caves in southwest Virginia during the Late Prehistoric Period (400–1000 years ago).

Another quite widespread activity in caves was drawing images on the walls and sometimes on the ceilings and floors. While prehistoric petroglyphs and pictographs can be found in Mammoth and Salts Caves, they are not common and seem to be isolated, individually inspired acts of expression (Fig. 2.10). The drawing of images, however—including

Fig. 2.10 Petroglyph in Mammoth Cave. Cross-hatching like this is a common design. *Note* that the prehistoric torch marks are superimposed on the drawing. Scale in centimeters. *Photograph* by Charles Swedlund

geometric, animal, human, and fantastical creatures—was an exclusive activity in some caves. One of the best-known sites is called Mud Glyph Cave because the images were drawn into soft mud that coats many of the walls (Faulkner 1986). The intensive use of caves as locales for creating rock and mud art is almost exclusively a late prehistoric phenomenon (ca. 400–1000 years ago).

The prehistoric use of caves as mortuary sites and art galleries is not evident in Mammoth or Salts Caves. Archeologists do not know why, but apparently they were not deemed suitable for these purposes by those Native Americans who continued to live in the region. Perhaps the history of extensive mineral mining and stories of the ill-fated miners who lost their lives underground made Mammoth and Salts Caves places to avoid. Even though archeologists have learned much about the world's greatest prehistoric cave explorers, the caves themselves still hold many secrets.

References

Crothers GM (2009) North American cave archaeology. In: Palmer AN, Palmer MV (eds) Caves and Karst of the USA. National speleological society, Huntsville, AL, pp 413–416

Crothers GM (2012) Early Woodland ritual use of caves in eastern North America. Am Antiq 77(3):524–541

Crothers GM, Faulkner CM, Simek J, Watson PJ, Willey PS (2002) Woodland cave archaeology in Eastern North America. In: Anderson DG, Mainfort RC Jr (eds) The Woodland Southeast. University of Alabama Press, Tuscaloosa, pp 504–525

Crothers GM, Willey PS (2009) Mortuary caves and sinkholes in the Interior Low Plateaus and Southern Appalachian Mountains of the Eastern United States. In: White WB (ed) Proceedings 15th international congress of speleology. International Union of Speleology, 1(1): 74–79

Faulkner CH (ed) (1986) The prehistoric Native American art of Mud Glyph Cave. University of Tennessee Press, Knoxville

Freeman JP, Smith GL, Poulson TL, Watson PJ, White WB (1973) Lee Cave, Mammoth Cave National Park, Kentucky. Bul of the Nat Speleological Soc 35(4):109–126

Marquardt WH, Watson PJ (eds) (2005) Archaeology of the Middle Green River Region. Florida museum of natural history, University of Florida, Gainesville, Kentucky

Meloy H (1977) Mummies of Mammoth Cave. Micron Publishing, Shelbyville, Ind

Nelson NC (1917) Contributions to the archaeology of Mammoth Cave and vicinity, Kentucky. Anthropological Papers of the American Museum of Natural History, 22(1)

Pond AW (1937) Lost John of Mummy Ledge. Nat Hist Mag 39:176–184

Robbins LM (1997) Prehistoric people of the Mammoth Cave area. In: Watson PJ (ed) Archaeology of the Mammoth Cave Area. Cave Books, St. Louis, pp 137–162

Sobolik KD, Kristen J. Gremillion KJ, Whitten PL, Watson PJ (1996) Sex determination of prehistoric human paleofeces. Amer J of Physical Anthro 101:283–290

Watson PJ (1969) The Prehistory of Salts Cave, Kentucky. Report of Investigations No. 16, Illinois State Museum, Springfield

Watson PJ (ed) (1997) Archaeology of the Mammoth Cave Area. Cave Books, St. Louis, Originally published 1974 by Academic Press, NY

Watson PJ, Kennedy MC, Willey P, Robbins LM, Wilson RC (2005) Prehistoric footprints in Jaguar Cave. Tennessee J Field Archaeol 30 (1):24–43

The History of Mammoth Cave

3

Colleen O'Connor Olson

Abstract

Mammoth Cave is a long cave with a long, colorful history. This history includes all kinds of interesting characters including the frontier entrepreneurs who ran a saltpeter mining operation in the cave and then entered the strange and risky new business of underground (literally) tourism at a time when tourism above ground was rare. It also includes the adventurous travelers who ventured into the cave by lantern light with guides who were slaves. The guides' lives differed from those of most slaves. Instead of working in the fields, they spent hours walking through the cave talking with wealthy travelers including scientists, writers, and European nobility, yet they still had to deal with the harsh realities of being slaves, including having their children sold away. The cave's uses include a tuberculosis hospital, scientific research, a sleep study in which the father of sleep science and a graduate student lived in the cave for a month, inspiration for art, early cave photography, and a national park.

3.1 The Early Days

Mammoth Cave was discovered at least twice. American Indians explored and used the cave as far back as 5000 years ago, but they did not leave us a written account of how they found it, so we must turn to the modern rediscovery. (For the story of the cave's first explorers, see Chap. 2.)

Passing through what is now Kentucky, French hunters traveled on the Green River near Mammoth Cave in search of game in the early and mid-1700s. By the 1760s, American long hunters, so named because their hunting trips lasted several months, came down Green River in quest for game.

Hunters sometimes ventured into caves to look for saltpeter, the main ingredient in black gunpowder, but the French and long hunters apparently missed Mammoth Cave in their search for game and saltpeter with which to shoot it. Supposedly a hunter eventually discovered the cave (a legend), and saltpeter miners came soon after (a true story).

The legend says that in 1809, a young hunter named John Houchin shot and wounded a bear that ran into a cave, and John followed it. The two of them discovered Mammoth Cave together, but John took all the credit. At least that is the most common version of the story. I have never heard the bear's family's side of the story.

Mining may have started as early as the 1790s. A 1799 survey of the property mentions two unnamed saltpeter caves, referring to Mammoth's Historic Entrance (Fig. 3.1) and nearby Dixon Cave, implying that someone was already mining or at least saw the potential (Warren County 1800).

Abridged instructions for how to mine saltpeter from Mammoth Cave:

1. Shovel saltpeter-rich dirt into wooden vats in the cave.
2. Pour water piped from the entrance waterfall onto the dirt to leach out the saltpeter. The water and saltpeter drains into troughs beneath the vats.
3. Pump the water and saltpeter through pipes to kettles outside the entrance.

C. O'Connor Olson (✉)
Cave Research Foundation, Cave City, KY, USA
e-mail: colleen@scrtc.com

H.H. Hobbs III et al. (eds.), *Mammoth Cave*, Cave and Karst Systems of the World,
DOI 10.1007/978-3-319-53718-4_3

Fig. 3.1 Historic Entrance. Prior to the 1920s, the Historic Entrance was the only known entrance to Mammoth Cave; today it is one of 27 entrances

4. Add potash (a source of potassium needed to make gunpowder) to the solution and boil the water away in kettles.
5. Ship the saltpeter to Delaware where the DuPont Company combined it with sulfur and charcoal to make gunpowder.

Important note: DuPont got out of the gunpowder business in the 1970s and mining is not allowed in National Parks, so do not follow these instructions.

Modern visitors to Mammoth Cave see what is left of the vats, pipes (Fig. 3.2), pumps, and spent dirt in cave rooms called the Rotunda and Booths Amphitheater, about one-third of a mile into the cave. The cave's oldest map, the "Eye-Draught," drawn before 1811, extends about two miles (3.2 km) from the entrance—way past the mining operation. Perhaps the owners anticipated eventually mining that far into the cave, or maybe they wanted to lead investors to believe the mining operation was more extensive than it was. DuPont's founder E.I. duPont and prominent physician and signer of the Declaration of Independence, Dr. Benjamin Rush had copies of the map (Hill and DePaepe 1979), and it appeared in the 1853 edition of Thomas Jefferson's *Notes on the State of Virginia* (Jefferson 1853). Jefferson died 27 years earlier, so it is questionable whether or not he had seen the map or knew of Mammoth Cave.

However, maybe he did know about the cave. Jefferson wrote to Pierre duPont de Nemours of the DuPont Company on February 12, 1806,

> …the supply of saltpeter which the Western country can furnish is immensely beyond what had been expected. A single cave is known which would supply us for the whole term of a war (Jefferson 1806).

Was Jefferson referring to Kentucky and Mammoth Cave? Kentucky had other large saltpeter caves, as did other states in the "Western country."

In 1813, Ebenezer Meriam visited the saltpeter mining operation and wrote about it in 1844. Memories get cloudy after 31 years, but he provided first-hand information found nowhere else. He wrote that 70 blacks (probably slaves, but he is not specific) worked by the light of lard-burning lamps. This was labor-intensive work: a bushel of dirt yielded only three to five pounds of saltpeter. By 1815, the operation ceased to be profitable. With the War of 1812 brought to a close, saltpeter from the East Indies became cheaper than saltpeter mined in the USA (Meriam 1844).

On April 20, 1810, the Richmond, Virginia *Enquirer* published "The Subterranean Voyage, or The Mammoth Cave, Partially Explored," the earliest known article on the cave and the first known use of the name Mammoth Cave. The anonymous writer mentioned the saltpeter operation, but he focused more on the natural cave. He called the Rotunda

> …one of the most sublimely beautiful and picturesque amphitheatres in the world…The most elaborate effort of the pencil would fail to do justice to the rich scenery and varied drapery with which the senses are delighted (Anonymous 1810).

At this time, real tourism had not started at the cave, but curious people already came wanting to see it. The writer from the *Enquirer*, and others like him, helped make Mammoth Cave famous; they introduced readers across the USA and abroad to a cave they otherwise never would have heard of. In addition to descriptions of large, spectacular chambers, many early articles mentioned something else that grabbed readers' attention—a mummy.

3.2 How a Mummy Helped Make Mammoth Cave Famous

About 1813, Mammoth Cave owner Charles Wilkins heard that saltpeter miners found a mummified infant at Short Cave, a saltpeter cave his brother-in-law, Peyton Short, owned a few miles from Mammoth. Wilkins wrote that he

> …hastened to the place; but, to my mortification, found that, upon its being exposed to the atmosphere, it had fallen into dust, and that its remains, except the skull, with all its clothing, had been thrown into the furnace. I regretted this much, and promised the labourers [sic] to reward them, if they would preserve the next subject for me. About a month afterwards, the present one was discovered, and information given to our agent at the Mammoth Cave, who sent immediately for it, and brought and placed it there, where it remained for twelve months (Wilkins 1817).

The workers found the mummy, an adult woman, seated in a stone coffin with many grave goods including "splendid head dresses," strings of beads, whistles, thread, horn and bone needles, and a necklace strung with fawn hoofs (Meriam 1844), from which she got her modern name—Fawn Hoof.

Fawn Hoof left Mammoth Cave after a year because Nahum Ward, a land speculator from Marietta, Ohio, came to the cave and with Wilkins' permission, and took Fawn Hoof and her "apparel, jewels, music, etc." with him (Ward 1819).

Ward wrote to the American Antiquarian Society in Worchester, Massachusetts about the mummy, and the society wrote to Wilkins for details on the find. After being displayed at the Antiquarian Society and elsewhere, Fawn Hoof went to The Smithsonian Institution in Washington D.C. in 1914, where her remains are today, though not on display.

Fig. 3.2 Saltpeter Pipes. Miners hollowed out tulip poplar logs to make water pipes for the saltpeter operation

Even with Fawn Hoof gone, cave visitors saw torches, gourds, and other fascinating ancient artifacts. Some visitors realized American Indians left these antiquities, but others could not believe that Indians had made such fine things. In her 1837 book *Society in America*, British sociologist Harriet Martineau wrote,

> It is supposed that this cave was made use of by that mysterious race which existed before the Indians, and of which so many curious traces remain in the middle States of the West; a race more civilized, to judge by the works of their hands than the Indians have ever been; but of which no tradition remains (Martineau 1837).

Today it is no mystery that American Indians used Mammoth and Short caves and left the artifacts that Martineau saw (for more archaeology, see Chap. 2).

3.3 The Show Cave

Early writers and Fawn Hoof gave Mammoth Cave publicity that led to a new use for Mammoth and eventually other caves—becoming a show cave.

To be fair, Mammoth Cave was not the world's, or even America's, first show cave. Virginia's Grand Caverns had guided tours in 1806, and Baumannshöhle Cave in Germany had its own tour guide in 1668. But prior to the Industrial Revolution in the early 1800s, few people could afford to travel. Around 1810, when curious people first came to see Mammoth Cave for fun rather than work, it was quite difficult to get there. Eventually, the country was soon to have better roads, railroads, steamboats, and (at least for some people) more money and leisure time, making tourism easier.

At first, tourism at the cave was slow. Even calling it "tourism" may be an exaggeration; early visitation consisted of curious travelers showing up, sleeping in saltpeter manager Archibald Miller's cabin, and having him show them around the cave.

In 1838, Franklin Gorin, a lawyer from Glasgow, Kentucky, saw potential to make money from tours and bought the cave. Gorin brought his slave Stephen Bishop to the cave and leased two others, Mat and Nick Bransford (more on all of them below), to be full-time guides. Mammoth Cave soon became a genuine tourist attraction.

Ralph Seymour Thompson's 1879 book, *The Sucker's Visit to Mammoth Cave,* gives modern readers a good (and amusing) look at post-Civil War tourism at the cave. But first, some useful trivia—Thompson was not implying that only gullible people paid to see the cave. In the 1800s, Thompson's home state of Illinois was nicknamed the Sucker State. The name may have come from the practice of sucking water through a reed from crayfish holes, or comparing migrant workers arriving by boat on the Mississippi River to sucker fish. Either way, it is clear why modern Illinoisans prefer "The Land of Lincoln."

Thompson paid $3.00 for a night in the hotel and four meals, $2.00 for the tour called the Short Route, and $3.00 for the Long Route (Thompson 1879). For comparison, in 1992 the Historic Tour, somewhat similar to the short route, cost $3.50.

The four-hour Short Route covered passages on the near side of Echo River (Hovey 1880). More energetic visitors could opt for the nine to twelve-hour Long Route that crossed Echo River by boat (Fig. 3.3), followed Silliman Avenue to the Snowball Room, and continued up gypsum-encrusted Cleveland Avenue to a pit called the Maelstrom, near what is today Carmichael Entrance (blasted in the 1930s, the Carmichael Entrance gives modern visitors quick, easy access to that side of the cave). According to *A Guide Manual to the Mammoth Cave of Kentucky* (Wright 1860), the walk to the Maelstrom was nine miles (14 km) from the entrance; it is actually closer to five miles (eight km), but the low lantern light, stooping, boating, and rough trails convinced visitors they had walked almost twice as far.

Early guides discussed geology, animals, and history with visitors, but they tended to focus more on the sights and the adventure. (You're in the bowels of the earth. It's eternally dark.) They pointed out many of the landmarks visitors see today: the remains of the saltpeter operation, the illusion of a starry sky at Star Chamber, and Bottomless Pit—which by lantern light really does look bottomless. Early visitors also mentioned shapes of figures in limestone that are lost to history—they are still there, we just do not know where. One 1852 visitor saw Napoleon on his horse and a large catamount (cougar) near Giant's Coffin. Somewhere in Gothic Avenue there is a flying Indian, an elephant's head, and a deer.

Early visitors saw the cave by lantern light. Old accounts often mention lanterns but do not give details about them. Because lanterns and similar lamps were common in the 1800s, writers probably did not think they were worth describing any more than a modern writer would describe the LED light bulbs currently used in the cave.

Fortunately, many of the old Mammoth Cave lanterns still exist in the cave, park curatorial, and private collections, as well as in old photographs. The basic design is typical of

Fig. 3.3 Boat on Echo River. Cave visitors listening to a horn echo on Echo River

Fig. 3.4 Lantern. This abandoned 1800s era lantern still sits in Mammoth Cave

lanterns and lamps going back to ancient times: a wick (or in the case of most cave lanterns, two wicks) comes out of a fuel vessel, is lit, and the fuel burns. Unlike most lanterns, Mammoth Cave lanterns had no globe. That posed a problem at the breezy entrance. But the risk of having the lantern blow out may have been a fair trade; glass was fragile, expensive, and susceptible to getting sooty. The specific design is apparently unique to Mammoth Cave. Rick Olson (park ecologist) has shown the lantern to historians, antique dealers, and tinsmiths, and has searched the Internet, but he has not found the exact design outside of the Mammoth Cave area (Fig. 3.4).

Legend has it that cave guides fueled the lanterns with chicken fat from the hotel kitchen, making the guides and their lanterns popular with the local dogs. We really do not know what fuel they burned. Whale oil, lard oil, and kerosene were common lantern fuels in the 1800s, but records do not indicate their use. The only references to fuel are records mentioning purchases from the American Cotton Oil Company in 1904 and 1905 and a bill for "J.V.I. Winter White Miners Oil" from the American Cotton Oil Company dated April 20, 1917 (Olson 2008). The park burned kerosene for most of the twentieth century and uses paraffin oil for lanterns today.

Fig. 3.5 Bottomless Pit. By lantern light, Bottomless Pit could be as deep as the guide wanted it to be. Modern cave visitors are harder to fool; electric lights and a laser measurer reveal the pit is only about 100 feet deep

Lanterns gave off enough light to see the trail and the immediate area, but the light only carried so far, certainly not down deep pits. Cave guide Mat Bransford used a different light at Side Saddle Pit in the 1860s.

> 'Careful! Pit on the right.'Of course we all crowded to the edge, and tried to pierce the thick darkness, but it was useless till Matt (sic) dropped one of his lights into it. The paper fell, blazing, whirling round, casting a lurid light on the rugged walls as it descended, until at last it rested – apparently but a taper – on the bottom. We involuntarily drew back a little from the edge. It looked a long way down to that little blaze.'How deep is it, Matt?''Ninety feet, and twenty across at the widest part' (Thompson 1879).

The tour continued to Bottomless Pit (Fig. 3.5), which Mat claimed to be 175 feet (53 m) deep as he dropped a burning paper down it. The pits are actually 42 ft (13 m) and 106 ft (32 m) deep, respectively, but with just lanterns and fiery paper to see with, they could be as deep as the guide wanted them to be—even bottomless.

3.4 What They Wore

Women arrived at the cave wearing long dresses and high button shoes. Men wore suits and ties, everyday attire for well-heeled Victorians. If visitors did not want to get their fine clothes dirty or felt they would not be comfortable, they could rent a cave costume for a dollar at the hotel (Thompson 1879). Journalist Nathanial Parker Willis wrote,

> There is an extraordinary uniform provided by the Hotel for visitors to the Cave. …I cannot say that the dress is becoming. A stuffed skull-cap of mustard-colored flannel is worn by the ladies to guard them from knocks on the head where the cave is low (Willis 1853).

Some visitors wore their own clothes into the cave. Victorian era clothing does not look comfortable for caving (or anything else) to this writer; apparently, at least two women agreed. In the Corkscrew, a rough climb up through breakdown not used on tours since the 1950s, a modern caver found two corsets.

Stereotyped as weak, frail, and helpless, women were expected to act properly in the 1800s, especially among the upper class. Climbing over rocks in a dark, dirty cave was not considered ladylike behavior, but Mammoth still attracted plenty of female visitors. Some guides and male visitors felt women should not climb Rocky Mountain near the far end of the Long Route (near the Carmichael Entrance today). In 1887, a German visitor wrote about the approach to Rocky Mountain,

> It is generally advisable for ladies not to hike farther along, since the next mile of the way is hardly passable for the tenderer sex (Zagel 1973).

Another German visitor in 1870 noticed women did not always comply with this advice.

> The guide now asks the women to stay behind; the way from here to the end might be too strenuous. This gracefully stated request is none the less vigorously rejected by the representatives of the fair sex. All declare most decidedly their desire to go along—as far as any *man*. We stumble, jump, creep, and teeter from one shard to another, one pile to the next, a dangerous progress in which each of us, in semi-darkness, has to find for himself the best route. Yet with laudable endurance the women cross every obstacle, and with the rest of the group, reach the end of the passageway, blocked by a sharply rising wall of stones (Kirchoff 1983).

Long-time guide Mat Bransford apparently did not try to discourage women. He and Ralph Seymour Thompson discussed the matter in the 1860s.

> 'Matt,' (sic) said I, 'do ladies ever climb this mountain?''Oh yes, sir. Most every day during the season some ladies go over the mountain.''Well, I don't see how they manage it…''Well, replied Matt, 'the air is good, and persons can stand an amount of fatigue they could not stand at ordinary times. And then there are almost always some young men along, and I never saw a young man so tired or so lazy he couldn't help a pretty girl up the Rocky Mountain.''…But how do the ugly girls do, Matt?''Well,' replied Matt, 'they most generally have to look out for themselves' (Thompson 1879).

W. Stump Forwood, learned in 1870 that women did more than manage to get through the trip.

> After having accomplished our first day's journey in the Cave, we remarked to one of the gentlemen connected with the hotel, that we supposed ladies must suffer extremely from fatigue in going through the Long Route. He replied that such was not the case; as a general rule, ladies endure the journey much better than men; and added that it was not an uncommon occurrence for ladies, after coming out in the evening, from a walk of eighteen miles, to enter the ball-room and dance until two o'clock in the morning. (Forwood 1870).

Even though plenty of women took cave tours, 1800s accounts do not mention children on cave trips, though they do not say children were forbidden.

3.5 Leaving Their Mark

In the early to mid-1800s, graffiti was popular, allowed, and maybe even encouraged in the cave. In the 1830s, guide George Slaughter Gatewood left his name and initials all over tour routes in the cave, and his handy work is still visible today. Maybe he was demonstrating the art of candle writing—smoking words onto rock with soot from candle smoke.

In 1838, Robert Montgomery Bird wrote that in the Register Room in Gothic Avenue, "many hundreds of such records of vanity are already to be seen deforming the ceiling" (Bird 1838).

In spite of the abundance of graffiti in the cave, most early accounts of it are negative.

In 1853, Nathanial Parker Willis wrote,

> Others more barbarous, or thoughtless, have hoisted candles upon sticks and smoked their names on the otherwise unblemished ceilings and walls, a disfiguration by which, in a very few years, the Mammoth Cave will have lost all its beauty—for those surfaces of delicate texture can never be cleaned. Stephen [Bishop] was eloquent upon this profanation, and doubtless puts in his protest, invariably; but a slave's remonstrances would not be much with the kind of white man that would thus immortalize his own bad taste (Willis 1853).

By 1888, cave management agreed. Vandalism and graffiti in Mammoth Cave became not only against the rules but against the law. *An Act to prevent trespass upon the Mammoth Cave in Kentucky* made vandalism punishable by a $50 fee (United States Circuit Court of Appeals, Sixth Circuit 1927a, b).

It is interesting that the guide Stephen Bishop, a slave, expressed disapproval of graffiti (though he wrote his name in the cave too). Did cave management discourage writing on the walls even before passing the law? Was it Stephen's idea to protect the cave? Even if Stephen was trying to enforce the boss's wishes, it was risky for a slave to correct a white man.

3.6 Who Owns This Cave?

I own Mammoth Cave. You do too, if you are a US citizen. We own it (and all other national parks), and the National Park Service manages it. But before the state of Kentucky began purchasing land to give the federal government to make Mammoth Cave a national park in the 1920s and 1930s, several private owners owned the cave.

In 1797, settlers could buy up to 200 acres (80 ha) of second-rate land (meaning bad farm land) south of Green River for $40 per hundred acres (40 ha) under *An act for encouraging and granting relief to Settlers* (Littell 1809). Valentine Simons (or Simmons, or Simon, depending on the source), a young man in his 20s, used the act to buy 200 acres (80 ha) that included "two petre caves" referring to Mammoth's Historic Entrance and nearby Dixon Cave, on September 14, 1798 (Warren County 1800).

A few years later, Simons sold the land to John Flatt for an unknown amount of money and, according to Simons' great-great-grandson, two mules (Warnell 1997).

Bartering the cave for draft animals supposedly continued around 1807 when Flatt sold the cave to brothers George, John, and Leonard McLean (sometimes spelled McClean) for a pony worth $40. Other stories say the cave was traded for a rifle (Shackleford 1920), and on another occasion, a farm on Highway 31W north of Bowling Green (Warnell 1997). Others claim the McLeans paid Flatt $400 for the cave. I like the pony story, but I am inclined to believe that the seller wanted money.

Charles Wilkins and Fleming Gatewood bought Mammoth Cave for $3000 from the McLeans in 1810. Wilkins and Gatewood both had experience dealing in saltpeter and may have done business with the McLeans before buying the cave. Along with dealing in saltpeter, Wilkins had owned a farm and a general store, founded the Kentucky Mutual Assurance Company, started a state lottery, and organized the Lexington Fire Department. Wilkins stuck to the business end while Gatewood moved near the cave to run the operation on site. Under their ownership, Mammoth Cave grew from a small-time saltpeter operation to one of the most productive saltpeter caves in the country and eventually to a new phenomenon—the tourist attraction.

Wilkins and Gatewood hired a young Irish immigrant named Archibald Miller Sr. as general manager, and Miller's brother-in-law, John Holton, to oversee the slaves that did most of the work (Shackleford 1920). Miller and his son, Archibald Jr., worked at the cave off and on until the 1830s.

Gatewood sold his half of the operation to Hyman Gratz on April 18, 1813, for $10,000. Nine months earlier, the cave's previous owners signed a deed documenting the sale of the land to each owner up to Gatewood and Wilkins. This paper led some people to believe the buyers made some fast, impulsive land deals that day. Actually, they were documenting past sales. When trading a cave for a gun or a pony, you may not bother with legal documents, but Gratz wanted to make sure there was no question about who legally owned Mammoth Cave. When Wilkins died circa 1827, Gratz bought Wilkins' share of the cave from his heirs for $200.

In 1838, Gratz sold the cave to Franklin Gorin (Gorin 1876). When Gorin arrived, Archibald Miller Jr. and Robinson Shackleford were leasing the cave and guiding tours for $1.00 a ticket. Miller continued to work for Gorin, who hired Alexander Harvey to help Miller manage the cave (Shackleford 1920).

A year later, Dr. John Croghan, a Louisville physician, bought the cave for $10,000. He believed this new business of charging people money to tour the cave would catch on; he paid twice as much for Mammoth Cave as Gorin had. He started planning ways to expand what he believed to be a promising tourist attraction and even a health resort.

Later, when Gorin had second thoughts about selling the cave, he wrote to Croghan asking him to make him a partner. Croghan thought highly of Gorin, but declined the offer, since he "always objected to partnerships of all kinds" (Croghan 1839).

Though visitors had been paying to see the cave for about 23 years, some people found the idea of making money from tourism unrealistic. Croghan's brother-in-law, General Thomas S. Jesup, told him,

> Your purchase, should we have war, will turn out to be a valuable investment from the quantity of salt petre it will produce; but apart from the satisfaction of being the proprietor of so great a curiosity, I do not think it can be made valuable in peace, unless the lands be converted into a stock farm, and then only to a man who would attend to all the details of the farm himself (Jesup 1839).

Not persuaded to bring in livestock, Croghan continued in the tourism business. It paid off; Croghan and his heirs profited from the cave (some years less than others) until the state bought it to be a national park in the 1920s and 1930s.

Croghan came from a wealthy, well-connected family. His great-uncle, George Rogers Clark, was a Revolutionary War hero. Another great-uncle, William Clark, became even more famous by co-leading the Lewis and Clark Expedition. Family friends included Presidents James Monroe, Andrew Jackson, Zachary Taylor, and ornithologist and painter John J. Audubon. Croghan lived most of his life at the family mansion, Locust Grove, which is open for tours.

Lore has it that Croghan first heard of Mammoth Cave while in Europe, where some world travelers told him about this magnificent cave that just happened to be in his home state. Actually, Croghan was familiar with the area long before he bought the cave; his family owned land in Edmonson County, and they knew Gorin. Croghan was not just familiar with the surface; he carved his name and the date 1812 into the limestone in Ganter Avenue in the Historic section of Mammoth Cave. My park service colleague Dave Sholar and I compared the cave signature to Croghan's signature on a document; in spite of limestone being a rough surface to write on, they were a perfect match. The signature is about a mile (1.6 km) from the entrance in an area not used to mine saltpeter, so 22-year-old John and his guide (maybe Archibald Miller Sr.) went there apparently just for the adventure.

After buying the cave, Croghan began improvements. He hired carpenters to build more guest rooms, a dining room, and a ballroom. One of the carpenters, George Tapscot, made sure cave visitors would remember him long after his buildings were gone; he smoked his name twice onto the ceiling in Mammoth Cave's Gothic Avenue where modern visitors still see it (Shackleford 1920).

Before travelers could enjoy the new lodging, they had to get there. In 1839, the only public road to the cave came from Bell's Tavern (now Park City). Croghan got an order to build new roads from Cave City, Dripping Springs, and Grayson Springs. Oliver Shackleford, whose family worked many years at the cave, called the road to Cave City "a splendid road through rough country," (Shackleford 1920) but French artist Albert Tissandier wrote, "Only the victims themselves can believe the number of dreadful bumps, holes, and ruts in the road" (Tissandier 1885). Splendid or dreadful, the road from Cave City brought many travelers to Mammoth Cave.

Croghan often complained of his poor health; he may have had tuberculosis, possibly catching it from his patients at the cave (more on that later). He died in January of 1849 at 58 years old. Croghan never married and had no children, so he left Mammoth Cave and the hotel to his nine nieces and nephews. They were the cave's last private owners.

The nieces and nephews received the profits from the cave, but had little to do with daily operations. In his will, Croghan appointed three trustees, Joseph Underwood, George Gwalthney, and William Bullock, to hire managers for the cave and hotel and distribute the profits to the heirs (Croghan 1849).

Eventually, a distant relative, Albert Covington Janin, did manage the cave. "Judge" Janin (not really a judge, but a lawyer) married Croghan's great-niece, Violet Blair. Violet's mother, Mary Jesup Blair, was one of Croghan's nine heirs.

Violet came from impressive stock on all sides—her grandfather Thomas Jesup (Croghan's brother-in-law) was the Quartermaster General of the US Army, and her father's family, the Blairs, were politically influential in Washington D.C. Their family home, the Blair House, is now used by dignitaries visiting the president. Violet saw herself as American nobility and expected to marry well. Albert Janin, in spite of being well educated, the son of a successful lawyer, and a lawyer himself, disappointed her. His somewhat successful law practice was overshadowed by debt from failed get-rich-quick schemes, including development of a canal between the Mississippi River and Lake Borgne, land speculation in Minnesota, ice machines, and patent medicines. Violet compared Albert to Colonel Sellers, a constantly broke but optimistic character from Mark Twain's novel *The Gilded Age* (Lass 1998). Albert finally found success at the cave in his middle and old age; he started handling Mammoth Cave Estate's legal matters in 1900 at age 56.

Albert's business failures made him a bad prospect to manage his wife's family's business. His knowledge of law, a family feud, and the lack of available men in the family may have led to Albert landing the job.

The heirs disagreed about whether or not Henry Ganter, manager of the cave from 1888 to 1902, was stealing money. The Blairs believed Ganter innocent and hired Albert to act as their lawyer. By 1904, Albert moved from being the family lawyer to a trustee of the Mammoth Cave Estate. Though most of the heirs were women, they usually left business decisions to their husbands. Ganter's supporters (Ganter was gone by this time, but the family was still divided) had run out of men, except Albert (Algeo 2008).

In the late nineteenth century, the cave still had not completely recovered from the Civil War, but under Albert's management, the cave and the hotel prospered, making family members on both sides of the feud happy. Albert ran the operation until mental illness made him unable to in 1923 (Laas 1998). He formally resigned in 1926 (Janin 1926).

Dr. Croghan's last surviving niece, Serena Rogers, died in California in 1926. In his will, Croghan stated the cave would be auctioned off after the death of his last heir. You and I soon became Mammoth Cave's owners.

3.7 Guides

With over 400 miles (644 km) of passages (as of 2015), total darkness, no sun or stars for direction, and no food, Mammoth Cave is a bad place in which to get lost, hence the importance of cave guides. Saltpeter manager Archibald

Miller Sr.'s son Archibald Miller Jr., early owner Fleming Gatewood's son George Slaughter Gatewood, Robinson Shackelford and his sons J.C. and Oliver were the first guides. Until the 1960s, men guided all the tours—almost. Interestingly, British writer Harriet Martineau mentioned a female employee on her 1834 visit to the cave in her influential book, *Society in America*.

> Our hostess was with us the whole time; and it was amusing to see in her the effect of custom. She trod the mazes of this cave just as people do the walks of their own garden (Martineau 1837).

Martineau's hostess may have been Nancy Miller, Archibald Miller Sr.'s wife. Martineau called the men who took her in the cave the previous day "guides." Did Mrs. Miller guide this tour without the title of guide because she was a woman, or did she just take a break from her hostess duties at the hotel to go on the cave trip?

Slaves too played an important role as early guides. When Franklin Gorin bought Mammoth Cave in 1838, he brought with him a young slave named Stephen Bishop (Fig. 3.6) to guide tours. Stephen's name became more intertwined with Mammoth Cave than anyone else's. Twenty-first-century cave visitors are fascinated with Stephen[1]; he's had modern books, magazine articles, and a play written about him. Nineteenth-century visitors loved him too. The famous author, Bayard Taylor, wrote about Stephen in his 1855 book, *At Home and Abroad*.

> Stephen, who has had a share in all the principal explorations and discoveries, is almost as widely known as the Cave itself.…I think no one can travel under his guidance without being interested in the man, and associating him in memory with the realm over which he is the chief ruler (Taylor 1855).Marianne Finch wrote in *A English Woman's Experience in America,*He is certainly the very prince of guides. …he seems more like the high-priest and expounder of its mysteries, than a hired guide, much less a slave (Finch 1853).

Perhaps the most impressive and surprising description of Stephen comes from his former master, Franklin Gorin in a letter to W. Stump Forwood.

> I placed a guide in the Cave, - the *celebrated* and *great* [emphasis his] Stephen, - and he aided in making the discoveries.

Fig. 3.6 Stephen Bishop, Mammoth Cave guide and explorer. Artwork by Bonnie Curnock

> Stephen was a self-educated man; he had a fine genius, a great fund of wit and humor, some little knowledge of Latin and Greek, and much knowledge of geology; but his great talent was a perfect knowledge of man (Forwood 1870).

On the rare occasion that slave owners praised slaves, it was usually for their physical strength, loyalty, or obedience; Franklin recognized Stephen for his intellect.

Dr. John Croghan bought the cave and Stephen in 1839. Croghan apparently also thought highly of Stephen, and he had him draw a map of the cave for the book, *Rambles in the Mammoth Cave,* published in 1845. Stephen received credit for his work: "By Stephen Bishop, One of the Guides" is printed on the map. Compared to modern cave surveys, the map is not accurate, but considering that Stephen drew it from memory without any survey equipment, it is impressive.

Stephen left no journals or letters—just some cave graffiti (Fig. 3.7). He wrote his first name and sometimes first and last name on the rocks along tour routes. But in an untoured passage now called Stephens Canyon just off of Ruins of Karnak, he took time to scratch in a little extra in fine handwriting.

[1]In the 1800s, people usually addressed white adults as "Mr. Bishop" and slaves, children, or close friends by their first name. I refer to Stephen Bishop as Stephen not out of disrespect, but partially because that is what he is called in old accounts. Also because Mammoth Cave guides and cavers feel camaraderie with Stephen; we feel like we know him on a first name basis even though we have never really met him.

Fig. 3.7 Stephen's signature. Stephen Bishop discouraged visitors from writing on the cave walls but left his signature in Pensico Avenue and other passages

Mr. Stephen Bishop

Mammoth Cave guide

June 12 1841

Maybe Stephen felt that while he was in a part of the cave known only to him, never seen by visitors or cave management, that he was not just Stephen, a slave; he was *Mr.* Stephen Bishop. Or maybe he just felt like writing more that day. (For more on Stephen's after hours work as a cave explorer, see Chap. 4.)

The 1856 Jefferson County Court Order and Minute Book # 21 states that Stephen, his wife Charlotte, and their 13-year-old son, Thomas, gained their freedom in February of 1856 under the will of John Croghan. Eight months later, Stephen bought 75 acres (30 ha) near Mammoth Cave (Lyons 2006).

Stephen did not have much time to enjoy his freedom or his land. He died in the summer of 1857; somewhere between 36 and 40 years of age (records listing Stephen's age differ because slaves usually lacked birth certificates).

Records do not indicate how Stephen died. Modern cave guides and explorers ponder the possibilities: illness, a tragic caving accident that management covered up to avoid scaring visitors? Maybe even murder?

A May 11, 1857, journal entry by Maria Mitchell, America's first female astronomy professor, partially solves our mystery. She wrote,

> I called the landlord as soon as we arrived at the Cave House, and asked if we could have Mat, who I had been told was the best guide now that Stephen is ill (Mitchell 1896).

Since Stephen died shortly after Mitchell's visit, he probably died from that illness.

Like many slaves and former slaves, Stephen was buried in an unmarked grave. Several years after his death, cave visitor James Mellon met Stephen's widow, Charlotte. When she told Mellon that Stephen had no tombstone, he promised to send one. Mellon bought a tombstone with a military motif and had Stephen's name and year of death inscribed (Galey 1951). Mellon got the date wrong; the stone says 1859 instead of 1857.

About the same time Stephen arrived at the cave, Gorin leased two other young slaves, Materson and Nicholas Bransford, to guide tours (Fig. 3.8). Mat and Nick were not related, but like many slaves, they took their master's last name. Their master was Mat's father, Thomas Bransford of Nashville. When Thomas Bransford Sr. died in 1853, Mat's half brother, Thomas L. Bransford, bought Mat and Nick and continued to lease them to Mammoth Cave. The idea of

Fig. 3.8 Mat and Nick Bransford at Entrance. Mat and Nick Bransford at the Historic Entrance in the 1860s

owning one's own son or brother appalls us today, but many slave owners had children with their slaves. Some of these children got preferential treatment—that may be why Mat was sent to guide tours at the cave rather than put to work in the fields—but they were still treated as property to use, lease, or sell at the master's convenience.

Mat Bransford was still a slave in 1863,[2] but an article in the August 20, 1863, edition of the *Louisville Daily Journal* makes it appear he had some degree of freedom and respect that few slaves had. He apparently was able to visit Louisville not as a slave attending his owner's needs, but for his own pleasure.

No one that has visited Mammoth Cave during the last quarter of a century has forgotten Mat, the colored guide, to whose attentions they have been indebted for most of their pleasurable remembrances of a visit to that great subterranean wonder… He is a native of Nashville, Tenn., and is owned by Thom. L. Bransford, late of Nashville, but at present a seeker after "his rights" in the South… Although by no means scientific, he is familiar with the geographical and chemical formations peculiar to the Cave, and discourses of all its wonders with an apparent knowledge of his subjects that would do credit to Prof. Silliman.[3] Mat arrived in this city yesterday, and is a guest of our friends at the Louisville Hotel. He will sit for his portrait to-day at Brown's daguerrean saloon, after which he will take a shy at whatever is worth looking at above ground hereabout, returning to the Cave to-morrow (Louisville Daily Journal).

[2]The Emancipation Proclamation issued by Lincoln in 1863 declared slaves in the Confederate states free, but did not apply to slaves in the border state of Kentucky, which was still part of the Union.

[3]Benjamin Silliman was a well-known biologist from Yale who estimated that there were millions of bats in Mammoth Cave in the mid-1800s.

The record of slaves that Thomas Bransford's estate sold to his son and other buyers lists Nick's price as $1115 and Mat's as $900, indicating Nick may have had a skill like carpentry or blacksmithing that made him more valuable (Anonymous 1853). Most visitor accounts from the mid-1800s say Stephen or Mat guided them in the cave; visitors did not mention Nick guiding tours until after Stephen's death. Perhaps Nick used his skill above ground until he took Stephen's place in the cave.

In his *Blackwood's Magazine* article, "Adventures Underground, The Mammoth Cave of Kentucky in 1863," F. J. Stevenson wrote that Nick bought his freedom by catching and selling eyeless fish to raise money (Stevenson 1932). Mat remained a slave until the 13th amendment set him free at the end of the Civil War.

Mat and his descendants guided tours for a century (Fig. 3.9). His sons, Henry and William, grandsons Lewis and Mathew, great-grandsons Arthur, Clifton, Eddie, Elzie, and George guided until the 1930s, when the National Park Service replaced them with white guides. Mat's descendant, Jerry Bransford, came back to continue the family guiding tradition in 2004.

Fig. 3.9 Mat with a basket. Mat Bransford apparently had some of his own money and took pride in his work. He had a studio portrait taken and included the tools of his trade in it

3.8 Views on Slavery and African-Americans

From the late 1830s to the end of the Civil War, slaves at Mammoth Cave took care of cave visitors' needs: cooking and serving their meals, shining their shoes, cleaning their hotel rooms—typical jobs that slaves held in hotels and plantations across the south. Their work often went unnoticed (unless they did something wrong). However, visitors could not help but notice the guides who led them through the cave. In articles, books, and letters, visitors not only described their guides, but often commented on the institution of slavery.

Mat Bransford and his wife Parthena had the unthinkable experience of having some of their children sold away. A farmer near Mammoth Cave owned Parthena and their children, so Mat and his brother/master had no control over their sale. Union officer, James Fowler Rusling, heard about the sale while he was visiting the cave during the Civil War. He said to Mat,

> Said I, 'Uncle Mat, I don't suppose you missed these children much? You colored people never do, they say.''Sho, Cap'n! Don't you b'lieve dat. Culled folks has feelins, jus de same as white folks! Corse I'se a man, and can bear sich things, do it went mighty hard at fust. But it most killed do old woman, dat's a fact! Sho went round kind o' crazy like, for long time; but what can niggahs do?' [sic] (Rusling 1864).

Rusling had little sympathy for the slaves' position, but many northern and foreign cave visitors expressed negative views of slavery. Federico Craveri, an Italian scientist who visited the cave in April 1859, called slavery "a bad gangrene corroding the State of Kentucky" (Cigna 1997).

British author Robert Ferguson wrote about a conversation he had with travelers and locals by the fire in the hotel lobby one evening during the Civil War.

> 'The condition of the slave is a happy one – he has no cares – he is clothed and fed, and that is all that he minds about, and when he is past work he is taken care of by his master.' 'Who,' said a hulking fellow, who seemed to me to have an uncommonly easy time of it, 'Who will take care of me when I get too old to work?' …when, in the course of the conversation, I felt constrained to say in a few words that I did not share in the opinions of the others about slavery, one of them came to me aside, with a kindly-meant caution, that it was not safe to say always what one thought in this country (Ferguson 1866).

Abolitionist Mattie Griffith wrote the anti-slavery novel, *Madge Vertner,* published as a serial in the *National Antislavery Standard* from July 1859 through May 1860, in which the main character, Madge, visits the cave. Griffith visited, or at least read about, Mammoth Cave—she called the novel's guides Stephen and Mat—but the cave owner in the novel had a fictional name. Maybe she feared her anti-slavery views would offend the management. She wrote,

> ...the most aristocratic ladies and gentlemen who visit the Cave seem to forget for a time those unnatural distinctions of race and caste, and associate with the colored guides in the most familiar manner...Truly, the Mammoth Cave should be a temple of abolitionism (Lockard 2002).

Dr. John Croghan often complained about his slaves, though I have never read any complaints specific to slaves at the cave. In letters to his brother-in-law, Thomas S. Jesup, he lamented, "Negroes are not now what they were five years ago," and "So much for the Darkies,—decidedly the most troublesome and worst property a man can have" (Croghan 1842, 1843a, b).

Croghan owned slaves his entire life, but in his will freed them after his death. His slave Isaac (who did not work at the cave) was freed immediately upon Croghan's death. His other slaves, including Stephen Bishop, were set free after being hired out for seven years. The money the slaves earned for the first four years went to Croghan's heirs. The slaves could keep and save the money they earned for the next three years

> ...so as to prepare them for freedom, and to provide the means for their support and removal to Liberia or elsewhere, and at the expiration of said three years, to emancipate the said slaves and all their increase (Croghan 1849).

Did Croghan at some level feel that treating human beings as property was wrong, but not so wrong as to give up the luxury of having slaves? This sentiment was common in many slave owners.

After the Civil War, African-Americans continued to make up most of the Mammoth Cave guide force and hotel staff until the National Park Service took over in the 1930s. Guides and other employees got paid and could come or go as they pleased, but racism continued.

Prior to the Civil War, only white travelers could afford to visit Mammoth Cave. In the early twentieth century, African-Americans began arriving at the cave not just as servants traveling with wealthy white visitors but as vacationers traveling on their own. To accommodate the rising number of black travelers, cave guide Matt Bransford (grandson of Mat Bransford) and his wife opened their home on Great Onyx Road as a hotel for black guests. They could not stay at the Mammoth Cave Hotel; cave manager Albert Janin insisted it was "exclusively for white people" (Janin 1908a, b). He allowed blacks in the cave, but enforced segregation. Janin wrote,

> I will not receive colored parties scheduled to arrive here on trains provided for our white visitors and expecting to be taken into the Cave at the time fixed for whites—that is at 9 A.M.; 1:30 P.M. and 7:P.M.—and to leave this place at the same time white visitors do.If I receive them at all, it will be in pursuance of a special agreement made for each separate party with a view to preventing conflict or contact between such party and any party of white people whom we may have to handle on the same day (Janin 1908a, b).

Segregation at Mammoth Cave continued many years after it became a national park. Mat Bransford's great-great-grandson, cave guide Jerry Bransford, told me that in the 1950s, his family would visit the cave and stop by the hotel to get ice cream cones. The Bransfords had to purchase their ice cream outside the back door; African-Americans still could not go inside the hotel.

3.9 The Civil War

When the Civil War began in 1861, Kentucky, along with Missouri, Maryland, and Delaware, chose not to join the other slave states in seceding from the Union. These border states had both Union and Confederate sympathizers. During the turmoil that the war created, Mammoth Cave remained a tourist get-away serving supporters of both sides.

In January 1862, Mr. Ousley, the proprietor of the cave and hotel (Fig. 3.10), heard that Confederate General Hindman planned to burn the Mammoth Cave Hotel. In hopes of saving what he could, Ousley had slaves hide the furniture and other valuables in the cave. Texas Ranger Frank Bachelor wrote to his wife about the raid.

> Our present camp being but 10 miles from Mammoth Cave some of the boys rode over there yesterday and found the Hotel closed and all valuables carried two or three miles into the Cave. The proprietor is a Union man & our men helped themselves freely to choice liquors, cutlery, bedding, cooking utensils, etc... (Rugeley 1864).

He likely exaggerated the distance into the cave.

The confederates destroyed furniture, carpets, looking glasses and other things too big to take (Underwood 1862).

The cave stayed open during the war, and the management did its best to replace damaged and stolen goods, but the raid and the war affected tourism. English visitor Robert Ferguson wrote,

> ...evil days had come upon the house, for the disturbed state of the district had thinned the stream of visitors and put an end to all its gay doings. ...the Confederates swooped down upon the place. ...took the things and spared the house. So this accounts for the beds being the hardest I ever slept upon in all my life (Ferguson 1866).On the bright side, the confederates did not burn down the hotel, after all.

Fig. 3.10 Hotel. The old Mammoth Cave Hotel survived the Civil War but burned down in 1916

3.10 Science

Scientists have been drawn to Mammoth Cave since the early 1800s, making it one of the most researched caves in the world. Constantine Rafinesque was probably the first scientist to visit the cave. Rafinesque was a genius whose thoughts were ahead of his time. He contributed to evolution (before Darwin), botany, zoology, anthropology, geology, and linguistics. (Though behavior like putting a curse on Transylvania University when the college fired him led to an eccentric reputation.) Rafinesque visited Mammoth and other Kentucky caves between 1818 and 1826. He wrote that Mammoth's passages

> …are supposed to extend under the bed of green River, which I doubt, as the whole cave appears to have been once the subterranean bed of a stream, which emptied into Green River, not far from the entrance, where the chasm leads and reaches the river (Rafinesque 1832).

Rafinesque was right. A large cave chamber called Rafinesque Hall and the Rafinesque big-eared bat are his namesakes at Mammoth Cave.

The fauna of the cave has been studied since the early 1800s. After River Styx and Echo River were discovered, Robert Davidson wrote the first report of Mammoth Cave's eyeless fish.

> We were informed that a species of *white fish* were found here without eyes, and the keeper of the hotel assured us he himself had seen them…

But…

> As for us, on our visit, we were not favoured with a sight of these natural curiosities… (Davidson 1840).

Davidson was a tourist writing about his fun cave tour, but scientists soon heard of the fish and became fascinated by them. Charles Darwin wondered how the fish and other eyeless cave animals benefited from losing their eyes. In *The Origin of Species* he wrote,

> As it is difficult to imagine that eyes, though useless, could be in any way injurious to animals living darkness, their loss may be attributed to disuse (Darwin 1909).

He was mistaken about the fish's lack of eyes resulting from disuse. Some believed the cave fish were deaf as well as blind. The August 27, 1867, *Louisville Morning Courier* reported that a Dr. H. Smith "discovered" this on

> …a visit of scientific research to that geological freak of nature, the Mammoth Cave…

he thought,

> that the auditory nerves only partially losing their stimulus – sound – will gradually become weakened …and in time become paralyzed and useless (Anonymous 1910).

Even though the fish do not have visible ears, they can hear; they just were not listening to Dr. Smith.

Some ideas about cave science were not just mistaken, but downright unscientific. Many visitors thought Mammoth Cave had special healing properties, a belief going back to

Fig. 3.11 Tb hut. Two of the patients' huts from the tuberculosis hospital still stand in the cave today

the saltpeter mining days during the War of 1812. Robert Montgomery Bird visited the cave and heard stories about the cave's effect on miners' health. He wrote,

> The nitre-diggers were a famously healthy set of men: it was common and humane practice to employ labourers of enfeeble constitutions, who were soon restored to health and strength, though kept at constant labour: and more joyous, merry fellows were never seen (Bird 1838).

The idea that the cave could cure whatever ailed you became so widespread that Dr. John Croghan, who purchased the cave in 1839, wanted to take advantage of the health benefits as well as run tours. He considered establishing a health resort in the cave.

> I have no doubt there is nowhere to be found a spot so desirable for persons laboring under pulmonary affections, chronic Rheumatism, diseases of the eye, etc. … for the restoration and preservation of health it stands unrivaled and in fine is worth all the Niagaras and watering places in the Union put together (Croghan 1839).

Instead of an underground resort, Dr. Croghan set up what he called an "invalids village" in the cave for tuberculosis patients in 1842 (Fig. 3.11). About 20 patients lived in the cave hoping it would cure them. By January 1843, two had died and five had left the cave. That month Dr. Croghan wrote,

> There are now from 15 to 20 invalids in the Cave. I am convinced they would all return to the land above with greatly improved health, but for three considerations. 1st, the want of attention to diet, 2, their indiscriminate use of medicines and lastly, smoke. Whenever the temperature within the Cave corresponds with that without, smoke collects about the chambers, [and] by irritating the lungs, destroys measurably the good resulting from the uniformity of the Cave climate and the peculiarity of its air. By sinking a shaft in the vicinity of the 'invalids village' this may be obviated (Croghan 1843).

He never made a chimney, no one got cured, and before the end of the year, the patients gave up and left the cave. But people still believed in the cave's healing properties.

In 1860, physician and chemistry professor Charles Wright touted the cave's healing powers,

> …short easy trips have been known to effect a cure in chronic dysentery and diarrhea, where all other measures had failed.It is not an uncommon occurrence for a person in delicate health to

Fig. 3.12 Martel. E.I. Martel, the father of speleology, at Salt's Cave Entrance

> accomplish a journey of twenty miles in the Cave, without suffering from fatigue, who could not be prevailed upon to walk a distance of three miles on the surface of the earth.

But Dr. Wright warned women,

> The only condition in which risk is incurred is during the menstrual period. Serious, and even fatal results have been the consequence of inattention to this fact (Wright 1860).

Before you rush to (or from) the cave for your health, be assured that modern scientists have determined that Wright was wrong.

Though scientists had been poking around Mammoth and other caves for years, it took a French lawyer who preferred caves to courtrooms to make studying caves a science in itself. Often called the father of speleology, Edouard Martel (Fig. 3.12) visited Mammoth Cave in October 1912 to research the formation of the cave's passages and their relation to each other (Shaw 2003).

Martel had a difficult visit to the cave. He felt the area lacked an important caving staple.

> I felt most irritated by the strict application of the anti-alcohol laws in the dry state of Kentucky. A horror to European speleologists! Without a bottle of rum from my personal luggage I could never have finished this very strenuous visit to Hovey's Cathedrals (Kliebhan 1999).

Modern cavers recommend leaving the rum behind.

Rocks and strange animals have long attracted geologists and biologists to caves, but in 1938, Mammoth Cave attracted the first scientist devoted completely to the study of sleep. Dr. Nathaniel Kleitman founded the world's first sleep laboratory at the University of Chicago and wrote *Sleep and Wakefulness,* regarded as the sleep Bible by somnologists (aka, sleep scientists). He also discovered rapid eye movements (aka, REM) that we have when we dream.

In addition, Kleitman studied how changing one's diurnal rhythm (observing a 28 hour "day" instead of a 24-hour day) could modify body temperature. Changes in light, noise, and temperature from day to night make it hard to adapt to the artificial day, making this a tough subject to research. Kleitman found the near-perfect place for his experiment; it had eternal darkness, little noise, and a steady temperature—Mammoth Cave's Rafinesque Hall.

Kleitman and University of Chicago graduate student, Bruce Richardson, lived in the cave from June 4 to July 6, 1938. The Mammoth Cave Hotel provided them with beds, a table, chairs, washstands, and platforms for their equipment. A hotel employee delivered meals (T-bone steak, hickory-smoked ham broiled country style, and southern fried chicken) (Mader 1938), mail, and newspapers once or twice a day. Kleitman and Richardson slept nine hours and stayed awake 19 h to make a 28-hour day. While awake, they took their temperatures every two hours.

Twenty-five-year-old Richardson adjusted well to the artificial cycle; forty-five-year-old Dr. Kleitman, less so. They concluded that some people adapt to change easier than others (Kleitman 1939).

3.11 The Pen and the Brush

In his novel *Moby Dick*, Herman Melville wrote about life at sea, whaling, obsession with revenge, and… Mammoth Cave.

> Let us now…cant over the sperm whale's head, so that it may lie bottom up; then, ascending by a ladder to the summit, have a peep down the mouth; …with a lantern we might descend into the great Kentucky Mammoth Cave of his stomach.

His whale-cave comparison is a stretch, but Melville was enamored with this cave he never visited; he mentioned Mammoth not only in his famous 1851 book *Moby Dick*, but also in two other novels. Melville and other popular writers helped spread the cave's fame.

French writer Jules Verne never visited the cave either, but he mentioned it in six novels, including *Journey to the Center of the Earth*.

> The immense Mammoth Cave in Kentucky is of gigantic proportions, since its vaulted roof rises five hundred feet above the level of an unfathomable lake and travelers have explored its ramifications to the extent of forty miles. But what were these cavities compared to that in which I stood?

Science fiction fans could not enter the center of the earth like Verne's characters did in his 1864 classic, but they could visit the Mammoth Cave to which he compared his fictional underground world.

Horror writer H.P. Lovecraft took advantage of three common cave fears: darkness, getting lost, and dangerous animals that lurk in caves. In his short story, *The Beast in the Cave* (1918), the main character experiences the worst possible result of leaving a Mammoth Cave tour when he gets lost and meets a scary cave beast. He kills the animal, takes a closer look, and sees what could be his own fate if no one finds him,

> The creature I had killed, the strange beast of the unfathomed cave was, or had at one time been, a MAN!!!

James Fenimore Cooper, Ralph Waldo Emerson, David Thoreau, John Muir, and L. Frank Baum also referred to Mammoth Cave in their writing.

Writers told readers about the cave, but artists showed viewers the cave—or at least what the artist wanted the cave to be.

French painter Regis Francois Gignoux, most famous for painting *Niagara, the Table Rock-Winter*, which hangs in the Senate wing of the US Capitol, was famous for painting landscapes more realistically than other mid-nineteenth-century artists, yet his 1843 painting *Interior of Mammoth Cave* is nothing like the real cave. The painting shows a magnificent entrance that surely made some viewers want to travel to the cave to see it for themselves, but the entrance in the painting only exists only on canvas. It's like a scene from a J.R.R. Tolkien novel.

Danish artist Joachim Ferdinand Richardt, whose subjects included Niagara Falls, Virginia's Natural Bridge, and Yosemite's waterfalls, spent a week sketching above and below ground at Mammoth Cave in the spring of 1857. He drew the cave's Echo River, Bottomless Pit, the hotel, and cave guides Mat and Nick Bransford. Richardt used his sketches to paint six paintings (Thompson 2002).

Albert Tissandier, a French artist, journalist, and balloonist, visited the cave in May 1885. Along with a regular cave tour, Tissandier wanted a private guide to take him into the cave to sketch. Cave manager Henry Ganter complied because he knew Tissandier's work showing the cave would bring publicity (Tissandier 1885). Several of his Mammoth Cave drawings belong to the Utah Museum of Fine Arts. They're not on display but can be seen on the UMFA Web site (Thompson 2006).

3.12 The (Almost) Birthplace of Cave Photography

Caves, being dark and difficult to photograph, remained a subject artists had all to themselves for a while. However, by the mid-1800s, painters faced a new competitor: the photographer. The world's first underground photographs were of the catacombs and sewers of Paris in 1861. In 1865, Alfred Brothers shot a photograph of a chandelier in Blue John Caverns, a British show cave (Howe 1989). About the time Brothers shot the Blue John Caverns photograph, Mammoth Cave manager Larkin J. Proctor became interested in using photographs to advertise Mammoth Cave. Though the Blue John Caverns photograph was a breakthrough in underground photography, the fuzzy-looking cave picture was not widely seen in the USA (or elsewhere), so it is unlikely that Proctor knew about it or had any idea how to light the cave for a photograph. He did know that photographs would attract customers. Proctor's nephew, John R. Proctor, and John's friend, John O'Shaughnessy, secured the photographic rights to the cave and got Cincinnati photographer Charles L. Waldack to take the pictures (Fig. 3.13). Enthusiastic about the project, Waldack said,

Fig. 3.13 Waldack. Charles Waldack found photographing the cave a challenge

> Ever since the introduction of magnesium light as a photographic agent, it has been one of my pet projects to test its capabilities as such in the celebrated Mammoth Cave (Waldack 1866).

But after photographing the cave Waldack said,

> You will agree with me that photographing in a cave is photographing in the worst conditions (Howe 1989).

Waldack arrived at the cave in 1866 with 120 lb of photography equipment, including about 200 magnesium tapers and tin reflectors to light the cave. The magnesium tapers lit the cave sufficiently, but they had an offensive odor, created ash that got all over the cave, the camera, and everything else, and made a great deal of smoke, which you see in some of Waldack's photographs. Only eight of twelve photographs printed successfully, and they were not great quality.

In spite of the difficulties, Waldack, Proctor, and O'Shaughnessy liked the results. Proctor and O'Shaughnessy formed the Mammoth Cave Photographic Company, registered the photographs (an early form of copyright that did not offer much protection), and started selling prints, mostly in the form of stereo cards for stereoscopes (3-D viewers similar to a View-Master) (Howe 1989).

Waldack sent the photographs and a written account of their cave expedition to the *Philadelphia Photographer*. The magazine printed the account and added, "These pictures now lie before us, and are the *most wonderful* ones we have ever seen...Oh! Is not photography a great power?" (Waldack 1866).

Waldack, Proctor, and O'Shaughnessy returned to take more photographs of the cave in July 1866. The expedition was expensive, the magnesium alone cost $500, but the trip resulted in 42 photographs of sites still popular with tourists today, including the Historic Entrance, Giants Coffin, and Bottomless Pit. Edward Wilson, the editor of *Philadelphia Photographer*, wrote, "Mr. Waldack deserves the thanks of the world of science and art, and we hope his views will sell immensely. It is *our duty* to buy them" [Emphasis his].

People did buy them. Stereo cards of the photographs made money for Waldack, Proctor, and O'Shaughnessy for at least six years (Howe 1989). Waldack's photographs also gave Mammoth Cave a needed post-Civil War boost in publicity, allowed people who had never visited a cave to see what one looked like, and opened the way for future underground photographers.

Books soon contained photographs of the cave. *Photographic Views of Some of the Important Parts of Mammoth Cave Situated in Edmondson co., Kentucky USA,* and *Pictorial Guide To The Mammoth Cave, Kentucky,* published around 1888, included photographs W.F. Sesser took in 1886 and 1887 (Thompson 2000).

Sesser also showed off his Mammoth photographs at Chicago's Central Music Hall in a lecture he called "100 Miles Underground, W.F. Sesser's Brilliant and Eloquent Lecture on the Mammoth Cave, Illustrated, Scenes From a Land Where the Sun Never Shines" (Thompson 2005).

Indiana photographer Ben Hains' photographs of Mammoth Cave, along with White and Ganter caves in what is now the national park, helped publicize the cave for years. His photographs copyrighted between 1889 and 1900 were used for stereo cards, postcards, and books. The Mammoth Cave Hotel published the postcards until about 1930 (Thompson 2000).

In June 1892, Demorest's Family Magazine published Frances Benjamin Johnston's photographs of Mammoth Cave, along with an article she wrote called "Mammoth Cave by Flash-light," the flashlight being magnesium. The magazine called Johnston's photographs, "one of the greatest successes in flash-light pictures ever achieved" (Johnston 1897).

Not a typical Victorian lady, Johnston started her career as a photographer in 1864, when photography (and most other professions) was the exclusive domain of men. Her subjects included Mark Twain and Theodore Roosevelt. Johnston understood that cave photography is not easy. She wrote,

> As to the difficulties, disasters, but ultimate triumph of the photographic campaign, when I sought to vanquish the arch-enemy darkness with flashpowder, it is too long a story (Johnston 1897).

Disasters did not prevent Johnston from having fun. Her guide, William Garvin, who she said was her "most trusted ally, and aided very materially in the ultimate success of the venture," not only led her to photographic subjects but also told her jokes. He showed her formations called the "Elephant Heads" in Gothic Avenue, which Johnston said looked real except they lacked trunks. Garvin explained, "Oh! They're checked!"

By 1905, Mammoth Cave had its own photographer, Harry M. Pinson. Pinson did not take photographs in the cave, but of cave visitors at a studio at the Mammoth Cave Hotel where "...the cave donkey stands daily for dozens to have their pictures made while reposing on his comfortable back" (Wilson 1909). The donkey had no real cave connection but added to the novelty of a visit to Mammoth Cave.

Visitors liked getting souvenir photographs of themselves at the cave. Joseph McDaniels took group photographs of tours at the Historic Entrance from 1919 to 1949, and W. Ray Scott continued the tradition from 1946 through 1967 (Thompson 2000). By then, the rise of inexpensive, easy-to-operate cameras eventually enabled visitors to take their own photographs.

In the early days, Mammoth Cave had little competition. In the mid- and late 1800s, a few other caves opened for tours to take advantage of travelers arriving by rail. By the 1920s, automobiles and better roads brought even more tourists to cave country; with more tourists came more competition, which led to the...

3.13 Cave Wars

Mammoth Cave's biggest rival in the cave wars was ... Mammoth Cave. George Morrison, an oil man from Ohio, came to the Mammoth Cave area to drill for oil in 1914. He soon discovered that the money underground in Kentucky was not in oil, but cave tourism. Morrison took several cave tours with guide Ed Bishop who told him visitors had offered him $50 to go out an entrance other than what we now call the Historic Entrance, the only entrance known at that time. Morrison correctly believed the cave extended past land owned by the Mammoth Cave Estate and decided if people were willing to pay to use another entrance to the cave, he would make one himself beyond the estate's boundary.

Morrison was open with the locals about wanting to create a new entrance to Mammoth Cave. He formed the Morrison Development Company, leased land, and drilled holes trying to find a good place to blast.

There was an underground side (literally and figuratively) to Morrison's plan. He paid Mammoth Cave guide Bob Lively to make an impression of the cave entrance key in a cake of soap; Morrison had his own key made in Louisville, giving him access to the cave. He also had Lively set off some dynamite in the cave that made an impression above ground on land Morrison claimed he had leased. They succeeded in making an entrance into Violet City. This illegal entrance was used for clandestine surveys so that Morrison could find other places to make entrances. Morrison made what is now called the Cox Entrance, but that land was owned by the Colossal Cavern Company, spoiling his plans again.

Morrison gave up temporarily, but came back five years later in 1921 with a new company name—the Mammoth Cave Development Company. He leased some land next to the Mammoth Cave Estate and blasted a new entrance that to this day is called the New Entrance.

This entrance to Mammoth Cave opened for tours in May 1922. Along with signs advertising the New Entrance, Morrison hired solicitors to encourage tourists to come see his section of cave. Mammoth Cave Estate and other show caves also had solicitors vying for tourist dollars; the competition was not friendly.

In a court transcript, former New Entrance solicitor Dave Estes stated that George Morrison and his partner William O'Neal who ran the New Entrance Hotel told him to tell tourists that they would see Echo River from the New Entrance (they did not). Morrison told Estes if he heard any old entrance solicitors say otherwise, he should "slap it out of them," and to tell Mammoth Cave Estate employees if they so much as entered the neighborhood around the De Luxe Inn where Morrison's men solicited, the "dead wagon would be following close behind." Estes followed instructions and punched an old entrance employee named Mr. France "so hard that it staggered him and caused him to lose his ring." Estes was arrested, Morrison paid the $5.00 fine and got him out of jail.

Estes admitted he "diverted to the New Entrance Cave many hundreds of thousands of tourists who told me…they wanted to see the old Mammoth Cave" (United States Circuit Court of Appeals 1927a, b). Since Estes worked for Morrison only a few months in 1925, that number may be exaggerated.

Understandably, the trustees at Mammoth Cave Estate objected to Morrison's men lying, stealing potential customers, and beating up their employees, so they sued him. The trustees did not want Morrison even to use the name Mammoth Cave. Several old entrance guides (including Bob Lively) signed a statement titled, "What the Mammoth Cave Guides Have to Say About the Pretended 'New Entrance' Four Miles East From Green River" (Wilson undated).

Though Morrison's tactics were less than honest, his New Entrance really was part of Mammoth Cave—it just was not part of the cave visitors had toured for over a century. Judge Chas. I. Dawson allowed Morrison to use the name Mammoth Cave with his New Entrance, but required him to print this "cautionary phrase":

> We do not show any of that part of the cave which, prior to 1907, was generally known to the public as 'Mammoth Cave.' That portion of the Cave can be seen only through the old Entrance (United States Circuit Court of Appeals 1927a, b).

3.14 The Move Toward National Park Status

The lies, fights, and lawsuits weren't useful for long. When Judge Dawson made his decree in 1926, The Mammoth Cave National Park Association had already formed two years earlier and began the process of making the cave and the land above it a national park.

The idea that the cave was worthy of being a national park was not new; it came up as early as the turn of the century. A 1936 article in the Louisville *Courier-Journal* said that Milton Smith, president of the L&N Railroad, had proposed making Mammoth Cave a national park thirty-five years earlier (Goode 1986). The L&N Railroad ran through Cave City and Glasgow Junction (now called Park City) carrying travelers destined for the cave, so the railroad's interest in increasing the cave's popularity is not surprising.

About 1905, former Attorney General of Kentucky, M. M. Logan, suggested to Congressman James M. Richardson that Mammoth Cave be made a national park. Richardson said Secretary of the Interior Ethan Hitchcock liked the idea (the National Parks are part of the Department of the Interior). If public interest could be drummed up, Richardson planned to introduce a bill to have the park created. Richardson was not re-elected, so the bill did not come to be (Goode 1986).

The concept of trying to make Mammoth Cave a national park caught on, though it took a while to get anywhere. In 1910, Richardson's successor, Congressman R.Y. Thomas, announced that he would introduce a bill in the next congress for the purchase of Mammoth Cave as a national park or forest reserve, but the bill failed (Anonymous 1910).

The push to make Mammoth Cave into a national park really started to move in 1924 with the creation of the Mammoth Cave National Park Committee, which became the Mammoth Cave National Park Association. The Association generated public interest with meetings, advertising, and a membership drive. An ad told readers:

> Your membership in the Mammoth Cave National Park Association will help to save Mammoth Cave for all our country… Its membership is composed of men and women from many states who believe that there should be established more national parks in the East—and especially a national park in the Mammoth Cave region, in order to preserve for all people for all time one of the greatest of the natural wonders of the world.Interested citizens of the United States are invited to become members of the Mammoth Cave National Park Association and, through their membership fee of $1.00, make possible the organized presentation of a plan for Congressional action (Goode 1986).

The Association's efforts paid off. In April 1926, Congressman M. L. Thatcher introduced legislation to make the cave a national park, and Senator Richard P. Ernst introduced the bill in the Senate. President Calvin Coolidge signed it that May (Goode 1986).

Not everyone was happy to see the creation of Mammoth Cave National Park. The legislation stated that at least 45,310 acres had to be acquired for the area to become a national park (Goode 1986). That meant the acquisition of not only the cave entrances and the land immediately around them controlled by Mammoth Cave Estate and George Morrison, but land belonging to nearly six hundred families in the area. Understandably, some people did not want to sell their land. In 1928, the Kentucky National Park Commission was set up with the power of eminent domain, which means the ability to purchase land for public use whether the owner wants to sell or not.

Dr. Croghan stated in his will in 1848 that when all nine of his nieces and nephews who inherited Mammoth Cave Estate had died, the property was to be put up for sale. Croghan's last niece, Serena Croghan Rogers, died in August 1926 (Trust Agreement 1926). The Mammoth Cave National Park Association was able to buy two-thirds interest in the Estate for $446,000, but one-third interest was controlled by an heir's offspring who did not want to sell. That third was purchased through eminent domain (Goode 1986). George Morrison's New Entrance to Mammoth Cave was purchased for $290,000 in 1932 (Barren County Courthouse 1932).

Most of the families who had to sell their land lived in Edmonson County. The Edmonson County News printed several negative articles about the creation of the new park. A February 26, 1931, article compared the plight of local landowners to that of Native Americans.

> To H—with the National Park… We find in the Kentucky History where Elsquawater and Ficheamsha, two great Indian chiefs and fighters and dear lovers of Old Kentucky, our happy hunting grounds. But they were driven out and had to leave their happy hunting ground. I wonder who will be to blame if we have to leave our happy hunting ground, it won't be Daniel Boone and Simon Kenton.

Today, the National Park Service has a good relationship with park neighbors, but understandably, some people still have negative or mixed feelings about the park. Edmonson County native George McCombs, who was a small child when his family had to sell their land for the creation of the park, shared his view.

> I'm crazy about Mammoth Cave, but another part of me still dislikes it. I was born at Joppa. There was an old high school across the road. My dad taught school up there. He told me stories about when they first got ready to buy the park that officials—I don't know if it was somebody local who was hired or appointed, or officials from Washington—would come to the schools and talk to all the little kids. They'd tell them what wonderful things they were going to do with Mammoth Cave. They were going to build this beautiful place, there was going to be recreation for everybody, there's going to be jobs for all your fathers, your brothers and your uncles, they'd make good money. Then they'd take up a collection from the little kids to help pay for the park. My dad hated that so much. Then when the park was formed, people from Edmonson County got pushed aside.

Mammoth Cave's transformation to a full-fledged national park was gradual. In 1936, the cave was given the status of a national park, but it received no government funding until 1941, when the National Park Service completely took responsibility for its administration and protection. In 1946, about forty-five years after talk first began, and the formal dedication took place (Goode 1986).

References

Algeo K (2008) The puzzling Mr. Janin and Mammoth Cave management, 1900–1910. In: Proceedings of the Max Kämper Centennial symposium and 9th science symposium: cultural history and research. Mammoth Cave, Kentucky, pp 11–21

Anonymous (1810) The subterranean voyage, or the Mammoth Cave partially explored, vol 6, Issue 109. The Enquirer. Richmond, Virginia, 20 April 1810

Anonymous (1853) List of sale of the slaves belonging to the estate of Thomas Bransford. Mammoth Cave National Park manuscript collection

Anonymous (1910) A neglected natural wonder. The Louisville Courier-Journal, 29 Oct 1910

Barren County Courthouse (1932) Book 91. pp 228–229

Bird RM (1838) Peter pilgrim: or a rambler's recollections. Lea & Blanchard, Philadelphia

Cigna AA (1997) An Italian traveler at Mammoth Cave in the 19th century. J Spelean Hist 31(3):59–61

Croghan J (1839) A letter to General T.S. Jesup. Library of Congress microfilm (R-2), 15 Nov 1839

Croghan J (1842) A letter to General T.S. Jesup. Library of congress microfilm (R-1), 2 Oct 1842

Croghan J (1843) A letter to General T. S. Jesup. Library of Congress microfilm (R-1), 13 Jan 1843

Croghan J (1843) A letter to General T.S. Jesup. Library of Congress microfilm (R-1), 20 Nov 1843

Croghan J (1849) Will Book Number 4, Jefferson County Clerk's Office Probated February 5, 1849, p 121

Darwin C (1909) Reprint. The origin of species. P.F. Collier & Son

Davidson R (1840) An excursion to the Mammoth Cave, and the Barrens of Kentucky. A.T. Skillman & Son, Lexington
Ferguson R (1866) America during and after the war. Lonmans, Green, Reader and Dyer, London, pp 138–139
Finch M (1853) An English woman's experience in America. Richard Bentley, London (Published in Ordinary to Her Majesty)
Forwood WSMD (1870) A historical and descriptive narrative of the Mammoth Cave of Kentucky. J.B. Lippincott & Co, Philadelphia
Galey TM (1951) A letter from Galey to Margaret Bridwell, 11 Mar 1951
Goode C (1986) World wonder saved, how Mammoth Cave became a national park. The Mammoth Cave National Park Association, Kentucky
Gorin F (1876) The times of long ago, Barren County, Kentucky. Reprinted 1992. Jett Press Quick Print, Glasgow, Kentucky
Hill CD, DePaepe D (1979) Saltpeter mining in Kentucky caves. The Register, 77 Kentucky Historical Society, pp 249–250
Hovey HC (1880) One hundred miles in Mammoth Cave-In 1880. Reprinted in 1982. Outbooks, Golden, Colorado
Howe C (1989) To photograph darkness. Alan Sutton, Gloucester, UK
Janin A (1908) Letter to S.L. Wood, 27 Sept 1908. Mammoth Cave National Park Manuscript Collection
Janin A (1908) Letter to William A. Russell, 3 Oct 1908. Mammoth Cave National Park Manuscript Collection
Janin A (1926) Letter of resignation, 11 Aug 1926. Mammoth Cave National Park Manuscript Collection
Jefferson T (1806) A letter to Pierre S. Dupont de Nemours, 12 Feb 1806. The Thomas Jefferson Papers Series 1. General Correspondence. Library of Congress
Jefferson T (1853) Notes on the state of Virginia, 1853rd edn. J.W. Randolph, Richmond, VA
Jesup TS (1839) A letter from Thomas S. Jesup to Dr. John Croghan. Library of Congress microfilm (R-4), 30 Nov 1839
Johnston FB (1897) What a woman can do with a camera. Ladies' Home Journal
Kirchoff T (1983) Mammoth Cave through a German's eyes: a visit by Theodor Kirchoff in 1870. Register of the Kentucky Historical Society, Reprint, pp 402–403
Kleitman N (1939) Sleep and wakefulness. The University of Chicago Press
Kliebhan B (1999) The contribution of E.A. Martel (1859–1938) to the development of caving technique. http://www.kliebhan.de/spelhist/mar/earn-contrib.html
Laas VJ (1998) Love and power in the nineteenth century. The University of Arkansas Press, Fayetteville, AK
Littell W (1809) The statute law of Kentucky, vol 1. William Hunter, Frankfort, Kentucky, pp 682–683
Lockard J (2002) "A light broke out over my mind" Mattie Griffith, Madge Vertner, and Kentucky Abolitionism. The Filson Club Hist Q 76(3):259
Lyons JM (2006) Making their mark, The signature of slavery at Mammoth Cave. Eastern National, Fort Washington, PA
Mader JH (1938) Captions for picture series on scientists in Mammoth Cave. Unpublished notes from the Mammoth Cave Hotel, Mammoth Cave, Kentucky
Martineau H (1837) Society in America. Saunders and Otley, London
Meriam E (1844) Mammoth Cave. Municipal Gazette, 21 Feb 1844, pp 317–319
Mitchell M (1896) Maria Mitchell, life, letters, and journals. Compiled by Phebe Mitchell Kendall
Olson RA (2008) The lamps that lit their way. In: Mammoth Cave National Park's Max Kämper centennial symposium and 9th science symposium: cultural history and research proceedings, pp 46–56
Rafinesque CS (1832) Atlantic journal and friend of knowledge. Philadelphia
Rugeley HJ, Batchelor-Turner H (1861–1864) Letters written by two of Terry's Texas Rangers
Rusling JF (1864) A trip to the Mammoth Cave. Self-published pamphlet, Nashville, TN
Shackleford OP (1920) An account of the history of Mammoth Cave during the middle nineteenth century by Oliver Shackleford. Unpublished interview
Shaw TR (2003) Martel's routes in Mammoth Cave, Kentucky, 1912. Intl J Speleol 32(1):33–36
Stevenson FJ (1932) Adventures underground. The Mammoth Cave of Kentucky in 1863. Blackwood's Magazine, pp 722–723
Taylor B (1855) At home and abroad. G.P. Putnum, New York
Thompson RS (1879) The sucker's visit to The Mammoth Cave. Springfield, Ohio. Live Patron Publishing Office. Reprinted 1970, Johnson Reprint Corporation, New York
Thompson B (2000) The Mammoth Cave of Kentucky, a catalogue of selected photographers and their images. J Spelean Hist 34(1):7
Thompson B (2002) Historic paintings of Mammoth Cave. J Spelean Hist
Thompson B (2005) W.F. Sesser's brilliant and eloquent lecture on the Mammoth Cave illustrated. J Spelean Hist 39(2):1
Thompson B (2006) Albert Tissandier's drawings of Mammoth Cave. J Spelean Hist 40(1):23–27
Tissandier A (1885) Six months in the United States. Translated by Mary F. Francey. Collection of the Utah Museum of Fine Arts
Trust Agreement from Mary Jesup Sitgreaves et al (1926) To Fidelity & Columbia Trust Co. Wm E. Wyatt and Violet Blair Janin as Trustees, Creating the Mammoth Cave Trust Estate, pp 3–4
Underwood JR (1862) Report of affairs at the 'Mammoth Cave' for the year 1861. J.R. Underwood, Trustee of Mammoth Cave Estate, 2 April 1862
United States Circuit Court of Appeals (1927) Sixth Circuit. Vol. I, Transcript of Record, 1927. Acts 1887–88, vol 3, p 563
United States Circuit Court of Appeals (1927) Sixth circuit, Transcript of Record, Wyatt and Janin vs. Mammoth Cave Development Company. p 7, 46–48, 86
Waldack C (1866) Photography in the Mammoth Cave by magnesium wire—1866. Reprinted in NSS News 26(12):200
Ward N (1819) The attributes of God displayed. Methodist Mag 2:418
Warnell N (1997) Mammoth Cave, forgotten stories of its people. Self published
Warren County Courthouse, Bowling Green, Kentucky, 3 June 1800. Document recording a September 3, 1799 survey for Valentine Simmons, Surveyor's Book A pp 268–269
Wilkins C (1817) A letter to Samuel M. Burnside, Corresponding Secretary of the American Antiquarian Society dated October 2, 1817
Willis NP (1853) Health trip to the tropics. Charles Scribner, New York
Wilson G (1909 The Mammoth Cave Excursion. From Gordon Wilson's diary
Wright CW (1860) A guide manual to the Mammoth Cave of Kentucky. Bradley & Gilbert, Louisville, KY
Wilson OJ et al (Undated) What the Mammoth Cave guides have to say about the pretended 'New Entrance' four miles east from Green River. Statement of the guides. Mammoth Cave National Park Manuscript Collection
Zagel H (1973) An excursion to Mammoth Cave in Kentucky. The Register of the Kentucky Historical Society 71(3):289

History of Exploration at Mammoth Cave

4

Stanley D. Sides

Abstract

Many forces have led humans to enter caves in the Mammoth Cave area over the past 5000 years. Caves have served as sources of adventure, social gathering, scientific discovery, and exploration. They have been valued for economic benefits including the mining of cave minerals (calcium nitrate from which to make saltpeter for gunpowder) and speleothems as well as their commercial potential as show caves. Caves were used as places of shelter from the elements of nature, for storing produce and meat and as refuge from criminal activity or war. Curiosity drove many prehistoric and historic explorers to enter caves seeking answers to many questions concerning their length, depth, and extent. The curiosity of modern explorers led to the connection of the Flint Ridge Cave System (FRCS) to Mammoth Cave in 1972. Since then, discoveries have continued and nobody alive today will hear of the end of the Mammoth Cave System.

4.1 Prehistory

Many forces have led humans to enter caves in the Mammoth Cave area over the past 5000 years. Caves have served as sources of adventure, social gathering, scientific discovery, and exploration. They have been valued as a place where one could mine cave minerals and speleothems for potential economic return from a show cave and for storing produce and meat. Caves were used as places of shelter from the elements of nature and as refuges from criminal activity or war. Passages became mines for the extraction of calcium nitrate from which to make saltpeter for gunpowder. Curiosity drove many explorers to enter caves seeking answers to many questions concerning their length, depth, and extent.

Archeologist Patty Jo Watson and researchers associated with the Cave Research Foundation (CRF) Archeological Project discovered that Late Archaic Indians explored Mammoth Cave, traveling far into its passages between 4000 and 5000 years ago. This early exploration was not associated with mining activity, and the motivation of these ancient cavers is not evident (see Chap. 2). What led these prehistoric people to risk venturing far underground is unknown. Whether it was simply for exploration or for cultural ritual cannot be determined today, but the Indians eventually assessed the caves as sources of minerals they wished to mine. Radiocarbon dating indicates they were going far from the entrance of Mammoth Cave around 5000 years ago in Salts Cave 3100 years ago (Carstens and Watson 1996). Some time after initial exploration, extensive mining of calcium sulfate, magnesium sulfate, and sodium sulfate began and extended far into side passages. Moving forward in time to the historic period, systematic looting of textiles, slippers, gourds, and other aboriginal artifacts for sale to the public has occurred over the past 150 years.

4.2 Early History

Long Hunters from Virginia explored the Green River Valley near Mammoth Cave in the 1780s. Settlers followed rapidly after the Revolutionary War and freedom from Great Britain. Sources of potassium nitrate (saltpeter) for production of gunpowder were necessary for survival, and the caves

S.D. Sides (✉)
Cave Research Foundation, 2014 Beth Drive,
Cape Girardeau, 63701, MO, USA
e-mail: ssidesmd@aol.com

H.H. Hobbs III et al. (eds.), *Mammoth Cave*, Cave and Karst Systems of the World,
DOI 10.1007/978-3-319-53718-4_4

were explored for the mineral. Caves containing large amounts of saltpeter had been reported in the Green River Valley by Gilbert Imlay in 1792 who wrote as follows: "Sulphur is found in several places in abundance; and nitre is made from earth which is collected from caves and other places to which wet has not penetrated. The making this salt, in this country, is so common, that many of the settlers manufacture their own gunpowder. This earth is discovered in greater plenty on the waters of Green River, than it is in any other part of Kentucky" (Imlay 1797).

It was no coincidence that Valentine Simons, the first landowner of Mammoth Cave, chose land containing two saltpeter caves. Initial exploration of the caves and assessment of the caves' potential value as saltpeter mines must have been made before he claimed the land. Simons was born in Maryland or Pennsylvania, probably around 1776. Pheltius Valentine Simons is likely his full name, as tax records used both names. On September 14, 1798, the 22-year-old Simons received a land grant for $80, paying $10 down for 200 acres (81 ha) in Warren County that included two saltpeter caves—later named Dixon Cave and Mammoth Cave. Simons' claim was entered in the Warren County survey book a year later, about the time he sold the property to John Flatt, probably a relative by marriage through Flatt's wife, Patience Logsdon. Simons most likely claimed the land on Flatt's behalf of Flatt to make saltpeter. The land boundary he selected included only the caves and hillsides with little suitable agricultural land. He soon moved to favorable agricultural land northwest of Brownsville, away from the supposedly unhealthy environment of a river valley. Simons married Annie Durbin on February 8, 1803 and became a prominent landowner in Edmonson County. Valentine and Annie Simons had ten children and became progenitors of the long line of Simons families of Sunfish, Kentucky (Sides 1997).

On March 14, 1811, Frederick Ridgely, brother-in-law of eventual Mammoth Cave owner Charles Wilkins, sent the Eye-Draught map of Mammoth Cave copy to Benjamin Rush in Philadelphia. The map was made prior to 1811. It is speculated that Mammoth Cave co-owner Fleming Gatewood was the originator of the map. This map confirmed all the main upper level passages of Mammoth Cave had been explored before that time. The Bogert Map from 1813 was believed to be a map drawn by Archibald Miller, manager of the saltpeter operation and was a working map for the saltpeter operation. The map showed lower level passages, especially Ganter Avenue, and indicated the presence of an aboriginal basket far out the passage from Wooden Bowl Room. This map confirmed cave owners or miners explored long distances in search of niter for successful operation of the cave as a saltpeter mine.

4.3 The Tourism Era

In the winter of 1834–1835, a Cincinnati civil engineer, Edmund F. Lee, made the first instrument survey and map, with a 30-page booklet on Mammoth Cave. He established that known cave passages consisted of about eight miles, and the longest path to the end of Symmes's Pit Branch was little more than two miles.

Robinson Shackelford and Archibald Miller, Jr. leased the cave from 1836 to 1837 for its tourist potential. Exploring from near the Methodist Church cave room, the main cave entry to the Corkscrew was discovered in 1837. However, no way down to Bandits Hall was found until William Garvin went through in 1871.

Archibald Miller, Jr. stayed as manager for the new cave owners, Franklin Gorin and A.A. Harvey. Gorin brought with him his slave, Stephen Bishop, who was about 17 years old. Gorin leased teenage slaves, Mat and Nick Bransford (not brothers) from Thomas Bransford, and the three young black men arrived at Mammoth Cave where Archibald Miller, Jr. and Joe Shackelford taught them the routes in the cave. Soon after, Bishop and tourist Hiram C. Stevenson got to the other side of Bottomless Pit in October of 1838, discovering Pensacola Avenue. This was the first major discovery of new passages since the days of saltpeter mining. A year later, on October 8, 1839, John Croghan, M.D. purchased the cave from Gorin. Bishop's continued exploration soon led to River Styx and Echo River. The latter was crossed in 1840, leading to many more miles of cave. Croghan's vision to develop a popular show cave attraction was advanced by Bishop's exploration and discoveries.

Bishop and Mat and Nick Bransford's discoveries contributed to the cave's fame and extended the known cave far beyond Echo River. However, exploration virtually ceased outside established tourist trails after Croghan's death in 1849. Passage surveys were suppressed to restrict knowledge of where the cave undoubtedly extended beyond Mammoth Cave estate's property boundary (Meloy 1975).

In 1856, David Dale Owen projected there were 150 miles (241 km) of cave in the area, and Mammoth Cave management promoted this speculation for a century. Some notable exploration did occur in Mammoth Cave after Croghan's death. Courtland Prentice descended the fearsome Maelstrom Pit at the end of Croghan Hall in August 1858. His highly publicized account stated the rope used to lower him caught on fire from friction over a log. Another visitor, F.J Stevenson from London visited the cave during the height of the Civil War and wrote that he found guides idle because of few visitors. Stevenson and Nick Bransford descended Gorin's Dome and supposedly explored "Stevensons Lost River" in a boat. Stevenson claimed he and Nick also explored Roaring River. Stevenson's grandiose account also stated that he, Nick, and guide Frank DeMonbrun also descended the Maelstrom (Meloy 1985).

Exploration beyond Pinsons Pass prior to 1860 resulted in the discovery of a passage draining the eastern side of Mammoth Cave Ridge, but the explorers have not been identified. Its location off Martel Avenue and Emily's Puzzle was apparently not shared among contemporary guides. Charles W. Wright's 1860 "Guidebook to Mammoth Cave" first mentioned Mystic River. In 1870, Wright's "A Guide Manual to the Mammoth Cave of Kentucky" (Wright 1870) mentions Mystic River, as well as W. Stump Forwood's, "An Historical and Descriptive Narrative of the Mammoth Cave of Kentucky" (Forwood 1875). In spite of these accounts, knowledge of the location of Mystic River was lost. Subsequent books by Reverend Horace C. Hovey, Richard E. Call, and others did not mention the stream. Mystic River was rediscovered in 1973, when CRF cavers Gary Eller, Walter Lipton, and C. Peterson found their way to this extensive stream complex.

During September 1879, guides noticed smoke coming from a passage off Serpent Hall in Silliman Avenue. Since fire is not normal in caves, they knew the smoke had to be coming from tour routes. Hovey wrote: "Going back to Serpent Hall they wormed their way through a series of extremely narrow crevices, finally emerging by Black Snake Avenue into the Wooden Bowl Room." This discovery and subsequent development of the passages, named "Welcome Avenue," allowed a safe escape route from beyond Echo River to the entrance if parties were trapped by rising river levels (Hovey 1882).

4.4 Exploration in Joppa Ridge and Flint Ridge Prior to Formation of the Park

The Proctor Cave entrance on Joppa Ridge was discovered by a slave, Jonathan Doyle. It led to narrow passageways and numerous scenic shafts. Mary and Larkin Proctor opened the small show cave in 1877. The Colossal Cavern Company and L&N Railroad bought the cave in 1901, later conveying the property to the developing national park. Seventy years later, Proctor Cave's lower levels were connected to Mammoth Cave.

Woodson-Adair Cave (Fig. 4.1) became the first show cave on Flint Ridge and was also purchased by L&N Railroad. In 1896, the railroad was enjoined from using the entrance and offered a reward of $150 for discovery of another way to provide access to the cave's Grand Avenue. It took only two days for Lute and Henry Lee to demonstrate a connection between Bedquilt Cave and Grand Avenue. Railroad surveyors Edgar Vaughn and W.L. Marshall began their survey that resulted in a map of the cave (Fig. 4.2) published by Horace (Hovey 1912). L&N Railroad blasted a new entrance, and Colossal Cavern was opened for visitor use in the summer of 1896 (Sides 1971).

In 1889, twelve-year-old Floyd Collins was selling Indian artifacts at Mammoth Cave and met a visitor from New York named Edmund Turner. Turner was 40 when he returned in 1910 and lived with the Collins family on Flint Ridge. Turner and Floyd explored and dug in many caves in the area. They surveyed the upper passages of Salts and Great Onyx caves. With Floyd Collins' help, Edmund opened Dossey Domes Cave across the river from Mammoth Cave ferry landing on Green River. On June 12, 1915, Turner opened a shallow pit leading to Great Onyx Cave (GOC) on the property of L.P. Edwards on Flint Ridge. Today, Blind Fish River off Ralphs River Trail in Unknown Cave portion of Mammoth Cave is tantalizingly close to Lucy kova River in GOC, but no human-size connection has been made to the Mammoth Cave System. Turner discovered two show caves and explored many local caves, but died penniless on Flint Ridge in May 1917, just before Floyd Collins discovered Floyd Collins Crystal Cave (FCCC).

Floyd Collins performed great feats of cave exploration while alone underground. He was 23 years old in 1910 when he opened Donkey Cave, or Floyds Cave, on a hillside he owned near the family home. He lived in a cabin built over the pit entrance to the cave and showed the cave to a few visitors. Floyd and his brother Homer noticed cool air coming out at the base of a sandstone outcrop and began enlarging this small opening near the Collins home. On December 17, 1917, Floyd crawled through breakdown to discover what we now realize is the highest passage stratigraphically in the Mammoth Cave System (Collins and Lehrberger 2001). Over several years, he and his family enlarged passageways, and in 1921, FCCC was opened to visitors. However, few visited the cave due to its remoteness.

4.5 Exploration in Mammoth Ridge Prior to Formation of the Park

Increased regional show cave competition may have induced the Mammoth Cave estate to enhance the image of Mammoth Cave. There were no new passages found in the cave to draw visitors back, and though it was supposedly 150 miles (241 km) long, this projection did not attract visitors. Nearby caves with attractive formations were being explored and developed. There was no news of new passages in Mammoth Cave, but this void was soon filled when exploration blossomed with very talented cave explorers, notably Benjamin Franklin Einbigler (3) and John Nelson Fig. 4.3.

Fig. 4.1 Quinque Dome, and Woodson-Adair Cave. Gary Berdeaux photograph

Nelson began guiding at Mammoth Cave prior to 1895, and Einbigler was a lawyer from New York when he visited Mammoth Cave in 1905 (DeCroix 1998). He began exploring off Cleveland Avenue with Nelson, and on May 15, 1905, Einbigler and Nelson were lowered on a rope down the Maelstrom. The men corresponded and became good friends. Einbigler returned in the summers to explore the cave with Nelson for many years.

Ben Einbigler, Ed Hawkins, and William Bransford made what they thought was the new discovery of rediscovered Cathedral Domes on May 15, 1907. They found it had been explored in 1845 by a mysterious J.A. Creighton whose name and the date are carved nearby. Creighton's identity remains an enigma, but his name was found, elsewhere, in Mammoth Cave: J A. Creighton, New Orleans, with dates, 1843 and 1848.

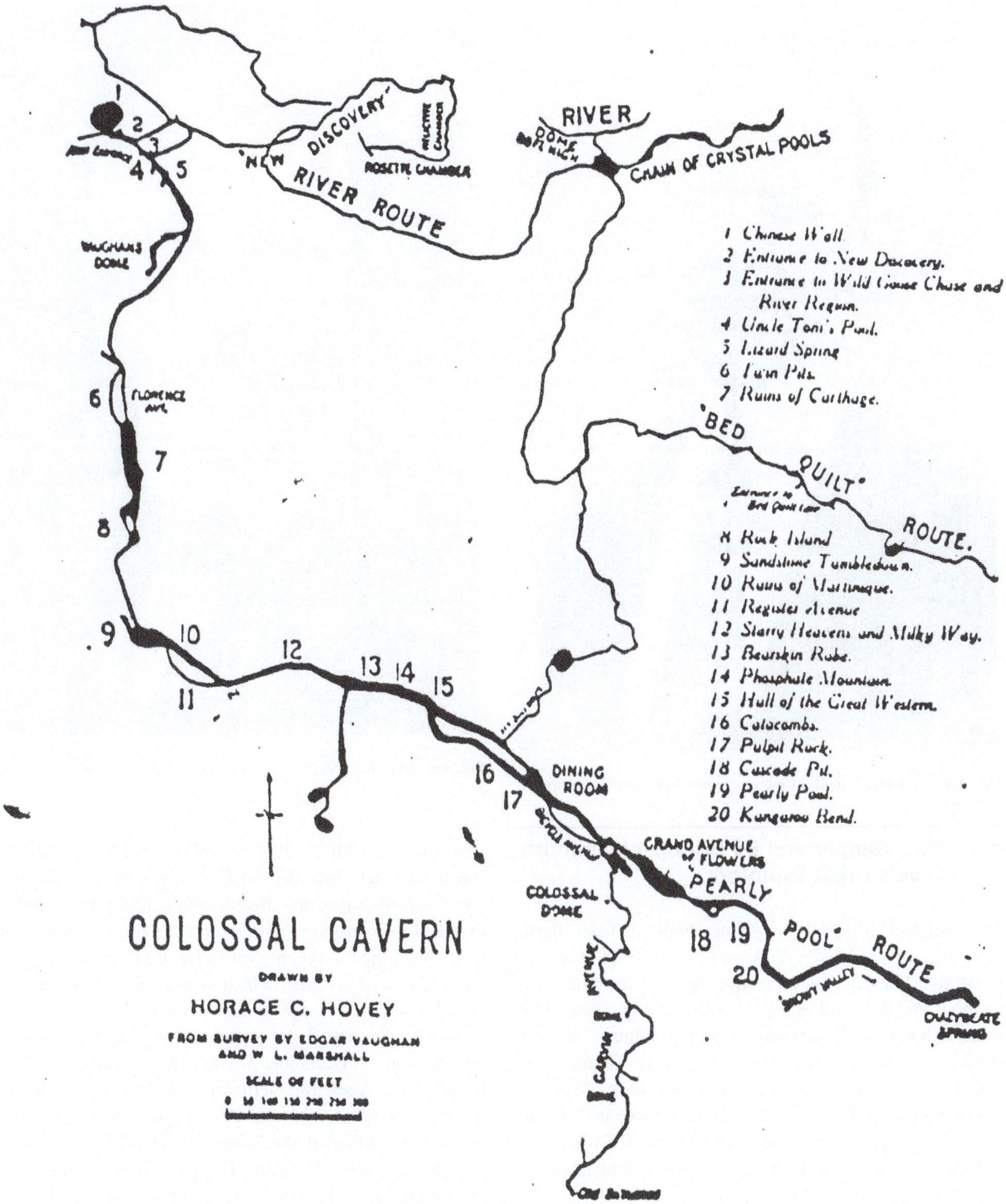

Fig. 4.2 Colossal Cave Map. H.C. Hovey, The Mammoth Cave of Kentucky, 1912

Fig. 4.3 Benjamin Franklin Einbigler. Joan Anderson Collection

Fig. 4.4 Max Kämper

4.6 Max Kämper and Ed Bishop—Mammoth Cave's Great Explorers

Max Kämper (Fig. 4.4), a young engineer from Berlin, Germany, arrived in New York on May 16, 1907, ostensibly to study American steel production and manufacturing methods and to learn English. He was initially employed by a Brooklyn firm, Lidgerwood Hoisting Engines and was active in the musical arts in New York City. German engineering students were frequently given a book by Max Eyth upon graduation. Eyth visited Mammoth Cave in 1866 and surveyed the cave. Max Kämper had a copy of Eyth's book, "In the Current of Our Time" in his personal collection and must have known about the cave before he arrived.

Max Kämper arrived at Mammoth Cave by train from Louisville on February 24, 1908. He became entranced by the extent of the cave and soon received permission to survey the cave's passages. Kämper and guide Ed Bishop penetrated beyond Ultima Thule, previously believed to be a dead end, on April 28, 1908 and discovered Violet City. His spectacular map included 36 miles (56 km) of passages in five levels displayed in different colors. Kämper and Bishop explored and surveyed far beyond the tourist routes until they ended their surveying on October 31, 1908. They discovered Grand Avenue, which was a major find and surveyed this large passage to Blairs Dome.

Kämper presented the completed map to estate trustee, Albert Janin, on December 3, 1908. On December 12, 1908, he left New York for Germany and departed the caving scene never to return to the United States. Kämper died in the trench warfare of the Somme Front in the First World War on November 10, 1916. His 1908 map was the cave's most complete and accurate survey up to that time, but was kept sealed away until 1963 when found and traced by Jim and Pat Quinlan and Roger Brucker. The Kämper map is still useful as a field map because of its cartographic excellence and accuracy. The names that Kämper applied to cave features are mostly those used today.

4.7 Exploration of Mammoth Cave Ridge Beyond the Estate Property

In March 1915, driller George Morrison (Fig. 4.5) arrived at Brownsville to drill for oil. He reportedly developed a strong interest in Route 4, the longest visitor route in Mammoth Cave. On July 6, 1915, he incorporated the Morrison Development Company to open a show cave beyond Mammoth Cave's boundary. In August 1913, he paid a former Mammoth Cave guide, Bob Lively, to make an impression of the gate key in a bar of soap. Morrison then had an illegal copy made in Louisville. Lively set off dynamite blasts that opened an entrance at Violet City. In March 1916, Morrison sneaked into Violet City to survey the cave. Three days later, the estate closed the unauthorized entrance. Morrison then used his drilling rig to try to locate cave passages while his employees listened in the cave. On April 9, 1916, Morrison's men were caught leaving the Corkscrew and exiting the cave after using the illegal key. Morrison was locked out of Mammoth Cave but, nevertheless, had surveyed enough caves to open the Cox Entrance outside the estate's property. In 1917, he was enjoined from using the Cox Entrance since the underground rights had been sold to the Colossal Cavern Company, but he continued to purchase land and underground rights beyond estate property. Morrison opened the New Entrance to Mammoth Cave (Fig. 4.5) at Doyles Big Break in May 1921, to explore and discover many miles of new passages. His workers explored a seemingly impenetrable breakdown pile at Grand Central Station near the New Entrance. Lute Lee, Roy Jaggers, and Earl Lee managed to work their way through the dangerous pile of rock on March 12, 1923 and discovered miles of passages leading to Frozen Niagara. Their discovery resulted in Morrison's Frozen Niagara Entrance to Mammoth Cave, having formations to display those were far superior to those in historic Mammoth Cave.

John D. Hackett of Tesnus, Texas, cave explorer and digger of note, arrived in the cave region in 1917. Hackett boarded with E.M. Doyel at Chaumont and first began exploring caves under Joppa Ridge. Hackett lived in Long Cave during the harsh winter of 1918, but still explored local caves and would walk daily to the Chaumont post office and store. His letters state he nearly froze to death walking to the store on March 4, 1918, when the temperature fell to 24° below zero. Hackett and George Morrison owned the land and/or underground rights to Mammoth Cave beyond the estate's boundary. They extended exploration southeast in Mammoth Cave Ridge. Their exploration and digging were extensive and dangerous both above and below ground. Harrison Logsdon, a 21-year-old working for Hackett, was killed in a rockfall in Hackett's Cave on December 31, 1921.

Visitation was slow at Mammoth Cave during the depression. However, new discoveries provided publicity for the fledgling park. In 1937, an organized exploration trip, with ten guides from Morrison's New Entrance and historic Mammoth Cave, visited many low-level passages of the cave. However, Mystic River was not found and remained a place of legend.

Mammoth Cave guides, Carl Hanson and Leo Hunt, explored passages off far upstream Roaring River. They found passages leading to unknown upper levels and recruited Carl's son, Pete, and Leo's cousin, Claude, to assist them in clearing a breakdown plug. After the blockage was removed, the crawlway led to magnificent virgin cave passages. They entered the trunk passages of New Discovery on October 10, 1938 (Lix 1946). Their exploration and discovery led to important publicity for the developing park. The New Discovery entrance was opened in 1940, but public tours were never given. In the Civilian Conservation Corps (CCC) days, the entrance did not lead to exploration beyond easily accessible passages. However, the tremendous effort

Fig. 4.5 George Morrison at New Entrance Hotel. R.T. Neville photograph

and tenacity shown by the guides to make this significant discovery off Roaring River is still legendary today.

4.8 Flint Ridge Exploration and the World's Longest Cave

The entrapment and death of Floyd Collins (Fig. 4.6) in Sand Cave brought worldwide attention to the Mammoth Cave region (Murray and Brucker 1979). He died underground in Mammoth Cave Ridge only four years after Harrison Logsdon had died in Hackett's Cave only a half-mile away. Floyd had discovered a passage in FCCC, Floyd's Lost Passage, which was rediscovered in 1941 by FCCC guides Harry Dennison and Ewing Hood. Later, Jim Dyer, Luther Miller, and Bill Austin explored a seemingly endless network of lower level passages beginning in 1947. Most of the passages they found had been explored prior to 1925 by Collins while caving alone.

Great Onyx Cave and FCCC were private show cave inholdings within the park. The NPS sponsored a trip to Great Salts Cave in July 1950 to publicize the need for Congress to allocate the money to buy GOC and FCCC.

Fig. 4.6 Floyd Collins. R.T. Neville collection photograph

Newspapermen, photographers, local businessmen, and a group of Louisville members of the National Speleological Society (NSS) were invited. The correspondents and photographers left the cave after a short introductory trip, but the Louisville cavers were allowed to continue exploration until nearly midnight. They discovered the huge size of passageways in Salts Cave and began making frequent unauthorized trips under the leadership of Charles Fort.

During the Spring of 1953, the NSS held their annual convention in Louisville, with field trips to FCCC on April 8, introducing cavers nationwide to the potential of exploration in Flint Ridge. On September 26, 1953, the NSS Board of Governors accepted an invitation from the owners of FCCC to explore the cave's lower levels. The NSS mounted a Himalayan style expedition in the lower levels during February 14–20, 1954. Few previously undiscovered passages were explored, but the expedition resulted in a core of talented cavers who continued to explore and survey, guided by the concepts of serious scientific study of karst advocated by Ohio caver Philip M. Smith. The story of the expedition, "The Caves Beyond," was published soon thereafter (Lawrence and Brucker 1955).

Five months after the NSS expedition in July 1954, Roger Brucker and Jack Lehrberger explored tight passages beyond those encountered during the expedition to discover fish hook crawl and Black Onyx Pit. On September 11, 1954, Bill Austin and Jack Lehrberger descended an overhung, exposed shaft in Unknown Cave on park property near FCCC based on rumors that Louisville cavers had discovered the way down to large passages. Austin and Lehrberger explored miles of virgin cave under the central part of Flint Ridge. On Thanksgiving 1954, FCCC explorers pushed beyond Black Onyx Pit to discover a huge shaft, the Overlook. Exploring further along a drain named storm sewer, they encountered a base-level stream, Eyeless Fish Trail.

The core group of Ohio cavers, named the Flint Ridge Reconnaissance of the NSS, requested permission on April 20, 1955 from the National Park Service (NPS) to map Salts Cave. Superintendent Perry Brown requested more information. On May 27, the superintendent refused their entry to Salts Cave and suggested Colossal Cavern instead. This permission was also subsequently denied. Cave survey, as an important aspect of the scientific study of caves, was not believed to be an important need for the park.

On September 17, 1955, Bill Austin and Jack Lehrberger entered from the entrance shaft of Unknown Cave down into Eyeless Fish Trail in Crystal Cave and then exited via the Crystal Entrance. Roger Brucker and Red Watson, coming from FCCC, had come very close to connecting with Unknown Cave from the Crystal side, but not quite. On December 27, the Flint Ridge Reconnaissance announced at the American Association for the Advancement of Science meeting that the Flint Ridge Cave System (FRCS) was the longest in the world at 23 miles (37 km). They were quickly dismayed to learn that Switzerland's Hölloch Höhle was 35 miles (56 km) long. On December 4, 1955, NPS rangers caught Louisville cavers returning to their cars from Unknown Cave. The NPS learned of the existence of Unknown Cave, and future access by the cavers was denied. Bill Austin and his uncle, E. R. Pohl, then decided to open the Austin Entrance on FCCC property to provide access to Unknown Cave. Austin completed construction of the entrance in June 1956, providing access to many miles of passages under Flint Ridge that would connect ultimately with Mammoth Cave.

Superintendent Brown insisted Mammoth Cave was 150 miles (241 km) long, as had been proclaimed for a century. Park Naturalist Ray L. Nelson compiled a Mammoth Cave map including the Walker Survey (see Chap. 5 for more on this survey) and New Discovery surveys in 1956. A total of 32.5 miles (52 km) of cave passages was shown. Adding other passages from the Kämper Map, James Quinlan estimated the cave was a total of 46 miles (74 km) long at most. Thereafter, the NPS no longer stated the cave was 150 miles (241 km) long. On August 13, 1957, the Flint Ridge Reconnaissance of the NSS became the CRF, a nonprofit scientific research organization under the leadership of Philip M. Smith. In October 1959, the NPS signed an agreement permitting CRF to undertake cooperative scientific research in the park on Flint Ridge, including Salts Cave, with cartography central to the scientific study of the caves.

Continued exploration revealed a route to Candlelight River in the southwestern part of Flint Ridge in November 1959. A CRF party consisting of Marlin "Spike" Werner, Jack Lehrberger, and David Deamer connected Salts Cave to Colossal Cavern on August 22, 1960. Next, Salts Cave was connected to Unknown Cave on August 21, 1961 by Deamer, Judy Powell, Spike Werner, and Robert Keller. This connected the major caves of Flint Ridge with the exception of nearby Great Onyx Cave. CRF exploration and survey continued through the 1960s with passages getting ever closer to Mammoth Cave Ridge. In March 1969, Flint Ridge system's surveyed length was 64.5 miles (104 km) and the longest in the world.

A CRF survey party in Big Avenue in New Discovery made a major breakthrough during the Thanksgiving expedition on November 28, 1970. Pete and Karen Lindsley, Ken Abschnikat, David Hanna, and John Wilcox were finishing their survey of Big Avenue at the breakdown end of the passage. Pete squeezed up through a crack to find a huge room and virgin cave passages leading onward. This breakthrough named "Discovery '70" resulted in miles of passages subsequently surveyed on many trips. All ended tantalizingly near the end of Lee Cave in Joppa Ridge.

Passages from the FRCS had been explored across Houchins Valley and extended under Mammoth Cave Ridge. However, the explorers were stopped by a sandstone breakdown and could not find a way through or around. During a trip on July 15, 1972, Pat Crowther squeezed through a tight 10-in. wide canyon to find a stream passage draining westward toward Mammoth Cave. On Saturday, August 26, 1972 Tom Brucker and Richard Zopf explored beyond the tight spot to find the stream leading away. Roger Brucker, who was in their party, could not fit through the tight spot. On Wednesday, August 30, 1972, John Wilcox, Pat Crowther, Richard Zopf, and Tom Brucker surveyed stream passages beyond the tight spot and discovered "PH" and "LH" written on a mud bank. The following Saturday, September 2, John Wilcox, Richard Zopf, and Gary Eller took Joe Davidson to Hansons Lost River. Davidson, a strong but tall caver, also could not fit through the tight spot. The others continued to explore side leads in Hansons lost River. Expedition participants were assembled, and CRF president Stanley Sides revealed the discovery of the initials to everyone (Fig. 4.7). John Wilcox began planning a connection attempt with Mammoth Cave.

On September 9, 1972, John Wilcox, Pat Crowther, Richard Zopf, Gary Eller, Steve Wells, and NPS ranger Cleve Pinnix connected the stream, Hansons Lost River, to Echo River at Minnehaha Island. With this connection, the known part of Mammoth Cave became 144.4 miles (230 km) long and reached global significance as the undisputed longest cave in the world—a national treasure protected by the NPS. Shortly thereafter, John Wilcox began surveying Salts Cave passages outside the park under Hamilton Valley on unpublicized exploration trips, with the goal of exploring Mammoth Cave to the east of the park (Brucker and Watson 1976).

CRF received permission to study Joppa Ridge caves in the late 1960s. In 1876, Thomas E. Lee had visited a 40-foot (12 m) pit on Joppa Ridge and left his name and date. CRF cavers Gordon and Judy Smith rediscovered the cave, which they named Lee Cave, in November 1968. CRF teams visited the cave in early 1969, descended a tight crack Lee had not explored and found a passage 70 ft. (21 m) wide by 25 ft. (7.6 m) high going nearly 7000 ft. (2134 m). Aboriginal cavers had visited the main trunk passage, and their artifacts had not been disturbed. The passage was named Marshall Avenue for Charles Marshall of the NPS, a strong advocate for research in the park. Surveying and geologic study revealed it was the westward continuation of Mammoth Cave's Cleveland Avenue and New Discovery's Big Avenue. Exploration and survey by CRF connected Lee Cave to another entrance in Carpenter Hollow, discovered Thanksgiving 1970. Lee Cave remains very near, but not connected, to New Discovery and Mammoth Cave.

Fig. 4.7 CRF members announcing discovery of LH and PH signatures. L to R: Stan Sides, Tom Brucker, Richard Zopf, John Wilcox, Pat Crowther, and Judy Smith. Kay Sides photograph

On May 30, 1970, Stanley Sides, Gordon Smith, and Norbert Welch pushed through a long low crawlway in Proctor Cave to rediscover Procter River, a stream reported in booklets and documents from the late 1800s but unknown to modern cavers. Tom Brucker, Richard Zopf, Steve Wells, and Bill Hawes later climbed above a pit in Procter River on August 30, 1973 to discover an extensive large passage. The Proctor trunk passage had numerous shafts, but initial exploration revealed no extensive lower level passages.

4.9 Exploration in the Eastern and Southern Area of the Park

University of Kentucky college student Jim Borden and fellow caver Jim Currens joined forces on April 4, 1975, to discover a theorized major cave system under ridges east of Mammoth Cave National Park. On April 2, 1976, they found Roppel Cave, the key to exploration of the system they named Toohey Ridge Cave System. Borden, Currens, and Bill Walter named their group the Central Kentucky Karst Coalition (CKKC). They chafed under the management constraints of CRF and the NPS, preferring to develop their own group free of organizational encumbrances possible since they were operating outside of the Park proper. The CKKC cavers found a trunk passage in Roppel Cave, now five miles (8 km) long and also a key to understanding the hydrology in southeast of the park.

Don Coons and Sherri Engler explored forgotten Morrison Cave just outside the southern park boundary in the summer of 1977. George Morrison had explored the cave in the 1930s after he sold the New Entrance of Mammoth Cave on January 5, 1931. Difficult exploration and survey by Coons, Engler, and Horse Cave dentist John Branstetter led to the discovery on December 2, 1977 of a very large underground stream they named Logsdon River.

Richard Zopf descended a deep pit in Proctor Cave west of Morrison Cave on a CRF survey trip on April 21, 1979 and also discovered a large river, named Hawkins River for the park superintendent, Amos Hawkins. Exploration suggested Logsdon and Hawkins rivers were the same streams but exploration from both ends were blocked by breakdown.

Morrison and Proctor caves were finally connected when a way through the breakdown was found on July 2, 1979 by Coons, Engler, and Art and Peg Palmer during a CRF expedition. The connection between Morrison and Proctor caves yielded greater understanding of the stream-level passage relationships with Mammoth Cave. On August 11, 1979, John Wilcox and Tom Gracinin pushed through a long, very low stream passage from Mammoth Cave's Cocklebur Loop to Logsdon River. The connection, dubbed the French Connection, tied Proctor and Morrison caves to the Mammoth Cave System. Surveying off the Cocklebur Loop in Mammoth Cave on New Years Day 1980, Tom Brucker and Richard Zopf found a new shaft entrance named the Ferguson Entrance near the park's southeastern edge. This important discovery would enhance exploration by speeding access to upstream Logsdon River.

John Wilcox continued to lead parties exploring Salts Cave outside the park's boundary to the east under Hamilton Valley toward Roppel Cave and beneath Toohey Ridge east of Flint Ridge. CKKC cavers discovered the far upstream portion of Logsdon River under Toohey Ridge on May 24, 1980. Continuing downstream toward Mammoth Cave, they encountered deep water and a sump halting exploration. Dave Weller led the CKKC cavers into opening a new Roppel Cave entrance on Halloween, 1981 after a year of digging. The Daleo Entrance provided easier access to central portions of Roppel Cave, vastly speeding exploration.

Meanwhile, further east of the park near Northtown, Peter Quick, Joe Saunders, and many cavers from the Detroit Urban Grotto of the NSS found a pit on Sunday, January 24, 1981 that has led to over 121 miles (195 km) of surveyed cave, named the Fisher Ridge Cave system. Passages run under the park's eastern boundary near Dennison Ferry, but no connection with Mammoth Cave has yet been made. During the week of July 18–24, 1981, the Eighth International Congress of Speleology was held at Bowling Green, Kentucky. A beneficial result of the congress was allowing the separate cave groups to share maps and ongoing exploration both inside and outside the park's boundary. On August 27, 1983 with drought conditions similar to those in 1972, CRF parties from Mammoth Cave explored Logsdon River, upstream 1800 ft. (0.55 km) beyond what had been a siphon, leading toward Roppel Cave. Two weeks later on September 10, 1983, two teams of cavers representing CRF, CKKC, and Quinlan's NPS researchers entered Mammoth Cave via the Ferguson Entrance and Daleo Entrance. They met underground and connected Roppel Cave to the Mammoth Cave System, yielding a 294.42 mile (474 km) system which was officially announced on October 8, 1983 (Borden and Brucker 2000).

A new entrance to Roppel Cave, named Kahn, (Fig. 4.8) was opened by Dave Weller and a crew of CKKC and CRF cavers on Labor Day weekend, 1989. Passages reached through the Khan entrance are very near passages of the Fisher Ridge Cave System, but those at each end are not at the same elevation. Weller next opened the Downey Avenue entrance on his own land southeast of the park in 1992. This protected entrance, in the center of Roppel Cave, has led to steady exploration and scientific understanding of the portions of Mammoth Cave outside the park's boundary to the east.

During Thanksgiving of 1990, Rick Olson found two hidden crawlways on the same day, which led to the Southern Highlands of Mammoth Cave off Logsdon River, leading to about five miles (8.1 km) of new cave. Exploring in this same area on June 29, 1991, Bob Osburn, Julie Sotsky, and Greg Sholley surveyed 1000 ft. (305 m) of gypsum crawlway that led to a large trunk passage named Kämper Avenue, in honor of Max Kämper. By April 1991, Mammoth Cave System was more than 330 miles (531 km) long.

Beginning in 1998, exploration and survey trips involving very exposed climbing in Roppel Cave led to a series of major discoveries by James Wells, Dick Market, Seamus Decker, Peter Zabrock, John Feil, Rick Olson, and others. Starting at the top of Green Eggs Dome, which overlies Logsdon River, a traverse led to several miles of new passages in an area named "Dixie" and a stream called "Denial River". The new find did not head south to the main part of Toohey Ridge as desired but ended connecting back north to known passages.

James Wells and Alan Canon found the entrance to Hoover Cave in September 2003, while evaluating surface features southeast of Roppel Cave. Over the next two years, arduous exploration by CKKC and CRF cavers led to more than two kilometers of crawlways and canyons. On March 19, 2005, a party led by Wells and including Feil, Market, and Alan Canon widened a tight crack to descend into walking passage that reached the top of Wildcat Dome in the eastern part of Roppel Cave. With this connection, the Mammoth Cave System became 362 miles (583 km) long.

4.10 Exploration in the Western Area of the Park

In 1976 Rick Schwartz, working with Dr. James Quinlan and the NPS Uplands Research Laboratory, found Whigpistle Cave immediately west of the park. Two years later Don Coons, Sherri Engler, and others found a huge trunk in the cave. Dye tracing established that Whigpistle's Red River was the downstream continuation of Hawkins River beyond the sump under Proctor Cave. Whigpistle Cave became the key to understanding the hydrology of much of the park south of Green River. Quinlan's assistants and Western Kentucky University cavers surveyed 23.5 miles (38 km) of cave passages in the following seven years.

In 1994, explorers that had studied James Cave near Park City began exploring Jackpot Cave just outside the park's

Fig. 4.8 Khan entrance to Toohey Ridge Cave System, August, 1989. L to R: Tom Brucker, Dave Black, and Jim Borden. Stan Sides photograph

southern boundary. Don Coons and Jim Quinlan's assistants had earlier visited the cave they named Natural Bridge Cave. The James Cavers proceeded through tight crawls and difficult canyon traverses. In August 1995, they discovered a beautifully decorated passage named the Celestial Borehole. Returning Labor Day weekend 1996, they found that Alan Glennon and Jon Jasper had connected Jackpot Cave to nearby Martin Ridge Cave on June 24, 1996. Glennon and Jasper had discovered and explored Martin Ridge Cave on April 11 of the same year. Alan Glennon, Jon Jasper, and Chris Groves proceeded to connect Martin Ridge Cave to Whigpistle Cave on August 24, 1996, resulting in the 33-mile (53 km) Whigpistle Cave System.

Joel Despain, Pat Kambesis, Aaron Bird, and cavers from Western Kentucky University continue to explore passages in the park across the valley from Proctor Cave using the Whigpistle Cave System entrances. The Whigpistle System now has 35 miles (56 km) surveyed and has passages less than a quarter mile from Proctor Cave (Kambesis 2012). Hydrologic connection to Mammoth Cave has been confirmed, but explorers have found no high level passages that cavers can traverse in order to connect with the nearby Proctor Cave section of Mammoth Cave.

4.11 Recent Exploration Involving Prehistoric Human Activity in Mammoth Cave

Many years of study by the CRF Archeological Project were hampered by the fact that all areas where Indians explored and mined minerals underground had been visited and altered by historic visitors and tour development. This changed dramatically on June 30, 1996 when Rick Olson, on a CRF surveying party, found undisturbed Indian torch material in the lower portion of Mammoth Cave off Vanderbilt Hall. Survey and exploration up the Corkscrew to Broadway resulted in more evidence that Indians had explored down the Corkscrew to the lower levels of the cave long before Stephen Bishop crossed Bottomless Pit in 1838.

Olson and Dick Market returned to the Corkscrew area on January 17, 1998, to find a walking passage that the Indians had mined but was thereafter undisturbed. This passage was later named Robbins Run to honor archeologist Louise Robbins. The following day, another outstanding archeological discovery was made. Olson and Rick Toomey entered a tight crawl off Bunker Hill in Rafinesque Hall to discover a major passage leading on, previously explored and mined by Indians, but not historic man. This passage, named "Watson Trace" in honor of Patty Jo and Red Watson, remained untouched until May 28, 2005 when systematic study and survey began.

4.12 Underwater Exploration

On September 1–2, 1956, divers from the Cincinnati Diving Club performed reconnaissance of Pike Spring on Green River below Cathedral Cave and FCCC. Dave Roebuck, with four additional divers, explored a large conduit feeding Pike Spring on September 29–30, 1956. Very low flow from the spring and silting made travel up the conduit very difficult and no air-filled passages were found.

CRF members Roger Miller and Frank Fogarty entered Echo Spring resurgence and traveled 3340 ft. (1018 m) in flooded conduit to emerge in an air-filled passage near the third landing in Echo River on January 15, 1981. The divers returned a week later and completed surveying the water-filled passage in a remarkable feat of cave diving and exploration.

Arthur T. Leithauser arrived at Mammoth Cave in October 1981 to study the endangered Kentucky Cave Shrimp, *Palaemonias ganteri*, for the NPS as part of his doctoral dissertation research. He studied numerous passages in Mammoth Cave and dived regional karst springs in the 1980s with SCUBA equipment in his shrimp research. On February 16, 1985, he performed a dive of the upstream Mystic River sump in Mammoth Cave. He studied many base-level streams both inside and outside the park during research extending over several years. The full extent of his dive exploration is incompletely documented because it was part of a broader biological inventory.

Attempts were made by CKKC to dive the Logsdon River sump upstream in Roppel Cave to follow the cave's base-level passages upstream southeast toward the interstate highway. In October 1984, Wes Skiles and Ron Simmons passed the first sump heading upstream. Diving on Labor Day weekend, 2013, they found no lengthy extension of the Logsdon River passage upstream. Sumps in upstream and downstream Logsdon River, Whigpistle's Red River, and Hawkins River remain areas of promising underwater exploration.

4.13 Recent Mammoth Cave Area Discoveries

Stan Sides led CRF cavers back to Floyd Collins' Donkey Cave on January 3, 2009 with the purpose of evaluating the extent of formation mining by Floyd Collins in his cave. Survey and documentation began during the 2009 Thanksgiving CRF expedition. Exploration revealed many passages that had been found when the cave was last entered in 1955 by the Flint Ridge Reconnaissance cavers. Joyce Hoffmaster crawled through a very tight shaft drain on a CRF survey trip on May 29, 2011. She climbed down into Pohl Avenue in Mammoth Cave, connecting Donkey Cave to Mammoth Cave.

Rick Olson led CRF personnel to explore a shaft near the end of Gallows Way behind Ruins of Karnak during the Labor Day weekend of 2009, having first seen the shaft drain lead in 1975. Survey of the lead continued to a stream named River Acheron, which is an important tributary of River Styx, coursing about one kilometer beneath the Historic Section. Promising exploration and survey continue off River Acheron and other areas of Mammoth Cave, in Roppel Cave, in the caves of Flint Ridge, and in the Whigpistle Cave system.

4.14 Conclusion

CRF, CKKC, the James Cavers, and the Fisher Ridge Cavers continue to discover miles of cave passages both inside and outside park boundaries. Passages of the Fisher Ridge Cave System and Whigpistle System are within reasonable proximity of the Mammoth Cave System. Cooperative research agreements and the policies of the NPS enhance exploration while resource management remains central to all activities. Western Kentucky University's Center for Cave and Karst Studies has an important role in the exploration of the Mammoth Cave area and promotes academic studies in the region. More than 400 miles (640 km) of the Mammoth Cave System have been surveyed, and the potential to discover more caves is great. When will current exploration of the large cave systems surrounding the park yield successful integration with the Mammoth Cave System? Modern computer-processed cartography allows rapid understanding of passage correlations, and Global Positioning Systems allow surveyors to determine what surface features are above underground survey stations. Sump diving between the downstream Proctor Cave section of Mammoth Cave and the Whigpistle System has strong potential to connect these caves. The Fisher Ridge Cave System explorers will eventually find low-level passages near Roppel Cave, making connection to the Mammoth Cave System likely.

In the final assessment, exploration and cave cartography are essential to karst research. Survey of cave passages coupled with explorers insatiable drive to learn what is around the next corner is key to having an 800-mile (~1300 km) cave system, or even longer.

Acknowledgements The author would especially like to thank Gary Berdeaux, Chuck DeCroix, Nick Crawford, Harold Meloy (deceased), Bill Napper, Kay Sides, David and Stephen Sides, Gordon Smith, Norman Warnell, and Richard A. and Patty Jo Watson. I thank all the students I have had the honor to instruct in my class, "Exploration of Mammoth Cave," for Mammoth Cave National Park and Western Kentucky University's "Summer in the Park" series. I thank all my many colleagues in the Cave Research Foundation with whom I have been caving since 1962, for adding so much meaning to my life. This is written as part of the author's research for Mammoth Cave National Park research project, MACA H1.

References

Borden JD, Brucker RW (2000) Beyond Mammoth Cave. Southern Illinois University Press, Carbondale and Edwardsville

Brucker RW, Watson RA (1976) The longest cave. Alfred A. Knopf, New York

Carstens KC, Watson PJ (eds) (1996) Of caves and shell mounds. The University of Alabama Press, Tuscaloosa and London

Collins H, Lehrberger JL (2001) The life and death of Floyd Collins. Cave Books, St. Louis

DeCroix CJ (1998) Benjamin F. Einbigler: Pioneer explorer of Mammoth Cave. J Spelean Hist 32(2):31–36

Forwood W Stump (1875) An historical and descriptive narrative of the Mammoth Cave of Kentucky, 4th edn. J. B. Lippincott & Co., Philadelphia

Hovey HC (1882) Celebrated American caverns. Johnson Reprint Corp., New York and London

Hovey HC (1912) The Mammoth Cave of Kentucky. John P. Morton & Co., Louisville

Imlay G (1797) A topographical description of the western territory of North America, 3rd edn. J. Debrett, London

Kambesis P (2012) Whigpistle Cave and exploration history in the Mammoth Cave region. NSS News 70(5):4–5

Lawrence J Jr, Brucker RW (1955) The caves beyond. Funk & Wagnalls Co, New York

Lix H (1946) Mammoth Cave's underground wilderness. Living Wilderness 11(19):4–9

Meloy H (1975) Historic maps of Mammoth Cave. J Spelean Hist 8 (3&4):26–31

Meloy H (1985) The Bransfords show Mammoth Cave. J Spelean Hist 19(1):4–8

Murray RK, Brucker RW (1979) Trapped. G. P. Putnam's Sons, New York

Sides SD (1971) Early cave exploration in Flint Ridge, Kentucky: Colossal Cave and the Colossal Cavern Company. J Spelean Hist 4(4):63–69, 74

Sides SD (1997) Pheltius Valentine Simons—pioneer owner of Mammoth Cave. J Spelean Hist 31(4):72–76

Wright CW (1870) A guide manual to the Mammoth Cave of Kentucky. Bradley & Gilbert, Louisville

A History of Map-Making at Mammoth Cave

5

Michael Sutton

Abstract

Documentation of Mammoth Cave as the longest known cave has a history dating to the early nineteenth century, not long after Caucasian settlers first encountered it. The progress of underground survey and cartography is followed from the earliest crude sketch maps. Progress was frequently not linear, with advances in cartography followed by regression prior to further advances, but with an ever-expanding geographic range. Recent efforts have continued that expansion but have also focused on increasing levels of detail and precision.

5.1 Introduction

118 years after Call's remark, we still do not have a complete map of Mammoth Cave. Call's pessimism was prompted by the intransigence of the Croghan heir managers who knew that an accurate map would show that the cave extended beyond the boundaries of the Estate. Today, the managers are willing partners and the problem is logistical. Almost all the known cave has been cartographically documented to some degree, but in a long cave being certain that every obscure lead has been found and pushed to its limit is nearly impossible and major finds continue to be made to this day.

The earliest account of Mammoth Cave's cartographic history is that of Horace Carter Hovey (Hovey 1883) which discusses the major maps of Lee, Bishop, and Blackall. Hovey's article of 1899 was more comprehensive and discussed most of the earlier efforts except for the earliest map, which was not widely known at that time. About the same time, Hovey's sometime colleague, sometime rival, Richard Ellsworth Call published his own history of Mammoth Cave maps (Call 1897) up to and including Hovey's maps.

A complete map of the cave … will always be impossible. (Richard Ellsworth Call).

M. Sutton (✉)
Cave Research Foundation, Cave City, KY, USA
e-mail: haganandsutton@gmail.com

There was a gap until 1965 before the next major history appeared, in the form of unpublished notes for a presentation by James Quinlan. Harold Meloy's *Early maps of Mammoth Cave* (Meloy 1968) goes into more detail, and Meloy's research still forms the basis for much of our understanding of the earliest maps. In unpublished notes for a lecture, Meloy expanded his coverage to discuss most of the maps produced up until that date. Later, Sutton (1990) gave a history of CRF survey and cartography at Mammoth Cave, George (1996) gave another account of the early maps, and Brucker (2008) summarized the more important maps, emphasizing the use of cartography to guide further exploration.

5.2 The First Map: The Eye-Draught

What constitutes a cave map? Must a map be founded on an instrumental survey, or does an approximate sketch count? Hovey (1899) asserted that sketches could not properly be called maps "no use of chain or compass being made," but the dictionary is less picky, and we will adopt that more liberal view that non-instrumental sketches count as maps. By the time European settlers encountered Mammoth Cave sometime in the late 1700s, cave cartography already had a toe-hold in the USA, with a 1782 map of Madison's Cave, Virginia, created by Thomas Jefferson.

Jefferson also figures peripherally in the story of the first Mammoth Cave map of around 1808–1809 (Meloy 1975),

H.H. Hobbs III et al. (eds.), *Mammoth Cave*, Cave and Karst Systems of the World,
DOI 10.1007/978-3-319-53718-4_5

which like the Madison Cave map is called an "eye-draught," meaning that these maps are sketches and not based on surveys. There is an interesting if implausible excuse for this attributed to Frederick Ridgely (1757–1824): "in whatever position of the cave the compass may be placed, the compass directly points to the … entrance"! (George 1996: 75). Passage junctions are in the correct order, but the layout of the passages is unrealistic and written distances are greatly exaggerated—Blue Spring Branch, for example (shown but not named), is "explored 3 miles"—its length is actually about a half mile.

The Eye-draught map exists in three known versions: (1) the "Ridgely" copy in the Benjamin Rush papers (Anon. 1811a)—this is in color and includes a lot of crudely written notations; (2) the "E. I. DuPont" version, traced from the Ridgely map (Anon. 1811b)—the text is almost identical but the lettering is somewhat neater (Fig. 5.1) and (3) a published version in the 1853 edition of Jefferson's *Notes on the State of Virginia*, one of many posthumous editions. The so-called Jefferson map is a much more skillful job of drafting. The orientation has been changed from south up (more or less) to east up. The inscriptions are transcribed word for word from the Ridgely version and on both this and the "DuPont" copy, the entrance is clearly labeled. The map covers all of Main Cave (Broadway) and all of what was later named Audubon Avenue. It also shows junctions with the major side passages.

All three versions of the map are clearly instruments of commerce; the notations go into some detail concerning the location of "Salt Petre" deposits (peter dirt or calcium nitrate), the principle industrial product from Mammoth Cave, and include productivity figures, e. g. "6 lbs of Salt Petre to every bushel" of cave dirt. The map also makes note of Glauber salts (mirabilite), which was used in medicine and fabric dyeing but would have been too scanty for economic exploitation. Water sources, such as the spring at the entrance "from whence the water may be conveyed in pipes," are also shown as is the position of "leeches"—peter dirt leeching vats—in the Narrows. The places the known copies wound up are also indicative of commerce: Rush was a professor of chemistry and had lectured on saltpeter production; DuPont was a major saltpeter manufacturer and a customer of the Mammoth Cave saltpeter operation, and with the War of 1812 looming, Jefferson would have had a keen interest in potential sources of scarce war supplies. Meloy (1968) is probably right in speculating that the motivation for the map was to tempt a potential buyer, perhaps the saltpeter merchant Charles Wilkins.

Maps also enshrine place names, and the eye-draught is responsible for preserving the earliest known place names from Mammoth Cave: The Narrows (later Houchins Narrows), the Haunted Chamber (Gothic Avenue), the Sick Room (Harvey Avenue), the Big Room (The Rotunda), and not least the name Mammoth Cave itself. The earlier name, Flatt's Cave, was dropped in favor of a more commercially appealing and evocative name to go along with the cave's upcoming new status as the center of a fairly major industrial enterprise. The identity of the map's author remains speculative. As proposed by Meloy (1969) and others, the most likely candidate is Fleming Gatewood Sr., who along with Wilkins purchased the cave in 1810.

5.3 More Early Sketch Maps: Bogert and Ward

The next known map, "Green River, or Mammoth Cave," c. 1813, is the so-called Bogert map, named for John G. Bogert, a New York attorney who forwarded a copy to Samuel Mitchill, editor of the *Medical Repository*. Hence, in 1815 this became the first published map (Anonymous 1815).

This map too is an "eye-draught" with little conformity to the actual layout of the cave. The coverage is slightly less than the earlier map; although there is some uncertainty in interpretation of the passages, the map appears to end at the Cataracts. But some of the side passages are shown in much more detail, notably the passages later named Ganter, Gothic, Gratz, and Blacksnake Avenues and, with somewhat less confidence, Fox Avenue and Flint Alley. The correlation to actual passages is far from precise, not only in matters of proportion but also in details of passage junctions, especially in the mazy lower levels. Nevertheless, the map shows some cartographic advances. The earlier eye-draught is strictly two-dimensional, but the Bogert map makes an attempt to show passage levels, with underlying passages indicated by thin dotted lines. There are far more names on this map, and there is a numbered key to named places. Many of the names can be correlated to modern replacements, and the map features the first names other than Mammoth Cave to survive, barely, to modern usage—Deserted Chamber (Wooden Bowl Room and Blacksnake Avenue on the Historic tour) and the Basket Room in Ganter Avenue. The author is unknown, but the most likely candidate is Archibald Miller, Wilkins' agent in charge of the saltpeter operation.

A short time later, a new map was published by Nahum Ward, a wealthy businessman and real-estate speculator (Ward 1816). This was the first map with a known author, and it too exists in at least three different versions, one published as a single-page broadside and others in a variety of periodicals together with Ward's account of the cave. The complicated chronology of Ward's publications is discussed in detail by Riggs (2007). The map was based on a lengthy trip taken by Ward in November of 1815.

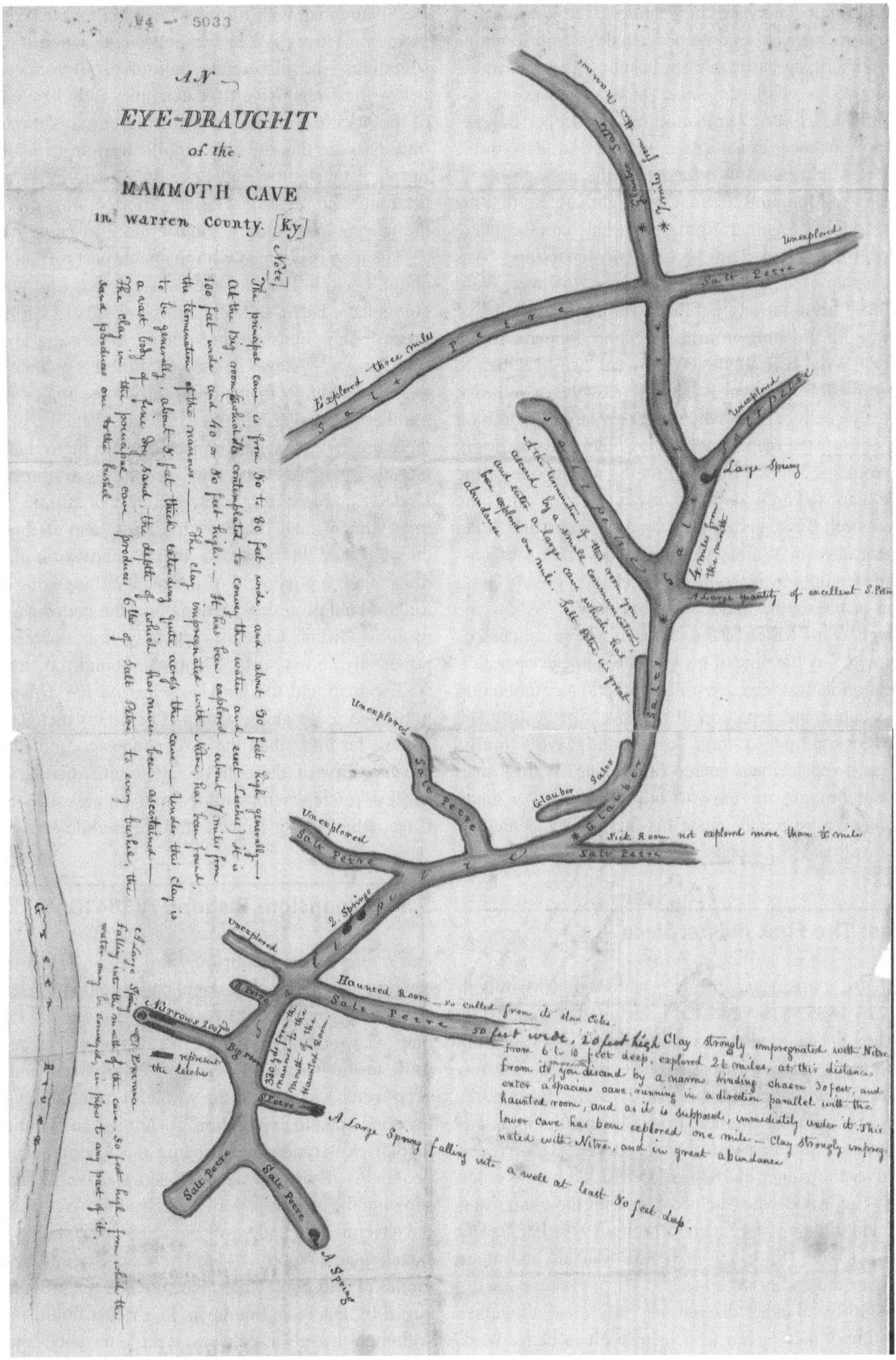

Fig. 5.1 "Du Pont" version of the eye-draught, the first known Mammoth Cave map. The approximate orientation is east upwards

The most immediately striking feature of the Ward map is that it shows passages extending beneath the Green River, an incorrect but not unreasonable proposition given the geological knowledge of the time and the lack of data on passage elevations. Closer examination reveals deeper flaws—were it not for occasional place names and descriptive notations, it would be hard to recognize the passage layout as belonging to Mammoth Cave. One might suspect that the map was cobbled together from somewhat hazy recollections, except that Ward states he used a compass and "was sketching a plan of the cave" during the trip. However, Ward may not have been entirely familiar with the functioning of the compass. In traveling from the "second hoppers" to the "chief city" (Wrights Rotunda), Ward went "west … then S. W." The direction of travel is actually east then southeast. This may also help explain why the arrow labeled south on some versions of the map actually points more or less north, and why some passages wound up on the wrong side of the river. The map is skimpy on place names—it even omits the name Mammoth Cave, giving the map the generic title "Plan of the great cave in Warren County." The naming of passages and features was soon to accelerate greatly with expansion of the tourist trade.

Although Ward writes of the "hoppers where saltpeter is manufactured," by the time of his visit the saltpeter operation was at best on its last legs, most likely due to a combination of price collapse and depletion of the cave's peter dirt. The owners were starting to look toward the cave's tourist potential as a replacement source of revenue. Ward's map and account brought the cave to the attention of a much wider audience, especially since his accounts appeared in British as well as US journals.

5.4 Lee: The First Masterpiece

By the 1830s, tourism was firmly established at Mammoth Cave, and by 1835 "a surveyor has been among its vaults; he has stretched his chain along its galleries, he has broken the heart of its mystery and, with cruel scale and protractor, has laid it down upon paper" (Anon. [Bird] 1838). Edmund F. Lee, a 24-year-old civil engineer from Cincinnati, Ohio, was hired to produce this first instrumental survey. He surveyed the cave over the winter of 1834–1835 assisted by the guide George S. Gatewood who "helped him with the chains and markers" and told him place names (Meloy 1975, 1977). The map that resulted (Figs. 5.2 and 5.3) was an enormous advance over the earlier sketches.

Hovey (1899) is rather dismissive: "it has had altogether too much importance given to it in relation to better work done since …" and criticizes the map for "serious errors." Call (1897), on the other hand, believes that Lee's layout of the passages is "absolutely accurate." Call also correctly credits the map with forming the basis for subsequent maps, including Hovey's. Lee was inaccurate when it came to pit soundings—he placed the bottom of Bottomless Pit well below the Green River, for example. But close examination of the map in comparison with current maps reveals that the only passage that deviates significantly from reality is Blue Spring Branch—by and large Lee's survey is remarkably accurate, and regardless of any shortcomings in the survey, the map is indisputably a masterpiece of cartography.

The map is oriented with magnetic east up and is drawn at about 1:3900. The passages are colored, and although the colors have faded, they are a pleasing mix of earth tones and greens. The colors are not related to passage levels; rather each major passage is colored differently than adjoining ones. In addition to including all of the major passages, a number of smaller side passages are sketched in. Although there are few passage details, there are many helpful annotations, such as "very low" or "heaps of broken stone." Underlying passages are traced faintly in outline, making the three-dimensional relationships clear, and if this were not enough, Lee introduces a major innovation in the third dimension: a good deal of the map surface is taken up with longitudinal profiles ("sections") of the major passages. The profile of Main Cave is especially impressive, and rather artistically shows surface features around the entrance.

The map and the booklet *Notes on the Mammoth Cave* published to go along with it (Lee 1835) include 125 place names, far more than the earlier maps combined. Although the majority of these have fallen into disuse, Lee is the earliest reference for still current names such as Solitary Cave, Blue Spring Branch, and Sidesaddle Pit.

5.5 Expansion: Bishop, 1838–1845

Lee's map shows about 12.9 km (eight miles) of passage. Three years later, the slave cave guide Stephen Bishop crossed, or more likely bypassed, Bottomless Pit, and the map was suddenly out of date. Bishop's feat opened up a route to the base-level Styx and Echo rivers, and once that deep water was passed, the whole of Mammoth Cave Ridge lay wide open to exploration. Bishop was a major participant in that exploration and became the leading expert on the geography of the passages beyond the rivers. The first map showing the new discoveries was by John Wood, who visited the cave in 1841 and wrote an article in which he included a sketch map (Wood 1841). The map extends beyond Bottomless Pit to Echo River. Passage and feature names are all keyed by letters and numbers in a rather confusing scheme. Although there is a scale and written distances, passage orientations and lengths are fairly random.

In 1842, following closely on the rush of exploration, the new owner John Croghan brought Bishop to his Louisville

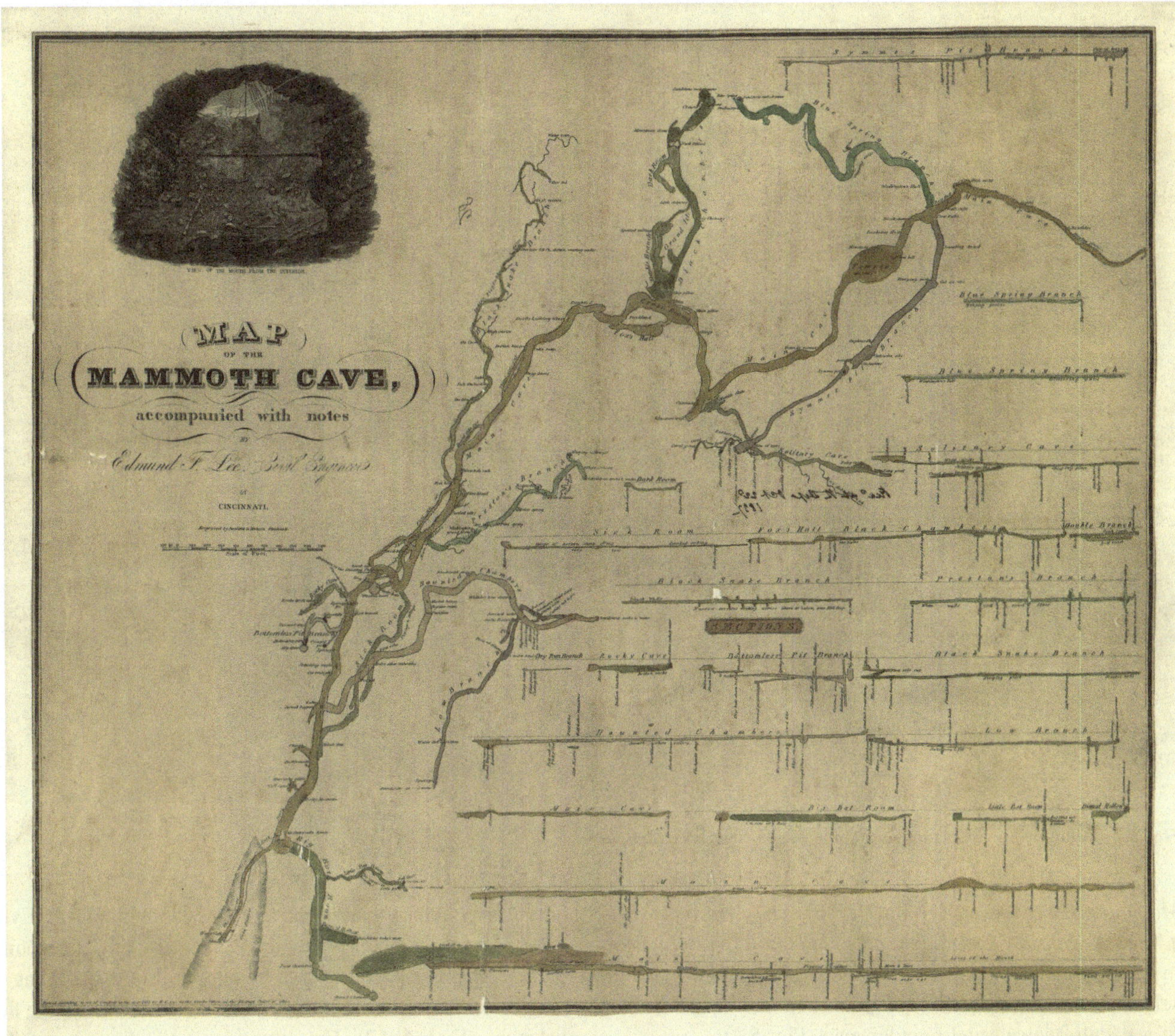

Fig. 5.2 1835 Edmund F. Lee map

mansion and had him draft a new map. According to Meloy (1975), Bishop drafted the map in pencil and Croghan's brother Colonel George inked and lettered it. The motivation was tourism potential; Croghan wanted to quickly open up the new discoveries for commercial tours. There is no scale or north arrow, but this map too is oriented east up. The old, well-known cave on Bishop's map is clearly based on Lee in that the passage relationships and general orientation are more or less correct, but Bishop did not trace the map directly from Lee, rather it is a crude sketch based loosely on the earlier map. There had been nowhere near enough time to commission an instrumental survey of the new discoveries, and that portion of the map is a rough sketch with greatly distorted distances and directions. Exploration, and the map, went as far as the end of Cleaveland Avenue and its side passages, and westward as far as "Harlan Avenue," probably present-day Boone Avenue. The map includes a generous portion of place names, and this assists greatly in correlating places on the map with geographic reality underground. But not all place names have obvious modern equivalents, so parts of the map simply cannot be resolved with confidence. Passage relationships are made more confusing by the treatment of overlaps; often the walls of the lower passage, but sometimes of the upper passage are omitted, and in most cases both passages are drawn with solid lines, making it impossible to know which is uppermost.

There are 100 place names on the map, and more than 30 of these are the first known occurrence of current names, including such well-known features as Giants Coffin and Echo River. Bishop's *Map of the Explored Parts of the Mammoth Cave of Ky.* was published in what became the standard guide book for several decades, *Rambles in the*

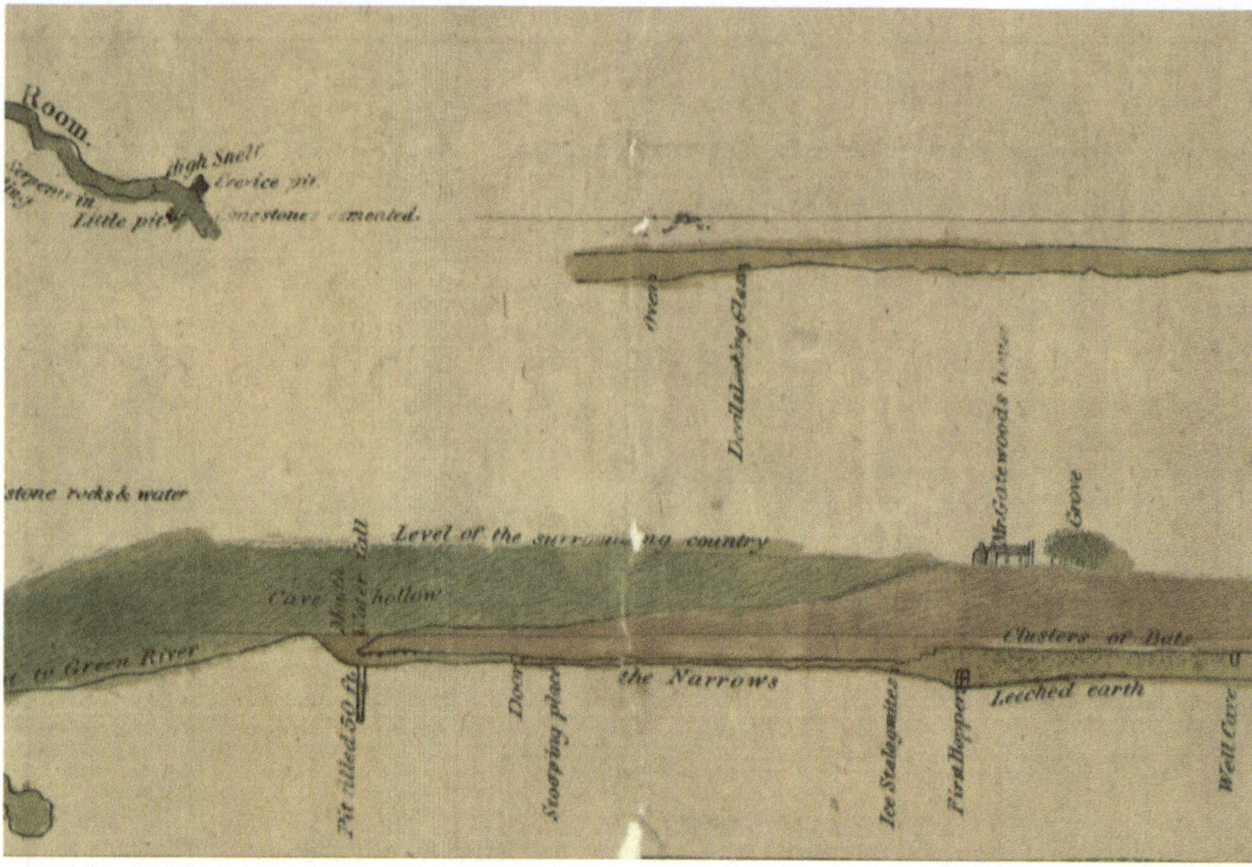

Fig. 5.3 Detail of Lee map showing profile view of entrance area

Mammoth Cave (Anon. [Bullitt] 1845). The map represents a huge advance in geographic coverage but a big regression in cartography. Still, it remains remarkable that a self-educated slave became the published and named author of a widely viewed and popular map (Fig. 5.4).

5.6 Secret Maps: Eyth and Blackall

Following Croghan's death in 1849, managers of the Estate became increasingly concerned over the implications of maps being publicly available. Any map including a scale would seriously undermine the advertising claim that the cave was 150 miles long (241 km). A more serious concern was that an accurate map would show that the cave extended well beyond the property boundaries and would be an invitation to open up entrances outside the control of the Estate. This fear later proved well founded. According to Meloy, even Bishop's map was suppressed after 1856, and in theory, no surveys of the new passages were permitted.

Nevertheless, at least one survey beyond Bottomless Pit took place. In 1866, the German engineer Max Eyth, famous for his innovations in agricultural machinery and a prolific author of popular travelogues, visited Mammoth Cave. Eyth was told of the prohibition against surveying but got around it by taking notes during two regular tourist trips using a compass borrowed from the only other visitor and pacing the distances (Eyth 1867). His rendition is the first to put the passages beyond Bottomless Pit on the long tour in a realistic perspective, although the relationship of the historic end of the cave to the new passages is badly skewed. Eyth's map includes an important innovation: while Lee had shown the third dimension in longitudinal profiles, Eyth shows passage cross sections (Fig. 5.5). The map is titled in German, but passage and feature names are in English. Although suppressed, the map somehow became available to the author R. Stump Forwood, who published an uncredited version in the 1875 edition of his Mammoth Cave guide book. The map is a close copy of the original but with the title in English and omitting Eyth's cross sections. The map was also featured in at least two later publications.

In 1870–1871, a would-be author of a new guide book, Dr. C.E. Blackall claims to have been allowed to survey. Whether or not the survey was sanctioned, the managing trustee, Joseph Underwood, suppressed the map, and it did not see the light of day until Hovey published it in Blackall

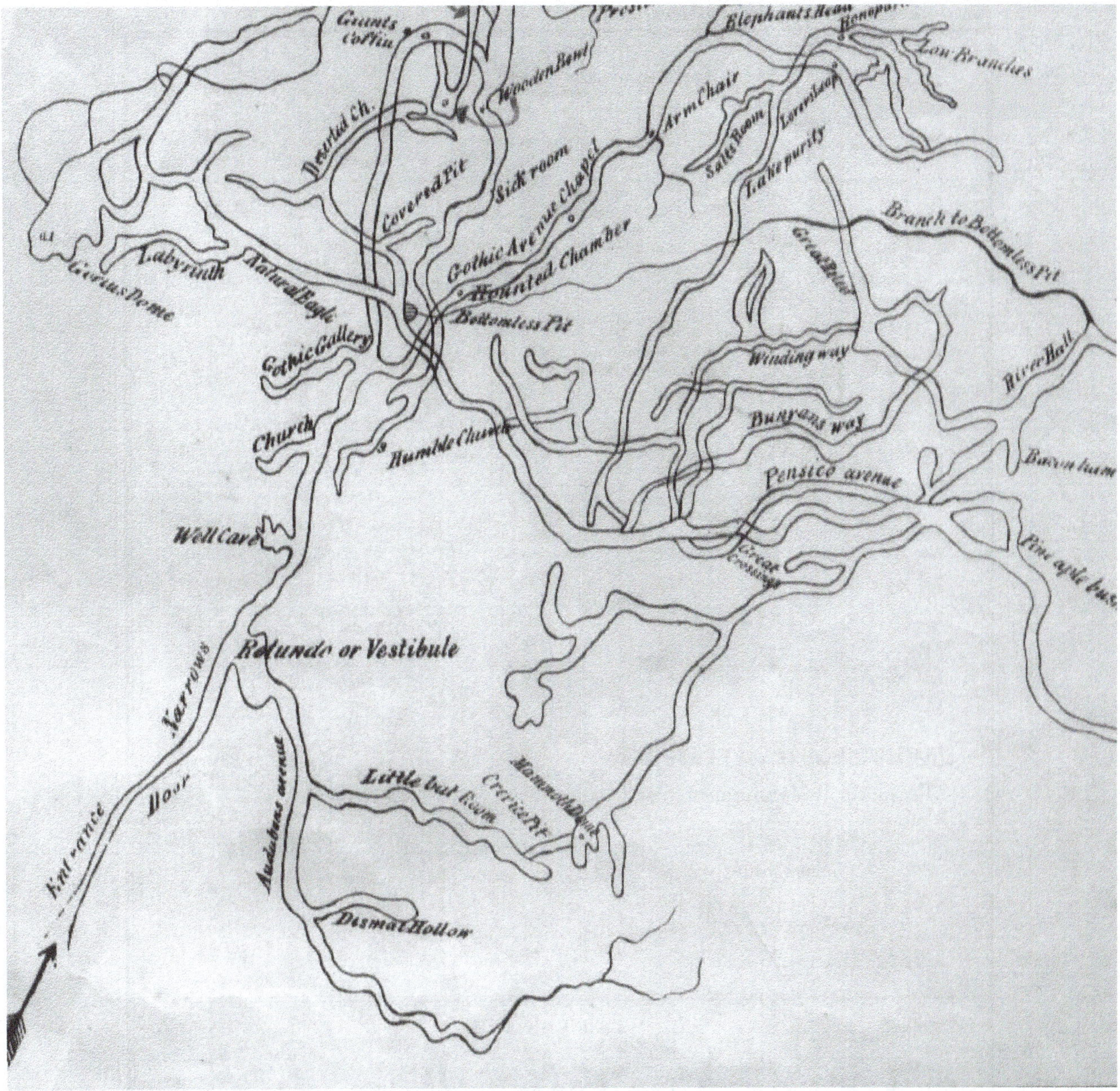

Fig. 5.4 Excerpt from the 1842 Bishop map showing the historic entrance area. This portion of the map is loosely based on Lee's map

1899. Blackall's map shows passages fairly realistically, although greatly simplified compared to Eyth. Unlike the Eyth map, Main Cave in the historic section is more or less in the right place. On a slightly different version of the map (Thompson 1909), Blackall states that the historic portion was based on Lee, but some passages are omitted, and there are some odd mistakes: e.g., the orientation of Solitary Cave is wrong, and the passage is mislabeled. Blackall's map is generous with place names, but few new names are added, and those few did not long survive. Hovey renamed Symme's Pit Branch in Blackall's honor.

5.7 The Hovey/Call Era

Horace Carter Hovey (1833–1914) was the most dedicated promoter of Mammoth Cave in the late nineteenth and early twentieth century. The numerous versions of Hovey's Mammoth Cave guide books, some of them co-authored with Richard Ellsworth Call (1856–1917), and dating from 1882 to 1912, included maps almost entirely synthesized from earlier maps. One of these was a map drawn in 1881 by Francis Klett, the cave manager at the time. A rendition of this map accompanies Hovey's story for children, *Brigham*

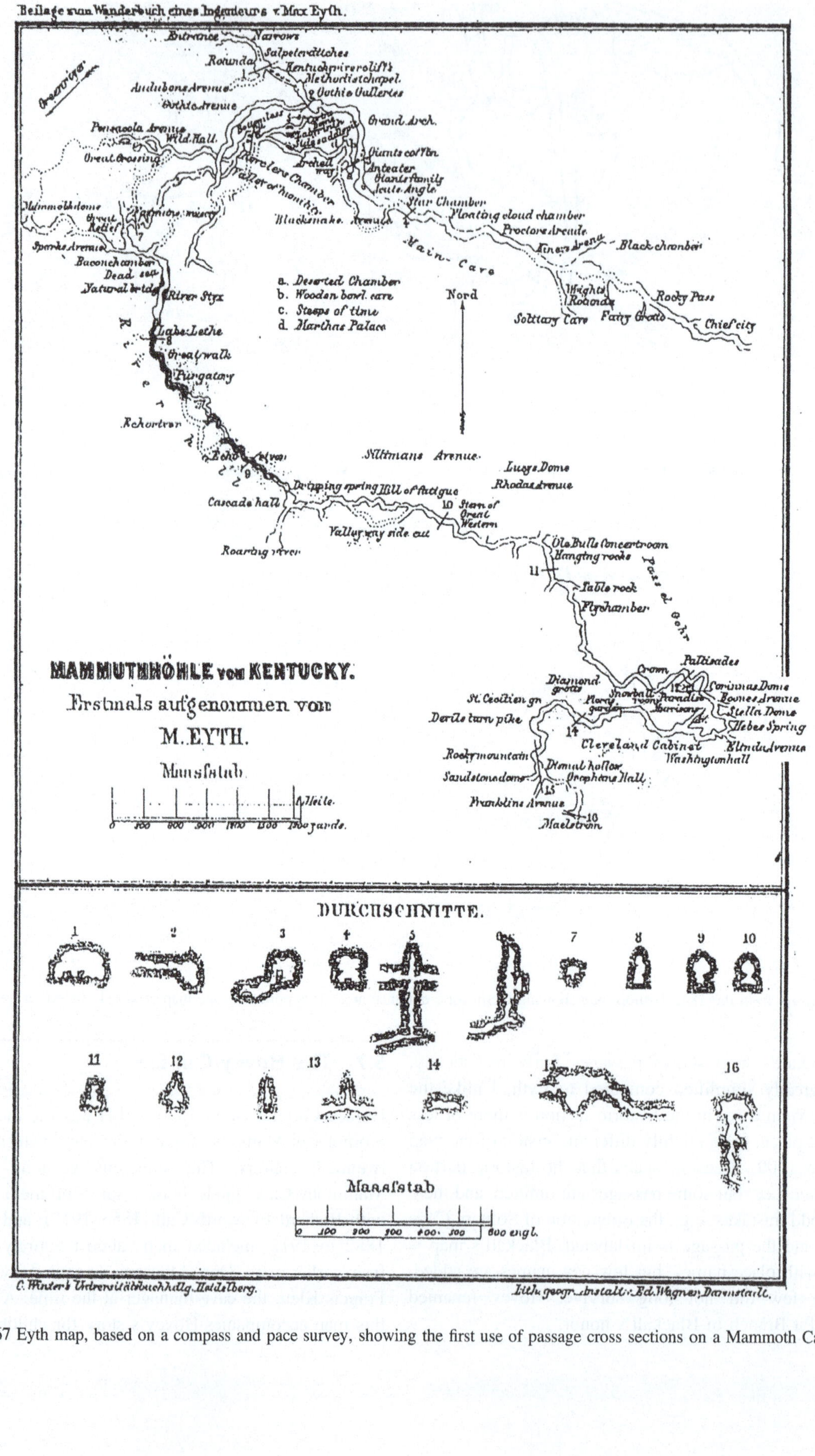

Fig. 5.5 1867 Eyth map, based on a compass and pace survey, showing the first use of passage cross sections on a Mammoth Cave map

the cave dog (Hovey 1882a). The map extends beyond Echo River only as far as Silliman Avenue, and it shows (but does not name) Welcome Avenue, a recently discovered alternative route from Ganter Avenue in the old cave to Silliman Avenue, bypassing Echo River. The map of the old cave is a survey not a sketch, with passages in reasonably correct orientations, but the rivers are closer to Bishop's sketch than to reality.

In the same year, Hovey published the first of his guide books (Hovey 1882b), incorporating the Klett map but relying on the Bishop sketch for passages beyond River Hall. The map was drafted by someone named "Bogart" of whom nothing is known. The map lacks a scale, and like most early maps, is oriented east upward. It is replete with place names but only six or so originate with Hovey. Simplified versions of the map appeared elsewhere, including the Encyclopaedia Britannica of 1883. In 1897, Call reworked the map with a few additions—e.g., Forts Way and Gallows Way are shown but not named, and parts of the Bishop sketch have been altered to bring them somewhat closer to reality. This rendition includes some half dozen new names still in use, including Dantes Gateway and Gratz Avenue. A minor but interesting map appears in Hovey's 1889 article on the domepits—both plan and profile views are given of the complex of pits around Bottomless Pit with depths and heights notated for each one, the depths measured by plumb line, the heights by balloon.

In 1903, there was another huge advance in geographic coverage, with the first survey to take place under Flint Ridge. In the late 1890s, Edgar Vaughan began mapping the connected Woodson-Adair and Bedquilt caves under the name Colossal Cavern, which an L & N Railroad Co. spin-off, the Colossal Cavern Company, was developing. The survey showed the potential for a more convenient entrance on property which the company promptly acquired (Sides 1971). The map by Vaughan and W.L. Marshall was published in Hovey 1903 (Fig. 5, Chap. 4). It shows both the new entrance and the Woodson-Adair entrance together with the Bedquilt Route from the Bedquilt Entrance. Passages are shown in black silhouette; there is a scale but no north arrow (orientation is north up). The map is the first source for many still current place names. In Hovey and Call 1912, a revised version was published showing additional passages, including the River Route and a fragment of what was later named Colossal River, in reality a small but extensive stream.

An unusual map appeared in 1904, published by a French visitor, Le Couppey de la Forest. The French language map, roughly translated as "a schematic map showing the respective positions of Mammoth Cave, Colossal Cave, etc. and of the Green River," gives a broad overview of Mammoth Cave Ridge, Flint Ridge and Joppa Ridge showing small-scale passages superimposed on the surface features. There is a scale and north arrow. Mammoth Cave is clearly based on the Hovey and Call maps, but the map also shows both Colossal and Salts caves on Flint Ridge and "Nectar Cave" (Proctor Cave) on Joppa Ridge, although none of these bears much similarity to the actual layout of passages. Interestingly, passages of Colossal Cave are shown extending under Houchins Valley toward Mammoth Cave. The map was compiled "from details given by M. Charlet," presumably Marty Charlet the Mammoth Cave Hotel manager. This map is a crude precursor of the much later map card series published by the Cave Research Foundation (CRF) showing the relationship of the surface to underlying caves.

5.8 Kämper 1908

Max Kämper's map eventually became the best known of all (Kämper 1908; Fig. 5.6). Kämper (1879–1916) had seen Max Eyth's map and was likely inspired by his famous predecessor (Kliebhan and Thomas 2008). Kämper was in the USA to study engineering and mining practices and visited the cave in early 1908. He became enamored with its vast size and complexity and proposed to carry out a comprehensive survey. Despite the earlier prohibition, the managing trustee Albert Janin approved Kämper's proposal, probably seeing the advantage of having an accurate map for his own use. He provided Kämper with board and lodging and the full-time services of guide Ed Bishop. Presumably Janin had no intention of making such a map available to the general public. Kämper stayed on at Mammoth Cave for eight months during which he and Bishop accomplished the most comprehensive and detailed cave survey conducted up until then.

Olson et al. (2013) show the comparative accuracy of Kämper's map compared to current surveys. Hovey (1909) writes that Kämper "used a good surveyor's compass … in the main cave and principal branches but relied on a pocket compass for the narrower passages and crawlways." He also wrote that the surveyors measured distances by pacing, but this method would have been seriously inadequate for surveying such a long and complex cave, especially since the many ups and downs would make the paced distance even more inaccurate. The precision achieved undoubtedly required a measuring chain (Olson et al. 2013). Kämper left a partial record of each day's activity, so we know that the survey began with his first new discovery, which he named Gertas Grotto for his distant relative and sweetheart Gerta Luyken (DeCroix 2008).

The resulting map lacks any apparent scale or orientation, but this information is in fact present in encrypted form (Olson et al. 2013). Kämper noted in a letter to Janin that the arrow showing the entrance points SE and a linear series of seven inconspicuous circles give a fixed scale (1:4000);

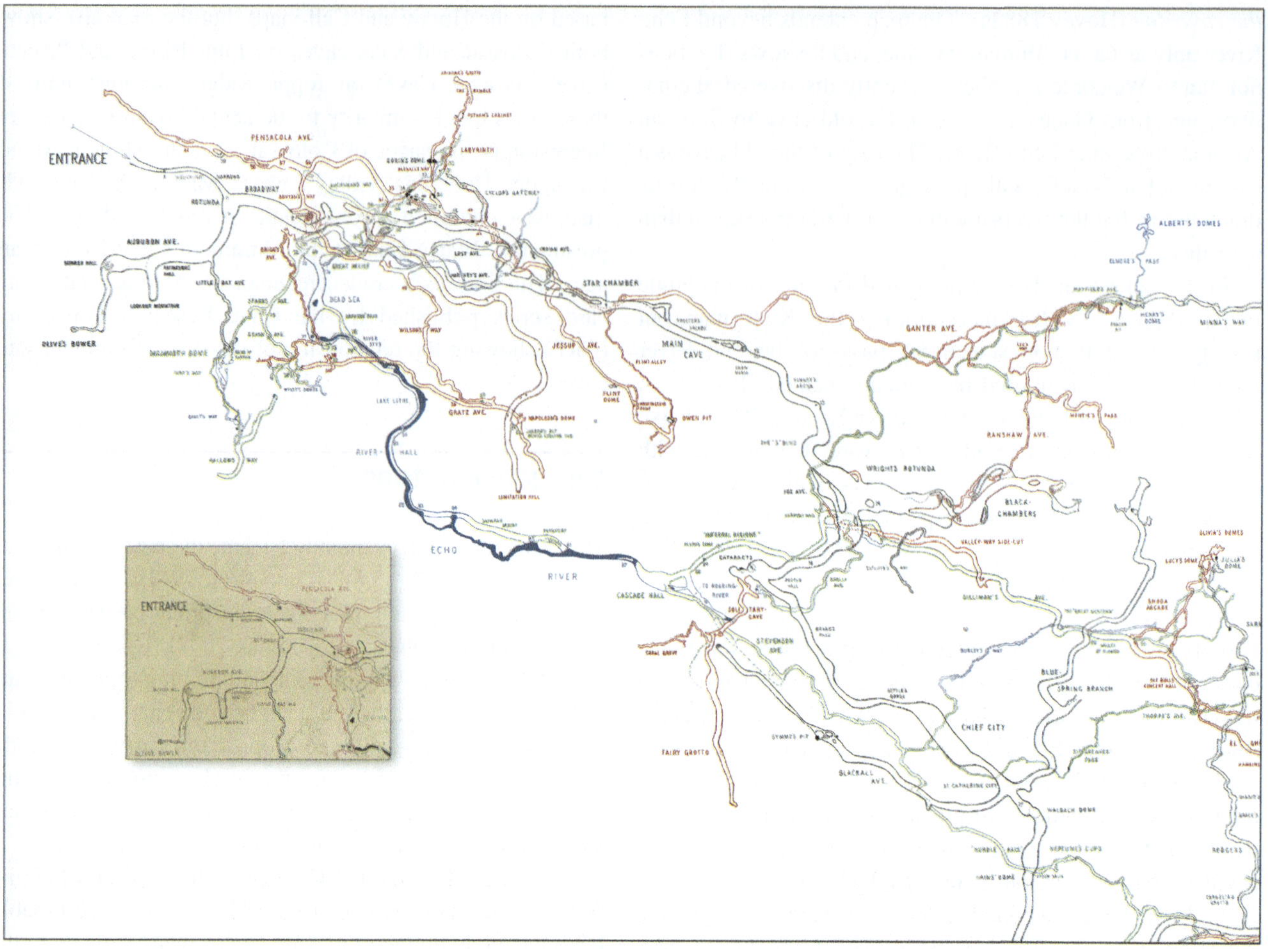

Fig. 5.6 Excerpt from the 1908 Kämper map as digitally restored by Seymour—*Inset* shows the unrestored original

moreover, one of these circles registers the map with a known point on the surface. A major innovation in Kämper's cartography is the use of color to show passages at different levels within the complex three-dimensional maze. The map divides the cave into five levels with the passage walls drawn with different colored inks. This important feature aside, the map is more noteworthy for the excellence of the cartography than for cartographic innovations. The map was drafted in pencil then colored in. In places, traces can be found of a lightly penciled-in line survey (Seymour 2013)—the main skeleton framework for any cave map.

The geographic coverage is greatly expanded. Earlier surveys were more or less confined to the Long Tour route, but Kämper carried his survey much further east, as far as Blairs Dome (later Aero Bridge Canyon) in Grand Avenue. Moreover, Kämper and Bishop pushed east into *terra incognita* on several different levels, from the upper level trunk of Grand Avenue to the near base-level Bransford Avenue. The density of mapped passages not surprisingly is highest close to the entrance and drops off progressively further east. At some point, Kämper must have realized that he was not going to get the entire cave mapped. The achievement is nevertheless impressive—the map shows approximately 36 miles of passage, all surveyed to an unprecedented precision.

The map is a major source for place names. Kämper uses a combination of names written on the map itself, and 112 names referenced with a numeric key for a total of 340. Well over 100 of these are first occurrences, in large part from passages that Kämper and Bishop discovered, including such well-known and still current names as Roses Pass, Emilys Avenue and Carlos Way. Swedlund and Crothers (2008) compared the names on the Kämper map with the Lee and Bishop maps and concluded that Kämper had used a far higher proportion of personal names, especially of local people. The original map is badly deteriorated, in large part due to earlier counterproductive attempts at preservation but Seymour (2016) produced a digital restoration, seeking to approach as close as possible to the original by processing computer images of the map.

5.9 Cave Wars

Even without knowing the scale, it would have been fairly obvious that Kämper's survey extended well to the east of property owned by the Mammoth Cave Estate—even parts of the existing Long Tour route ventured beyond the Estate boundary. Kämper's map was, like most of its predecessors, suppressed although late versions of the Hovey and Call compilations incorporated Kämper's discoveries of Violet City and the passages around Cathedral Domes—Martel Avenue, Bransford Avenue, etc. These are roughly sketched in, not traced directly from Kämper, and there is no scale. Hovey also derived accurate tour trail maps for the last edition of his guidebook.

Around 1915, George Morrison, an oil prospector, moved to the area and became convinced of the possibility of opening another entrance in competition with the Mammoth Cave Estate. To that end, he leased 2428 ha (6000 acres) on eastern Mammoth Cave Ridge and drilled many holes to try to locate the cave. This strategy failed, and Morrison determined to make a survey to locate a potential entrance beyond the Estate's property. In the annals of clandestine cave surveys, this was a classic. Back in 1908, Kämper and Bishop had tried to connect Violet City to Sandstone Avenue at the end of the long tour route but had instead blasted almost to the surface. Morrison probably heard of this from Henry Ganter, a disaffected former manager. In August 1915, Morrison had Bob Lively, a former Mammoth Cave guide, make an impression of the key to the entrance gate. With the copied key, Lively entered the cave and set a charge at Violet City. This allowed Morrison's crew to locate the point on the surface and excavate an illicit entrance on Estate property. Lively then guided a five-man survey crew led by Bruce Huffman on a 72-h survey trip. Soon after this marathon, the entrance was discovered and blocked.

In March 1916, Morrison's crews resumed nighttime surveys via the main entrance using the bootleg key. This came to an end soon after when the crew was caught red-handed while exiting at the Corkscrew, but the surveys had advanced far enough that Morrison was able to narrow the search for a new entrance beyond the Estate's property. The resulting Cox entrance was far more convenient for continuing the survey on eastward. Unfortunately for Morrison, the cave passages under the new entrance had been leased by the Colossal Cavern Company, and the Estate prevailed upon the company to bar Morrison from entry. But Morrison persevered, and in 1921, his workers blasted a new route into the cave on property he controlled. Morrison's New Entrance opened for business, and a map of the newly opened section by Roger Parrish, a local civil engineer, was published in Helen Randolph's 1924 guidebook.

The map pointedly gives prominence to the property boundaries and shows passages connecting new and old entrances—a fact that was vigorously contested by the Mammoth Cave Estate—and it includes a scale and orientation. Although good maps of the historic end of the cave were available to Morrison, the map shows it greatly simplified and distorted, presumably in order to downplay the Estate's tourist offerings. The New Entrance holdings on the other hand are shown realistically and with reasonable accuracy. The map is the principal source for place names in Morrison's part of the cave. Huffman also drew a map from his illegal surveys of the Estate's end of the cave, and this was published together with the Parrish map in Randolph (1924). The map purports to be "the first actual underground survey" a considerable exaggeration. It is fairly crude and imprecise, but passage layouts in the older section are at least recognizable and bear little relation to Parrish's wildly inaccurate version.

Major hostilities in the cave wars were brought to an end by 1936 when the rival properties were acquired by the Mammoth Cave Association for inclusion in Mammoth Cave National Park, although skirmishes continued into the 1950s and played an important role in later cartographic developments.

5.10 Mammoth Cave National Park

The Park had been in planning since the mid-1920s. As part of this process, Armin Lobeck was commissioned to write a monograph on the cave's geology (Lobeck 1928). Lobeck included an unusual map, which gives an oblique view of cave features as far as Echo River in relation to the surface, using cutaways to show the interiors of passages and domepits.

In 1935, a year before the property officially became a National Park (see Chap. 3), the US Geological Survey commissioned an accurate survey of the main trunk passages. The survey, led by H. D. Walker, used a 30-s transit to run survey lines from the Historic Entrance to Violet City and via Echo River to the Frozen Niagara Entrance with a branch to the new Carmichael Entrance. Another branch took the lower level route via Cathedral Domes to rejoin the main survey at Marys Vineyard. The survey was calibrated with Polaris sightings at each entrance. The surveyors also measured passage widths and heights at each station, using helium balloons for the higher ceilings. Accurate elevations were then obtained with a leveling survey, critical data for many purposes, not least the correlation of the levels of unconnected passages. Land survey practices had to be modified substantially to cope with cave conditions—the main constraints were the narrowness of many passages, the frequently short lines of sight, and the difficulty of bringing adequate light to bear. Permanent brass caps were set at entrances and major passage junctions, 37 in all (Mason 1950). The Walker data were later refined and corrected by Robert Hosley, and the benchmarks later formed a solid framework for the CRF surveys.

The most important spin-off from the Walker survey was a 1956 map by Mammoth Cave guide Raymond Nelson, the first map to show passages accurately all the way from the Historic Entrance to the Frozen Niagara Entrance—the eastern limit of the upper level trunk. The compilation map shows a good deal more than the trunk passages surveyed by Walker, although Nelson seems not to have had access to the Kämper map—for example, the low-level Belfry Avenue which was mapped by Kämper is merely estimated with dashed lines. There are also many passages discovered since Kämper, notably the passages in and around the New Entrance, including the long, low-level Cocklebur Avenue loop and the four-mile long New Discovery section. The latter had been discovered in 1938 by guides Carl Hanson, Pete Hanson, Leo Hunt, and Claude Hunt, and mapped under the direction of the NPS engineer, Paul McG. Miller. Miller's survey was accurate enough to permit a drilled shaft to intersect the passage. Nelson's map does not add to the nomenclature as it relies on earlier maps, especially Hovey. The map was revised in 1964 by Quinlan and Quinlan and again in 1972 by Pat Crowther, both unpublished. These versions change the orientation to north up and change some place names. Between Kämper and Nelson, about 74 km (46 miles) of passage were now mapped under Mammoth Cave Ridge. After the Nelson map, the traditional 150-mile (241 km) claim was quietly dropped.

5.11 The Longest Cave

Over on neighboring Flint Ridge was Floyd Collins' Crystal Cave, a struggling tourist cave left as an inholding within the Park. Around 1950, the owner commissioned an "accurate instrumental survey" of the tour route for a promotional brochure. Meanwhile, Bill Austin, a civil engineer and the owner's grandson, was exploring and mapping the extensive network of canyons below the tour trails. Austin plotted his surveys by a "latitude and departure" technique, whereby station positions are calculated by trigonometry from distances and bearings. Austin realized that he had a major project on his hands and began recruiting outside help from National Speleological Society (NSS) cavers from Ohio. This led to the Collins Crystal Cave expedition of 1954 to explore and map the cave's lower levels, an expedition most notable for its demonstration of how not to organize a survey of a large, complicated cave. The pace of exploration and cartographic development now became rapid. The story has been told in detail (Lawrence and Brucker 1955; Brucker and Watson 1976; Sutton 1990; Borden and Brucker 2000) and will be briefly summarized here.

The Ohio cavers conducted clandestine surveys due to Austin's unwillingness to share his maps. Austin eventually relented, showed line plots of surveys carried out illegally within the park, and involved the NSS cavers in illicit surveys in Unknown Cave; Austin and his co-conspirators had discovered that this obscure park cave was large and extensive. This led in 1955 to the connection of Unknown and Crystal caves. The NSS cavers started collecting a limited amount of data on the nature of the passages, generally by drawing occasional passage cross sections and rough plan views; these sketches were free-form and not to scale. Instrumentation consisted primarily of steel tapes and hand-held Brunton compasses, although transit surveys were conducted in some of the large upper level trunks (Brucker and Burns 1964). Vertical angles were measured only when deemed to be large—the built-in Brunton clinometer is imprecise when hand-held.

In 1957, the Flint Ridge cavers incorporated as the CRF with Roger Brucker as chief cartographer. In 1959, CRF negotiated an agreement with Mammoth Cave National Park and could now survey legally within the park, initiating a return to cartography by skilled volunteers rather than NPS professionals. This soon led to mapping in Colossal Cave, southeast of the Crystal-Unknown system, leading to the second major connection, between Colossal Cave and Salts Cave. By 1960, CRF had produced a series of 100 ft./1 in. maps of Crystal Cave, which by now had been absorbed within the park. The maps included surface contours that Brucker had scaled up from topographic maps and were drawn on 8½ in. by 11 in. sheets—small enough to be taken into the cave (Brucker et al. 1955). The cartography was an advance over Austin's line plots—passage outlines and topographic overlays were shown—but data processing regressed as Austin's laborious trigonometry was abandoned in favor of the faster but less accurate ruler and protractor. Later that year, a route was surveyed from the Lower Crouchway in Crystal-Unknown to Indian Avenue in Salts-Colossal, resulting in one vast integrated Flint Ridge cave system.

The collection of map sheets expanded to cover the whole of Flint Ridge, but survey was outpacing cartography. In 1964, Denver Burns took over as chief cartographer and along with Brucker brought the map sheets to completion. A professional artist was hired to draft the final version, and the maps were published in 1966. The Flint Ridge Folio of 30 sheets at about 250 ft./1 in. shows passage outlines and widths, surface topography and features, and occasional passage contents (Brucker and Burns 1964; Fig. 5.7). The Flint Ridge System now had 84 km (52 miles) of mapped passages and had overtaken Switzerland's Hölloch as the world's longest known cave.

In 1969, CRF's boundaries expanded as surveys were begun in neighboring Joppa Ridge and in Mammoth Cave

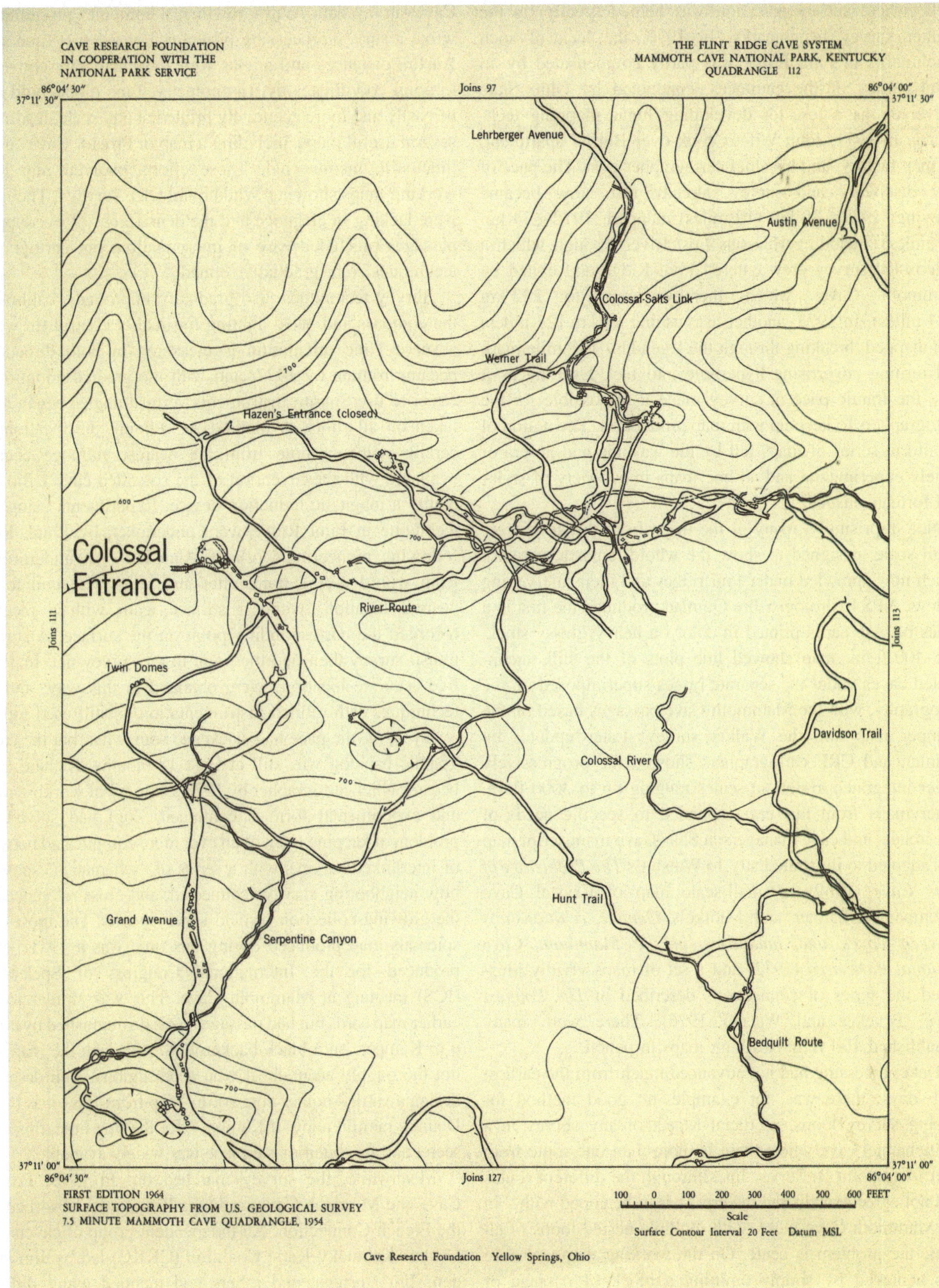

Fig. 5.7 Page from the Flint Ridge map folio, Brucker and Burns 1964

itself, where survey precision was helped greatly by the Walker survey benchmarks. In 1970, the lack of such benchmarks in Flint Ridge was partly compensated by an aerial survey of the entrances, conducted by Ohio State University as a test for developing lunar mapping techniques. In 1971, John Wilcox took over Burns' apartment, his map factory, and his chief cartographer post. The pace of new discovery and survey, already fast, now became dizzying: 1971—30 km (19 miles) mapped, 1972—34 km (21 miles), including Hansons Lost River through which a Wilcox-led survey crew entered Flint Ridge and exited by Mammoth Cave, which thereby grew to 232 km (144 miles). In 1973, another astonishing 43 km (27 miles) was mapped, breaking through the legendary 150-mile mark and turning advertising hyperbole into fact. Exciting as it was, the frantic pace of survey made it impossible for the cartographers to keep up with map production. The notion of a uniform series of maps fell by the wayside and an era of largely experimental and ad hoc maps in a variety of styles and formats ensued.

Not surprisingly, many of the maps from this time were small scale, designed to show the whole fantastic network which now sprawled under four ridges and their intervening valleys. NPS hydrologist Jim Quinlan produced the first in a series of map cards printed in color on heavy glossy stock. The 400 ft./in. map showed line plots of the still unconnected caves under the separate ridges superimposed on the topography, with the Mammoth Cave passages based on the Kämper map and the Walker survey. Later updates by Quinlan and CRF cartographers showed the progressively larger integrated system at scales ranging up to 3000 ft./in. Other maps from this era responded to specific needs of researchers and cave managers: a Salts Cave manuscript map was adapted to illustrate Patty Jo Watson's *The Prehistory of Salts Cave* (1969), a small-scale map of Crystal Cave accompanied Palmer and Miotke's *Genetic relationships between caves and landforms in the Mammoth Cave National Park area* (1972) and a set of maps vividly illustrated the series of connections described in *The Longest Cave* (Brucker and Watson 1976). There were many unpublished 100 ft./in. working maps in pencil.

Data processing had not advanced much from the earliest CRF days; there was, for example, no good method for closing survey loops, a crucial aspect of any survey in a multichannel cave where there is more than one route from point A to point B. Survey lines through the different routes will not agree exactly and the error has to be coped with—in the Mammoth Cave maze, with multiple nested loops of all sizes, the problem is acute. On the working maps, closure was achieved by simply omitting a block of passage or adding imaginary passage as needed to make the survey lines meet. This all started to change profoundly with the pioneering application of computer processing to Mammoth Cave survey data. Will Crowther, a computer programmer, wrote a survey-processing program. For the first time since Austin, distances and angles were systematically converted to point coordinates by trigonometry. Line plots could now be easily and more accurately printed at any scale, leading to several useful maps, including a map of Proctor Cave, at that time still unconnected. These efforts evolved into large working maps showing Walker and CRF surveys. The maps were lacking in elegance and the dense areas of overlapping passages precluded ease of interpretation, but cartographic techniques were gradually refined.

Survey techniques also gradually improved—calibrating the compass and back-sighting for quality control (in which a survey line orientation is measured in both directions) became routine by 1977, and with the gradual adoption of easier to use Suunto instruments vertical angles began to be taken on all shots. A method for refining the locations of certain points remote from the nearest entrance became available with the invention of the so-called cave radio—in reality a magnetic induction device. Experiments using this technique in Flint Ridge were undertaken by Frank Reid, following pioneering work by Alan Hill. An underground team would take a transmitter and large wire coil to the desired location, while a surface team with a receiver recorded the corresponding point on the surface. A surface transit survey then tied the point to the survey net. In 1973, five accurate locations were obtained in this way, and the technique, with refined instrumentation, is still used today.

By 1977, the pace had slackened somewhat, but the cartographic backlog was still at least 18 months. Richard Zopf became chief cartographer by 1978 as the era of working maps and experimental formats continued. Zopf had solved the problem of keeping track of an ever more complicated network of interlinked surveys with a series of "schematics" showing how neighboring surveys connected, and cross-referenced to the growing collection of field survey books. The most conspicuous map produced during this time was a 1981 poster produced for the International Congress of Speleology (ICS) meeting at Mammoth Cave. This was similar to the earlier map cards but had passage levels distinguished by color, *a la* Kämper, on a black background. Although the map was not thoroughly accurate, it won a prestigious, well-deserved design award. Another noteworthy map from 1981 was Diana Daunt's faithful copy of Kämper—for the first time, this once secret and forgotten masterpiece was widely available.

Meanwhile, the survey marched on. In 1979, Proctor Cave and Morrison Cave on Joppa Ridge were absorbed via the French Connection. An independent group of cavers, the Central Kentucky Karst Coalition (CKKC) led by Jim Borden, Jim Currens, and others had mapped many difficult miles in Roppel Cave under neighboring Toohey Ridge. In 1983, the CKKC became de facto Mammoth Cave mappers when a link through Logsdon River was surveyed by a joint

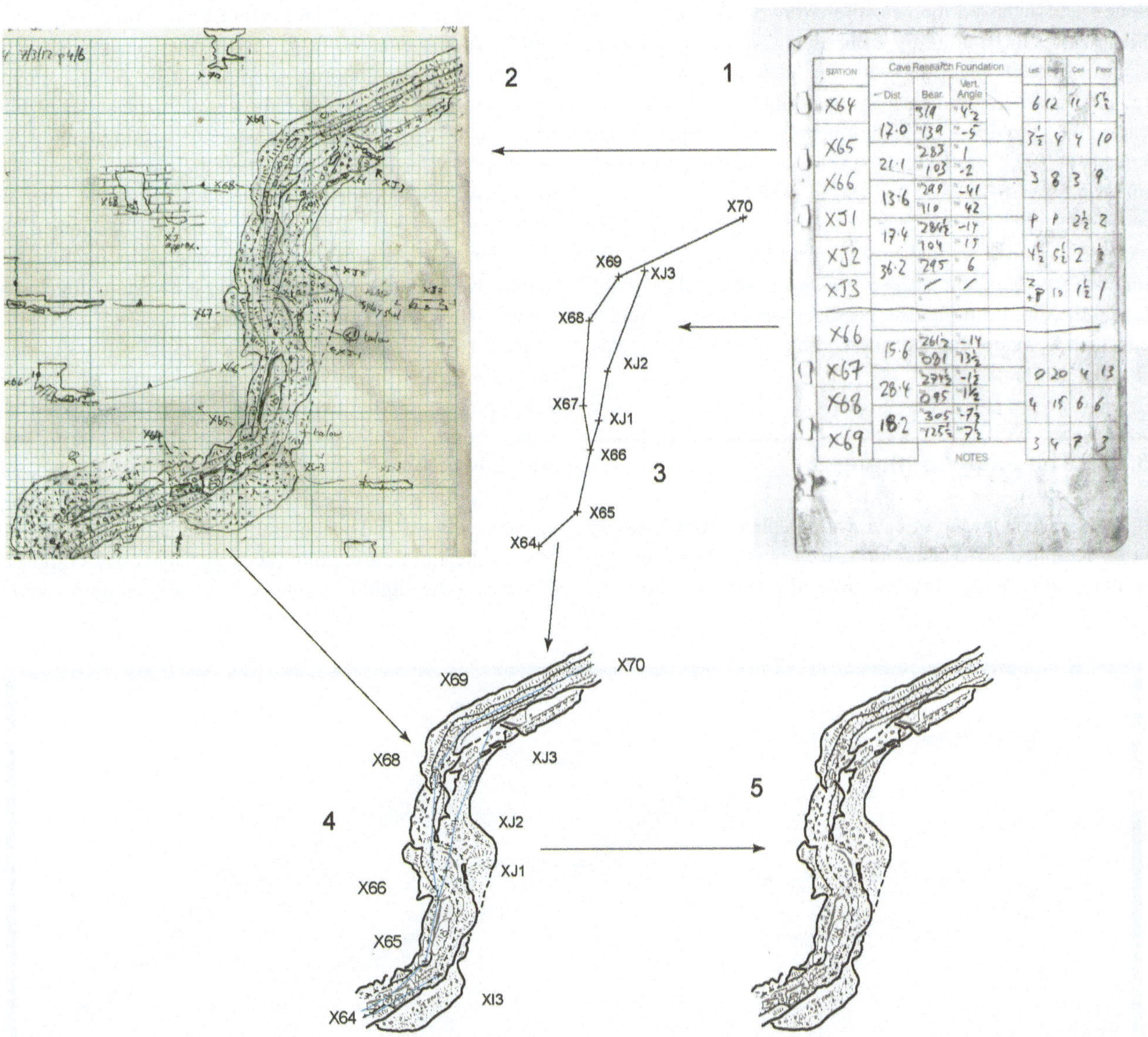

Fig. 5.8 Current process for making a map: *1* data collected in the field includes distance between survey stations, bearing and inclination with backsights; *2* these data are plotted to scale in the field survey book using ruler and protractor, and passage features, also to scale, are sketched around the survey line; *3* survey data are entered into a data reduction program which calculates precise station coordinates and generates a line plot; *4* the line plot and the field sketch are both imported into a drafting program and the sketch is manipulated one pair of stations at a time to match precisely the calculated survey plot, passage features are then traced over the sketch; *5* finally, the field sketch is deleted and the survey line and survey station names are hidden

CRF/CKKC crew, and an era of close cooperation between the two groups began, with Borden handling most of the Roppel Cave cartography. Mammoth Cave now extended well beyond the National Park. The immediate cartographic result was a revised map card showing the whole 480 km (300 miles) of combined CRF and CKKC surveys.

The elevation problem had, with the exception of the Walker survey, been pushed to the back burner. In 1967, several leveling surveys were conducted in Crystal and Colossal caves to begin to correct this major oversight. A more systematic effort was started in the late 1960s by geologists and cartographers Art and Peg Palmer. They experimented with water tubes but settled on a much easier system. In large passages, they used a tripod-mounted engineer's auto-level, which provided very precise elevations; elsewhere they used a hand level and vertical rods. By alternating foresights and backsights, surprisingly high precision could be achieved—a loop through Colossal and Bedquilt caves gave a misclosure of less than a foot. Nor was this a fluke. The Palmers applied the system to a comprehensive study of all major passage and stratigraphic levels throughout Crystal Cave—the first such study ever done—and achieved loop closure errors of less than 0.01%. Meanwhile, for routine cave surveying, it was possible to achieve enough precision with Suunto clinometers to get reliable passage gradients by combining foresights and backsights provided the instruments were properly calibrated. In 1973, the leveling survey was extended to other major passages.

The accurate measurement of stratigraphic layers, passage elevations, and low gradients allowed the Palmers to distinguish low gradient vadose passages, which closely followed the dip, from phreatic passages which did not, and to construct a model for the geomorphic history of the cave (Palmer 1981).

5.12 The Saga Continues

In 1984, a fresh group of cave cartographers from Missouri were recruited to Mammoth Cave. Scott House experimented with survey sketches detailed enough to allow the drafting of a map of Mather Avenue and Turner Avenue in Unknown Cave at 50 ft./in., twice the usual scale. This quickly led to neighboring map sheets by other cartographers and gradually evolved into a systematic series of Flint Ridge map sheets, each 40 in. wide, of varying lengths and oriented N–S or E–W as best fitted the passage layout. The map series was extended to Mammoth Cave Ridge when the NPS commissioned CRF to produce a series of detailed tour trail maps—these became the jumping off point for comprehensive maps showing all passages, a system directed by House, who had become chief cartographer.

Raw survey data were processed initially using a hand calculator, then by a variety of programs as software development rapidly evolved, eventually settling on the Walls program. Point coordinates for survey stations were originally transferred to gridded drafting film by hand, and survey book sketches were drawn freehand in pencil around the survey lines. The pencil draft was then traced in ink on a second sheet of drafting film. Only one sheet was completed this way—the digital revolution was accelerating rapidly,

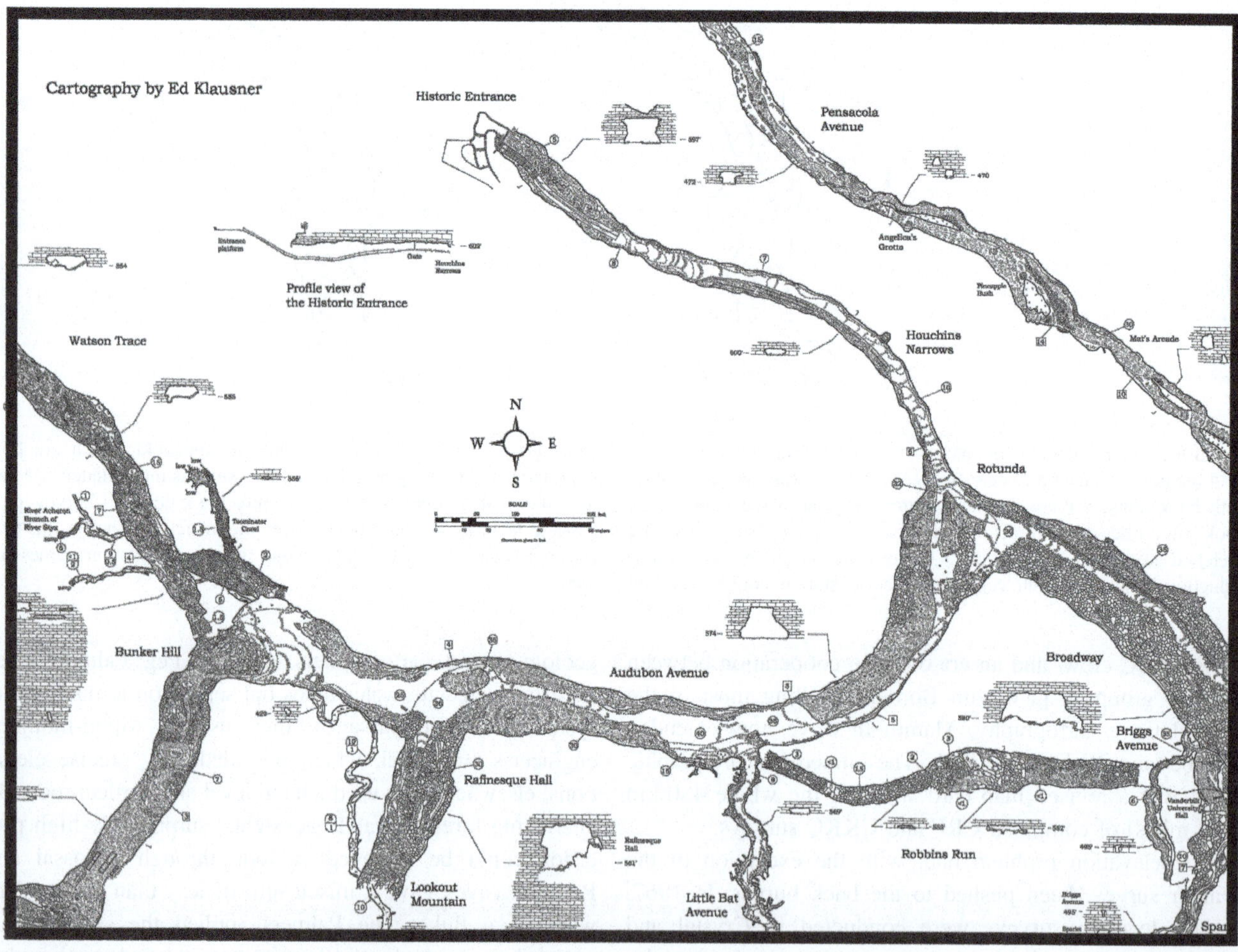

Fig. 5.9 CRF Historic area map sheet—detail showing historic entrance. *Note* Watson Trace—this passage was entered by aboriginal explorers but was lost to human knowledge for at least 2000 years until rediscovered by CRF surveyors in 2002

and by the early 2000s, computer drafting had become powerful and affordable. Computer drafting afforded enormous advantages, both in speed—a preliminary draft was no longer needed—and in flexibility for editing and updating (Fig. 5.8). About 60 map sheets, digitized versions of hand-drawn maps or maps drawn entirely electronically, are in various stages of completion (Fig. 5.9). In 1996, Bob Osburn took over management of the program. Osburn, with Aaron Addison and others, focused on tightening up the survey net by further standardization of data processing and by obtaining high precision Global Positioning System locations for entrances and a few other key points.

Showing the whole cave at such a large scale is no small undertaking. At the 2007 NSS convention in Marengo, Indiana, the entire collection of map sheets, from simple line plots to complete sheets with full passage detail, were printed with a heavy plastic coating and laid out in a 60 ft. by 60 ft. display on the floor of a gymnasium: visitors could walk around and over the giant map. An updated collection was similarly displayed at the 2009 ICS meeting in Kerrville, TX (Fig. 5.10). Another landmark was reached in early 2013 with a joint CRF/ NPS announcement that Mammoth Cave had passed the 400-mile (644 km) mark. With survey added since, mainly filling in detail in known areas, the length as of 2016 is 405 miles (652 km).

The miles of passage discovered on Flint Ridge, Joppa Ridge and Roppel Cave led to a veritable explosion in place names, many of them now formalized as names on maps. In 1956, about 1800 place names, the majority of them now obsolete, had been used at one time or another. The gazetteer of place names now has more than 2800 names, about 1300 of which are more or less current. Roppel place names such as the Lunatic Fringe or Canadian Mist Hike tend to be more colorful and imaginative than the generally more conservative names appropriate to a National Park.

Recent advances in instrumentation have greatly aided in producing accurate field sketches, most notably the adoption of laser rangefinders for measuring distances.

Fig. 5.10 Floor map: the entire Mammoth Cave survey at 50 ft./1 in.

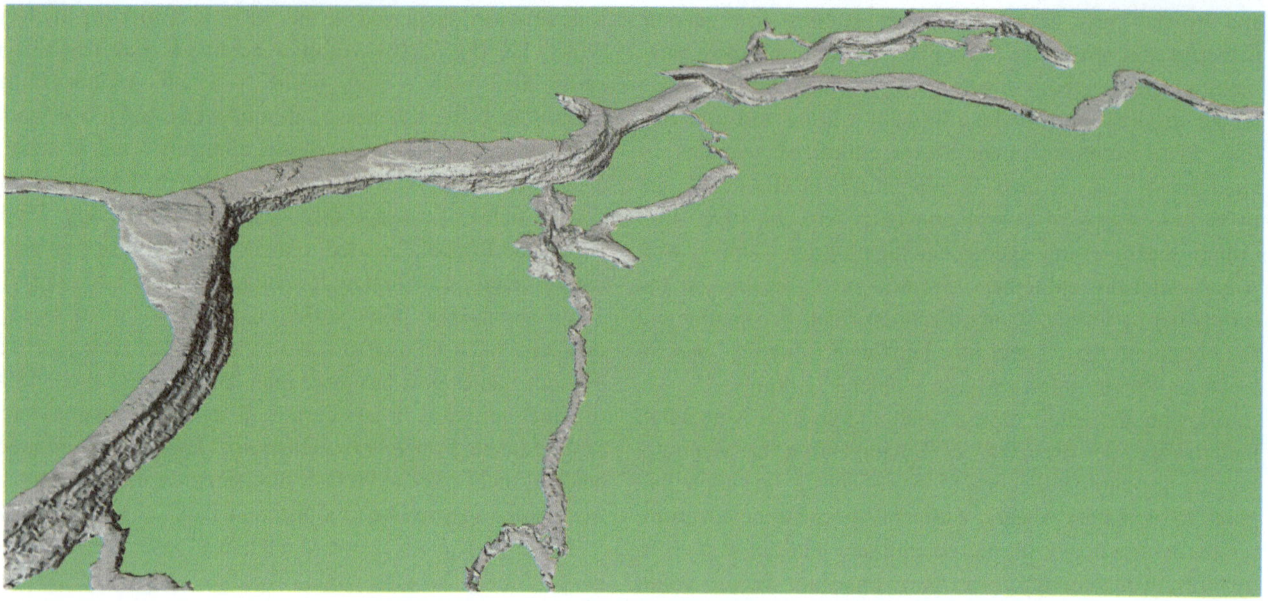

Fig. 5.11 LIDAR image of the historic entrance area, Osburn et al. (2013)

These instruments make it easy to routinely measure distances to walls, ceiling and floor, parameters which were previously either estimated or laboriously taped. A variant of the rangefinder, the DistoX, provides compass and inclinometer data along with the distance in a single instrument, speeding up the survey process.

The most recent advance has been the use of lidar ("light radar") which produces photographic-grade images using a point-cloud illuminating walls, floor, ceiling, and other features. The point-clouds can be processed into 3-D images viewed from outside the passage (Fig. 5.11) or views of passage interiors. In principle, this could make all earlier techniques redundant, but there are serious technical obstacles. The instruments are expensive and delicate, unsuited to many passages; moreover, the point-cloud files are large and difficult to process—it is not easy to generate a plan view showing floor detail and even more difficult to include ceiling features, such as is routinely done with orthodox drafting. Undoubtedly, these obstacles will gradually lessen.

In parallel with the late twentieth to twenty-first century mapping of Mammoth Cave, a variety of groups have surveyed hundreds of miles of passage in the near vicinity. Fisher Ridge Cave, which closely approaches Mammoth Cave passages and extends under the park, has 202 km (125 miles) of mapped passage; Whigpistle Cave, hydrologically connected to Mammoth Cave, has 56 km (35 miles) mapped. Sooner or later these and others are likely to become part of the Mammoth Cave cartography story. Richard Ellsworth Call's assertion that a complete map is impossible may eventually be disproven, but he will not be contradicted in the near future.

Acknowledgements Thanks to Roger Brucker, Sue Hagan, Scott House, Rick Olson, Colleen O'Connor Olson, Bob Osburn, Art and Peggy Palmer and Stan Sides for valuable critiques and comments.

References

Anonymous (1811a) An eye-draught of the Mammoth Cave in Warren County, Kentucky. Manuscript, Benjamin Rush papers, American Philosophical Society, Philadelphia. Reprinted in Meloy, 1968

Anonymous (1815) Green River, or Mammoth Cave, Henderson County, Kentucky. Medical repository of original essays and intelligence, relative to physic, surgery, chemistry, and natural history, 17(New Series, 2) (4):391–393

Anonymous (Alexander C. Bullitt) (1845) Rambles in the Mammoth Cave during the year 1844. Morton and Griswald, Reprinted Johnson Reprint Corporation, 1973

Anonymous (c.1811b) An eye-draught of the Mammoth Cave in Warren County, Kentucky. E. I. DuPont papers in Eleutherian Mills Historical Library, Wilmington, DE: Winterthur collection 1, group 4

Anonymous (Robert M. Bird) (1838) Peter pilgrim: or a rambler's recollections, vol 2, pp 47–162. Lea and Blanchard, Philadelphia

Blackall CR (1899) Map of Mammoth Cave, Kentucky. In: Hovey HC (ed) Mapping the Mammoth Cave. Sci Amer Supp 48 (1229):19707–19708

Borden J, Brucker R (2000) Beyond Mammoth Cave: A tale of obsession in the world's longest cave. Southern Illinois University Press

Brucker R (2008) Mapping of Mammoth Cave: How cartography fueled discoveries. In: Max Kämper centennial symposium and 9th science symposium, Mammoth Cave National Park:2–10

Brucker R, Burns D (1964; published 1966) The Flint Ridge cave system (map folio). Cave Research Foundation

Brucker R, Watson R (1976) The longest cave. Alfred A. Knopf. Reprint 1986, Southern Illinois University Press

Brucker R, Smith PM, Lawrence Jr J, Jones DB (1955) Some new approaches to speleology. Origins and the first twelve years 1957–1968: 49–61. Cave Research Foundation

Call RE (1897) The evolution of the map of Mammoth Cave. Proc Indiana Acad Sci 1896 1896(6):46–52

DeCroix C (2008) Max Kämper's explorations at Mammoth Cave. In: Max Kämper centennial symposium and 9th science symposium, Mammoth Cave National Park: 37

Eyth M (1867) Wanderbuch eines Ingenieurs-In Briefen von Max Eyth-Heidelberg 1871, vol 2, letter 7: "New Orleans, den 8 Janner 1867"

Forwood RS (1875) An historical and descriptive narrative of the Mammoth Cave of Kentucky, 4th edn. J. B. Lippincott, Philadelphia

George A (1996) Ante bellum cave surveys in America. J of Spelean Hist 30(3):72–84

Hovey HC (1882a) Brigham the cave dog. St. Nicholas Magazine 9 (6):425–431. Reprinted 1969. J of Spelean Hist 2(4):88–91

Hovey HC (1882b) Guide book to the Mammoth Cave of Kentucky. Robert Clarke

Hovey HC (1883) Subterranean map-making. Proc Am Assoc Adv Sci 31(2):345–348

Hovey HC (1889) The pits and domes of Mammoth Cave. Sci Amer 61 (16):247–248

Hovey HC (1899) Mapping the Mammoth Cave. Sci Amer Supp 48 (1229):19707–19708

Hovey HC (1903) Colossal Cavern of Kentucky. Sci Amer Supp 56:23316–23319

Hovey HC (1909) Kaemper's discoveries in the Mammoth Cave. Sci Amer 100(21):388–390

Hovey HC, Call RE (1912) Mammoth Cave of Kentucky with an account of Colossal Cavern. John P. Morton and Co

Jefferson T (1853) An eye-draught of the Mammoth Cave in Warren County, Kentucky. In: Notes on the State of Virginia, 1853 edition. Reprinted in J of Spelean Hist 1968, 1(1):1

Kaemper M (1908) Map of the Mammoth Cave, Kentucky. Unpublished MS map, published copy by Daunt 1981, Cave Research Foundation

Kliebhan B, Thomas N (2008) Searching for Max. In: Max Kämper centennial symposium and 9th science symposium, Mammoth Cave National Park:23–35

Lawrence J, Brucker R (1955) The caves beyond. Funk and Wagnalls, Reprinted 1975 Zephyrus Press

Lee EF (1835) Notes on the Mammoth Cave, to accompany a map. James and Gazley, Cincinnati

Lobeck A (1928) The geology and physiography of the Mammoth Cave National Park. Kentucky geological survey, sixth series, pamphlet 21

Mason CR (1950) Surveying in Mammoth Cave. Surv Mapp 10(2): 92–97

Meloy H (1968) Historic maps of Mammoth Cave. J Spelean Hist 1 (3):49–57

Meloy H (1969) The gatewoods at Mammoth Cave. J Spelean Hist 2 (3):51–62

Meloy H (1975) Historic maps of Mammoth Cave. J Spelean Hist 8(3–4): 36–31

Meloy H (1977) The legend of Stephen Bishop. J Spelean Hist 10(1): 5–7

Miotke FD, Palmer A (1972) Genetic relationships between caves and landforms in the Mammoth Cave National Park area. Böhler Verlag

Olson R, Kliebhan B, Toomey R (2013) How did Max Kämper and Ed Bishop map Mammoth Cave? In: Proceedings, Mammoth Cave national park's 10th research symposium, pp 69–77

Osburn GR, Addison A (2013) Unpublished map.

Palmer A (1981) A geological guide to Mammoth Cave National Park. Zephyrus Press

Quinlan J (1965) The history of the evolution of the map of Mammoth Cave, Kentucky (abstract). Bull Natl Speleol Soc 27(2):61

Randolph H (1924) Mammoth Cave and the cave region of Kentucky. Standard Printing Co., Louisville

Riggs PT (2007) Nahum Ward's 1816 account of Mammoth Cave. J Spelean Hist 41(1):6–15

Seymour T (2013) Restoring the Kämper map for the 21st century. In: Proceedings, Mammoth Cave National Park's 10th research symposium, pp 126–132

Seymour T (2016) The Max Kämper map of Mammoth Cave: an NPS centennial restoration. In: Proceedings, Mammoth Cave National Park's 11th research symposium paper 7

Sides S (1971) Early cave exploration in Flint Ridge, Kentucky: Colossal cave and the Colossal Cavern company. J Spelean Hist 4 (4):63–69

Sutton M (1990) Thirty years of mapping by the Cave Research Foundation. NSS Bull 52(1):1–15

Swedlund C, Crothers G (2008) Max Kämper's unique selection of place names for his 1908 map of Mammoth Cave. In: Max Kämper centennial symposium and 9th science symposium, Mammoth Cave National Park, pp 39–42

Thompson J (1909) Mammoth Cave, Kentucky: an historical sketch. Times Publishing Co., Smiths Grove, KY

Vaughan E, Marshall WL (1912) Colossal Cavern. In: Hovey and Call 1912. Reprinted 1971, J Spelean Hist 4(4):1

Ward N (1816) Plan and description of the great cave in Warren County, Kentucky. Massachusetts spy for May 8, 15 and 29, 1816. Reprinted in Riggs 2007

Watson PJ (1969) The prehistory of Salts Cave. Reports of investigations #16, Illinois State Museum, Springfield, IL

Wood JS (1841) Mammoth Caves of Kentucky. American magazine and repository of useful literature 1(6)184–190. Map republished 1974: Chart of the Mammoth Cave. J Spelean Hist 7(4):1,62

Geology of Mammoth Cave

6

Arthur N. Palmer

Abstract

Mammoth Cave owes its origin to the rock called limestone, which readily dissolves in water. Underground water passing through thick limestone layers toward the deep valley of the Green River has formed and enlarged the cave over the past several million years. The rocks are 330–340 million years old, but the cave is much younger. A cap of resistant sandstone overlying the limestone has protected the main parts of the cave from surface erosion, thus preserving many clues to past events. Passage types reflect the nature of the water that formed them. Below, at, and slightly above the local Green River level, all openings in the ground are filled with water. Above this zone, underground water drains downward by gravity along the steepest available cracks in the rock. Where water follows the narrow fissures between rock layers, it forms tall canyon-like passages. Well-shaped vertical shafts form where water is able to descend straight downward along fractures that cut across the layers. Below the water table, underground water follows the most efficient paths of flow, along the widest openings, to form tube-shaped passages. Former positions of the water table can be determined from the change in passage shape, which provides a history of groundwater flow in the region.

6.1 Introduction

Mammoth Cave, like most large caves, was formed by the dissolving of limestone bedrock by underground water. Its many passages indicate the paths of underground streams, many of which are still active today (Fig. 6.1). Some have been inactive for several million years. By preserving this record, the cave preserves the history of past changes in regional landscape, surface river patterns, and climates.

The cave streams are fed by runoff from rain and snowmelt that drains underground through depressions called sinkholes, which collect water and deliver it underground. Some sinkholes provide entrances to caves, but most are clogged by soil and rock so that only water can pass through. At Mammoth Cave, the underground water reappears at the surface through springs in the Green River valley. Water-filled cave passages below river level deliver their water upward through roughly circular "rise pools" along the banks of the river. This type of region, which contains caves, sinkholes, sinking streams, and large springs, is called a **karst** landscape, named for a high plateau in Slovenia where the first European studies of the subject were made.

6.2 Geology of the Mammoth Cave Region

Mammoth Cave has benefited from three essential ingredients: a large flow of underground water, bedrock that dissolves easily, and plenty of time. The humid climate and a large catchment area provide a great amount of underground stream flow. Although the limestone formations here are not as thick as in many other parts of the world, they cover an enormous area. They are also very soluble and contain more caves than any other limestone layer in the USA. Finally, the landscape in this region is very stable, and underground water has flowed through these same rock layers throughout

A.N. Palmer (✉)
State University of NY, Oneonta, NY, USA
e-mail: arthur.palmer@oneonta.edu

H.H. Hobbs III et al. (eds.), *Mammoth Cave*, Cave and Karst Systems of the World,
DOI 10.1007/978-3-319-53718-4_6

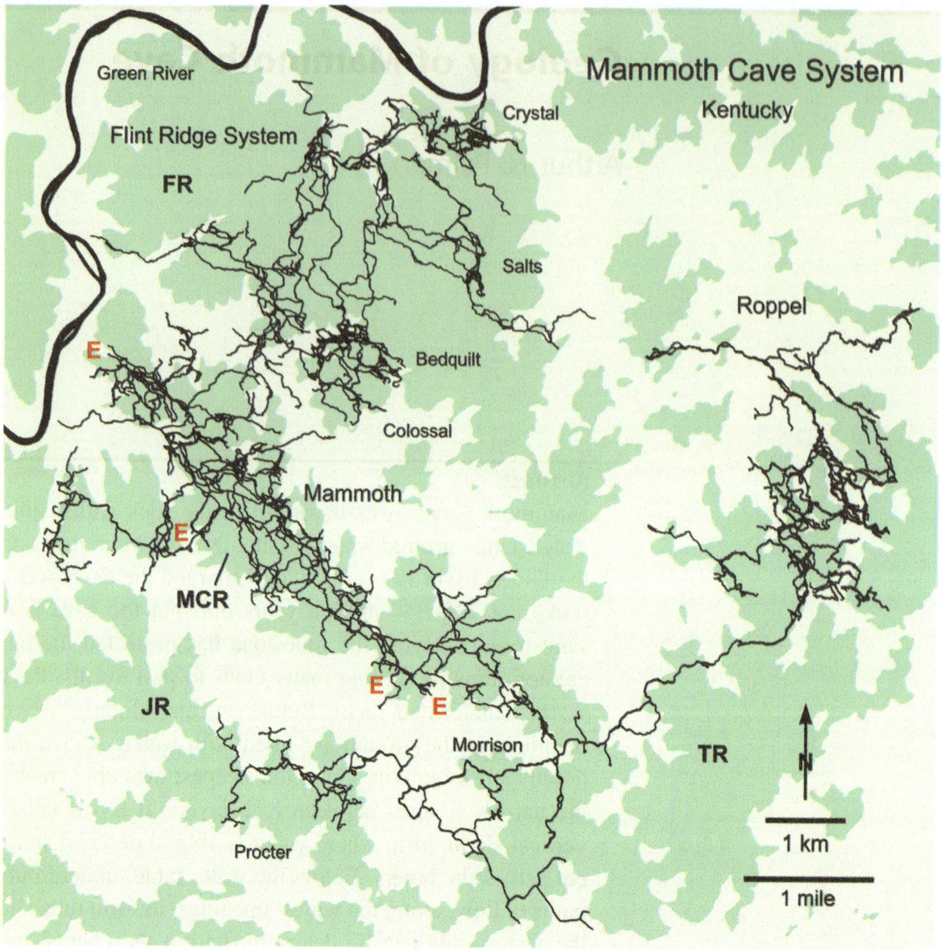

Fig. 6.1 Map of the Mammoth Cave System and its relation to the Green River and local topography, simplified from map by the Cave Research Foundation. Map of the Roppel Cave section by the Central Kentucky Karst Coalition. Names of major sections of the system are shown. *Green* sandstone-capped ridges; *yellow* limestone-floored valleys. *MCR* Mammoth Cave Ridge, *JR* Joppa Ridge, *FR* Flint Ridge, *TR* Toohey Ridge. *E* major entrances to tour routes of Mammoth Cave

the entire geologic history of the Ohio River basin, which predates the glacial ice ages.

The cave is located in rock layers that were deposited about 340–330 million years ago during the Mississippian Period of geologic time (Fig. 6.2; Pohl 1970). Sea level at that time was higher than it is today, and most of what is now the central USA was covered with shallow seawater. The various rock layers (beds) were deposited on the sea floor as loose sediment, which was converted to hard rock by compaction and internal crystal growth. Evidence preserved in the rocks, such as ancient ripple marks produced by flowing water, tell us that the depth of the continental sea averaged only a few feet or tens of feet.

At that time, the area that is now the southern USA was located near the equator, so the type of material deposited in the area was similar to that in warm climates today. Most of the deposits were **limestone**, composed of calcium carbonate ($CaCO_3$) from the hard parts of plants and animals, as well as direct precipitation from the seawater. Some of the beds contain both calcium and magnesium ($CaMg(CO_3)_2$), known as **dolomite**, which resembles limestone but dissolves more slowly. Another rock type that accumulated in small quantities was **chert**, also known as flint. This is a very hard material composed of silicon dioxide (SiO_2) that forms gray or black beds and nodules that project outward as the surrounding rock dissolves or erodes. It is deposited from water that has accumulated silicon-rich compounds from sources such as volcanic activity.

Throughout this time, a large river drained into the shallow sea from land that today is mainly Canada (Swann 1964). Geologists call this the Michigan River, an ancestor of today's Mississippi, and, like the Mississippi, its sediment load was deposited as a large delta where the river emptied into the shallow sea. Swift-moving water deposited sand and gravel in sheets and finger-like patterns like those in the Mississippi delta at New Orleans. Slow-moving water deposited only fine-grained sediment such as mud. The sand and gravel became cemented into hard **sandstone** and **conglomerate**. The mud (composed mainly of tiny particles of clay) compacted into the soft rock called **shale**. Beyond their limits, only chemical deposits such as limestone accumulated, and these are the rocks that contain Mammoth Cave. With time, the Michigan River delta grew southward and buried the older limestone layers with the insoluble rocks. The result is a gradual upward transition in rock type from nearly continuous limestone to almost none at all. The

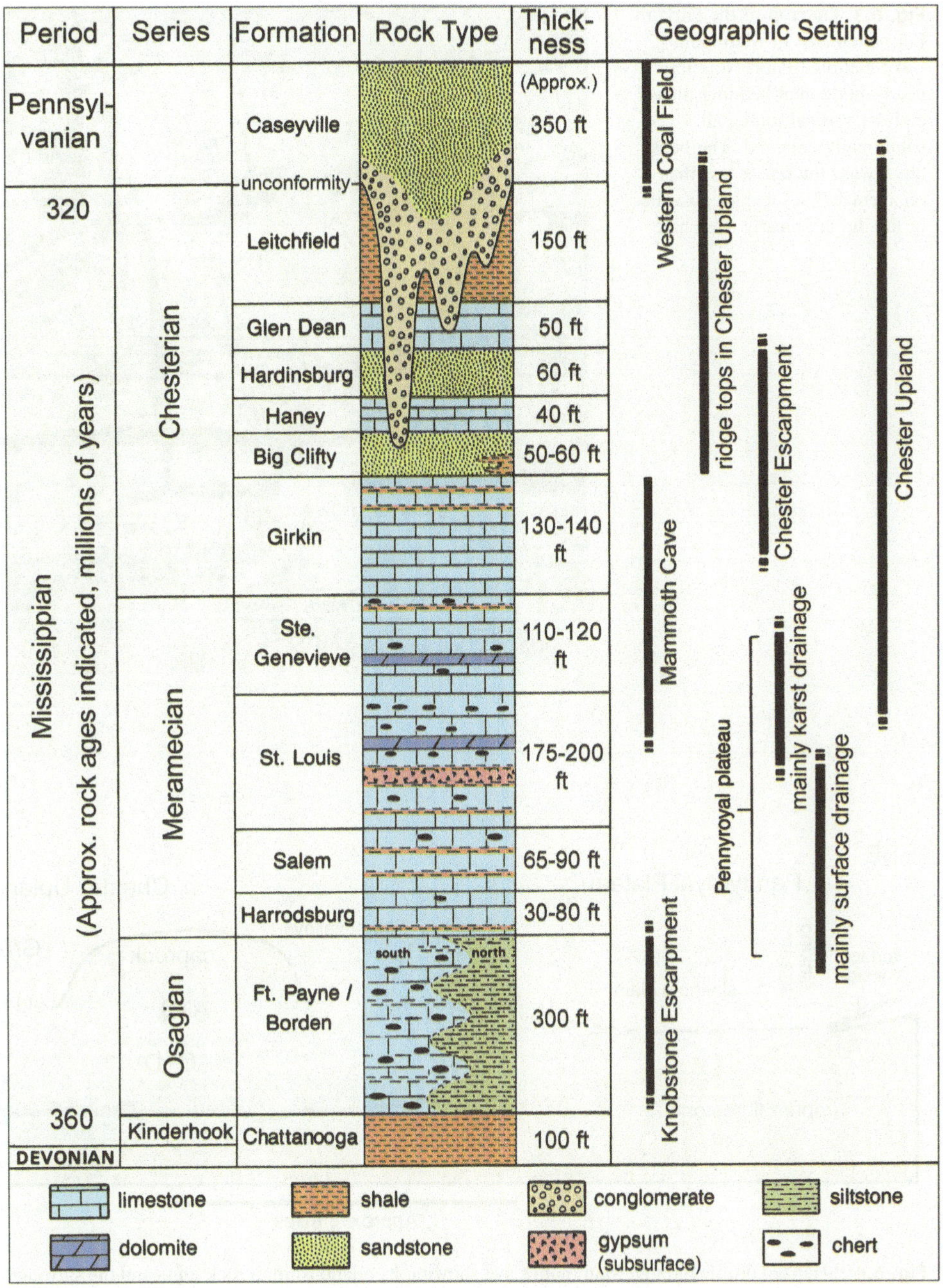

Fig. 6.2 Rock units in the Mammoth Cave area and their relation to the surrounding landscapes. Most of the rock units above the Hardinsburg have been removed by erosion in the Mammoth Cave area

limestones of Mammoth Cave are capped by thick beds of sandstone and shale, with sparse deposits of conglomerate.

These rocks were deposited as nearly flat-lying beds, but they have been tilted in various directions by slow rises, falls, and warping of the continent. At Mammoth Cave, the beds slope toward the northwest at about one degree or less —an angle too small to be detected by eye. The tilt of the rock layers is called the **dip**. Irregularities in the sedimentary deposits also provide local differences in dip angle and direction (Fig. 6.3). As shown below, the dip has a great effect on the shape of the land, as well as on the patterns of cave passages.

6.3 The Landscape

Where the resistant sandstone and conglomerate layers are exposed at the surface, they form plateaus that stand prominently above the surrounding landscape (Fig. 6.4). They once covered the entire region, but stream erosion has removed them from many areas. Because of the northwesterly dip of the rocks, they were entirely removed southeast of where Mammoth Cave lies today, the way a layer of paint is easily sanded away from the high parts of a board. The underlying limestone was exposed, and it dissolved and eroded downward more rapidly than the sandstone and

Fig. 6.3 Outcrop of the Girkin Formation east of Mammoth Cave National Park. Note the nearly horizontal bedding planes and the vertical joints, all solutionally enlarged. The beds dip toward the *left* at less than 1° on average, but local irregularities in the dip are clearly seen here

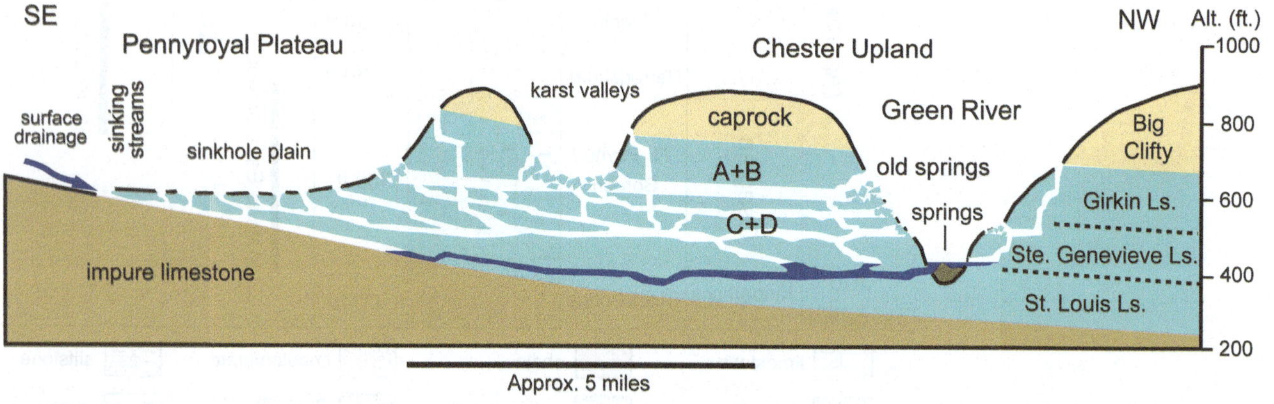

Fig. 6.4 Idealized profile through Mammoth Cave showing its relationship to rock units and the surrounding landscape. *A, B, C, D* major passage levels

conglomerate, to form a low flat region called the **Pennyroyal Plateau**. At first, the Pennyroyal stood only a small distance above the local rivers, but since then, the rivers have cut into the plateau about 200 ft (60 m), so that much of the drainage is now underground, and many sinkholes, sinking streams, and caves have developed within it. This is one of the most extensive karst areas of North America, and Mammoth Cave lies at its northwestern edge where it can receive much underground drainage from areas that lie farther southeast (Palmer and Palmer 2009; Quinlan and Ray 1981).

Northwest of Mammoth Cave, the insoluble rocks are still present and form a rugged hilly region called the **Chester Upland**. Its southeastern boundary is a steep slope that drops about 200 ft (60 m) to the Pennyroyal Plateau. Along the border, the elevation of the Chester Upland is about 800–950 ft (240–290 m) above sea level, and the Pennyroyal is at about 600–750 ft (180–230 m). In the vicinity of Mammoth Cave, the Green River has carved a deep channel through the insoluble rocks into the limestone below, and tributaries of the river have eroded the upland into many irregular ridges composed mainly of limestone but capped by sandstone. The protective caprock has allowed the limestone to stand almost 300 ft (90 m) above the river. This is one reason why the cave is so large.

6.4 Rock Layers in Mammoth Cave

Mammoth Cave extends through three major limestone layers. From top to bottom (youngest to oldest, the order usually encountered in one's descent into the cave), these are the Girkin Formation, the Ste. Genevieve Limestone, and the St. Louis Limestone (Fig. 6.2; Palmer 1981, 2007). Geologists call the Girkin a "formation" (a general-purpose name) because elsewhere it contains several rock types, including limestone, sandstone, and shale.

The Girkin consists of several limestone layers 10–20 ft thick (3–6 m), separated by soft beds of shaly limestone that are less resistant and produce recessed niches in the cave walls. Its total thickness is about 130 ft (40 m). These shaly beds represent the first stages of the Michigan River delta extending into the area. In Mammoth Cave, these rocks are best seen at the beginning and end of the Historic Tour (Fig. 6.5) and along the spectacular descending staircase of the Frozen Niagara Tour. The Girkin is capped by the resistant sandstone of the Big Clifty Formation, which forms the top of Mammoth Cave Ridge where the Visitor Center is located. The Big Clifty can be seen in small outcrops along the trail leading to the Historic Entrance.

Beneath the Girkin, the Ste. Genevieve Limestone includes a variety of thin-bedded limestones and dolomites with a few thin shaly beds that produce recessed niches (Fig. 6.5). The total thickness of the Ste. Genevieve is about 115 ft (35 m). Most of Mammoth Cave is located in it. Good exposures are seen between Gothic Avenue and River Hall on the Historic Tour, as well as on the Travertine Tour, and nearly all of the Grand Avenue and Cleaveland Avenue Tours.

The St. Louis Limestone also contains dolomite, dark shaly beds, and many beds and nodules of chert. With such a variety, the St. Louis forms rather irregular walls with many projecting and recessed ledges. The chert beds are highly irregular and stand out as odd-shaped nodules, shelves, and balls. Many of them are composed of fossils and include the hardened burrows of invertebrate animals. The overall thickness of the St. Louis is about 175–200 ft (53–60 m), but only the upper 130 ft (40 m) is exposed in Mammoth Cave, and less than half of that is seen on the tour routes. It is best seen in the lower passages of the Wild Cave Tour.

Above the main limestones, sandwiched between the Big Clifty and another higher sandstone layer, is the thin Haney Limestone. It is only about 30 ft (9 m) thick but contains many small stream caves at the contact with the underlying Big Clifty. These caves feed hillside springs, which were once used as the water supply for the National Park. By draining underground water to the sides of the ridges, the Haney diminishes the amount of seepage into the underlying passages of Mammoth Cave and determines the types of mineral deposits in the cave, as shown later. Most of the water from the Haney springs runs downhill across the eroded edge of the Big Clifty Formation and into sinkholes

Fig. 6.5 Broadway, on the Historic Route, extends through the lower Gikin Formation and upper Ste. Genevieve Limestone. The contact between them is located about half-way up the right wall. Note the wooden pipes, which were used to convey fluids during saltpeter mining in the early nineteenth century (see Chap. 7)

in the major limestones below, where it forms steeply descending passages along the fringes of Mammoth Cave. So the same water is able to form caves in two entirely separate limestone bodies.

6.5 Underground Water

The limestones at Mammoth Cave are very compact, with only tiny pores, so very little water seeps through the rock itself. Instead, most of it flows through cracks between layers and fractures that cut across the beds. **Joints** are fractures that cut across the beds with no perceptible movement between the adjacent blocks. **Faults** are cracks along which the blocks have slipped past each other. Major joints and faults tend to enlarge into narrow fissure-like passages (Fig. 6.6). However, in Mammoth Cave the most important paths of flow are **bedding-plane partings**, where beds have split apart by stresses in the ground (Figs. 6.5 and 6.6). The beds are irregular, so they remain tightly compressed in some places but have gaps between them elsewhere that can be enlarged by water.

Where water first enters the ground it drains downward by gravity and follows the steepest available openings,

Fig. 6.6 Leopard's Arch is a fissure in Sparks Avenue, on the Historic Route, that formed along a prominent joint in the base of the Ste. Genevieve Limestone, where water rose up a short distance from a lower level into a higher one

vertical ones if possible. This region is called the **vadose zone** (Fig. 6.7a). Where these openings become too narrow to transmit all the water, some, or all, of the water overflows into less steeply inclined openings, to form passages that follow the rock layers. Where it encounters a fracture that cuts across the beds, or intersects a lower-level passage, the water follows the steepest path downward, forming waterfalls in places (Fig. 6.7a). Eventually, however, the water reaches the **phreatic zone**, where all underlying openings are filled with water (Fig. 6.7b). Its top surface is the **water table**, which slopes in the direction of the nearest available outlets. In phreatic passages, the pull of gravity is offset by the downward increase in water pressure, so instead of following the steepest paths, the passages form along the most efficient routes to springs in nearby valleys. The largest and shortest openings are most efficient in transmitting water, and it is along these paths that most cave passages form. More information on groundwater in the Mammoth Cave area is given by White and White (1989) (see also Chap. 8 this volume).

6.6 How Limestone Dissolves

Limestone dissolves readily in fresh water supplied by rain or snowmelt. The speed of the reaction and the amount that can be dissolved are increased greatly if the water contains acid. The most abundant natural acid is carbonic acid produced by carbon dioxide absorbed from the atmosphere, and even more from the soil. This is the kind of water that forms most caves, including those of the Mammoth Cave region.

The main chemical reactions in the dissolving of limestone are:

$$\text{Water} + \text{carbon dioxide} \rightarrow \text{carbonic acid: } H_2O + CO_2 \rightarrow H_2CO_3 \rightarrow H^+ + HCO_3^-$$

Carbonic acid + limestone (calcium carbonate) → dissolved limestone carried away by water:

$$H^+ + HCO_3^- + CaCO_3 \rightarrow Ca^{2+} + 2HCO_3^-$$

Dolomite dissolves in a similar way.

If carbon dioxide is lost from the water, the reactions reverse and calcium carbonate is deposited as speleothems (cave formations) such as stalactites and stalagmites.

A constant flow of fresh water can dissolve limestone at a rate of a few hundredths of an inch per year (roughly 0.01–0.1 cm/year; Palmer 2007). In geological terms, this is rather fast. Enlargement rates are greatest during periods of high flow (Meiman and Groves 1997). But the solution rate decreases greatly as the dissolved limestone accumulates in the water. Most cave streams already contain enough to

Fig. 6.7 **a** Example of vadose water: a waterfall pouring out of a canyon passage into a larger passage below. **b** The water table in the Echo River passage, at the present level of the Green River. This is the *top* of the phreatic zone, in which every opening is filled with water. During high flow, the water in this passage reaches the ceiling. Also see Fig. 7.2

lower the solution rate tens or hundreds of times. A cave therefore requires a very long time to form, even if its streams run continually year-round.

6.7 Cave Passages

Mammoth Cave is composed of a vast number of individual cave passages, each formed by a discrete stream. Today only a minority of passages still contain the streams that formed them, and this was also typical in the past. The cave was not produced by a widespread seepage of water that filled all the openings at once, as though in a sponge. To understand the development of Mammoth Cave, it helps to know what controls the shape and direction of each passage.

The shape of a cave passage depends on the nature of the water flow and the structure of the bedrock through which it moves. Above the water table, water descends along the steepest available openings, and in Mammoth Cave these are mostly down the dip of bedding-plane partings. The water dissolves and erodes the floor to form tall, narrow **canyons** (Fig. 6.8a). These have sinuous patterns like surface streams. Many canyons enlarge in a stepwise manner where waterfalls or rapids cut through entire beds, or several beds, at a time and enlarge the channel as the waterfalls work their way upstream.

If vadose water encounters a vertical opening, such as fault or joint, it enlarges the opening into a **shaft**, which looks like the interior of a well or silo. The tallest known shaft in Mammoth Cave is Mammoth Dome, which is almost 200 ft (60 m) high—or deep, depending on which way you are looking (Fig. 6.8b). Most shafts in Mammoth Cave are still enlarging by active drips, which become significant waterfalls after heavy rains. Water enters their tops, usually through small canyons or vadose tubes, and drains out at the bottom through similar passages. Shafts deepen with time as their water diverts through fractures in their floors, or simply dissolves downward from bed to bed. As they deepen, new drains develop at lower levels. Each outflow tends to follow a different path, so a shaft can provide access to several passages at entirely different levels. Shafts are typically much larger in diameter than their infeeders or drains because the water spray reaches most or all of the shaft walls, especially during high water, while the water in the inputs and outlets tends to cover just the passage floor, which is commonly armored by relatively insoluble beds or sediment.

Phreatic passages form at or below the water table, and they follow the most efficient paths through the ground, rather than the steepest. These are the routes that offer the least resistance to flow, and thus depend mainly on the original width of the crack and the length of the flow path. Water flow is extremely sensitive to the crack width, and less so to the length of flow—so where several alternate paths are available, it is usually the widest that transmits the most

Fig. 6.8 **a** Pass of El Ghor, in the Ste. Genevieve Limestone, is a typical high, narrow, sinuous passage formed by a vadose stream above the water table. **b** Mammoth Dome is one of the tallest known shafts in Mammoth Cave. It extends through the entire Ste. Genevieve Limestone and about 30 ft (9 m) into the St. Louis Limestone (well-bedded rocks below trail level)

Fig. 6.9 Cleaveland Avenue is a fine example of a tube formed by phreatic water. Note the lens-shaped cross section elongated along the bedding. ("Cleaveland" is spelled correctly)

water and enlarges into a cave passage, regardless of the total distance.

Phreatic passages enlarge over their entire perimeters, so they form tunnels with rounded contours, called **tubes** (Fig. 6.9). These usually have lens-shaped or elliptical cross sections elongated along the bedding. Some have irregular cross sections that show the local effects of resistant or nonresistant layers in the bedrock. A few phreatic passages follow vertical or steeply inclined fractures that develop into straight **fissures** with high, narrow cross sections (Fig. 6.6). Phreatic passages tend to have very low overall slopes, but they can contain sections that loop up and down at various angles, like a garden hose draped over obstacles. Although some sections of phreatic passages can slope uphill in places, the surface inlets for water are invariably higher than the spring outlets, so the overall pattern of water flow is downward.

6.8 Passage Patterns

The Mammoth Cave map shows many complex passage relationships. Most passages lie over or under each other without connecting. A connection from one to another that is just slightly above or below may require roundabout trips of many hours. The main reason is that the local limestone beds are numerous but thin, so that water can easily divert to progressively lower beds, each following an entirely different path.

Adding to the complexity, passages tend to meander in a seemingly random way. But the bends are not at all random. Directions of underground water flow, and therefore, the paths of the resulting passages, are strictly controlled by the dip direction and angle of the bedding-plane partings. Irregularities in the openings cause the water flow, and thus the resulting cave passages, to follow equally irregular paths (Palmer 1981, 2007). Their slopes are so subtle in Mammoth Cave that precise surveys are needed to detect them.

The largest dip variations are caused by broad domes, basins, and warps that are large enough to appear on geologic maps. These features are produced mainly by stresses within the Earth. Most have dimensions that range from thousands of feet to hundreds of miles, and they influence the orientation of entire caves or cave passages. For example, the overall trend of many cave passages in the Mammoth Cave region is toward the northwest, into a broad dish-shaped depression in the beds called the Illinois Basin, in which the deepest point is located in southern Illinois.

Variations in local dip at the scale of tens to hundreds of feet are caused mainly by the irregular thickness of sedimentary deposits, and by differences in how much they were compacted as they turned into hard rock. These structures control the local trends of passage segments. Minor irregularities along bedding planes, with dimensions of typically a few feet, are the result of local conditions when the beds were deposited, such as currents and waves. These irregularities control the small-scale sinuosity that is typical of narrow canyon passages. Large structures in the beds also affect the flow direction, so small meanders are superimposed on broader ones (Fig. 6.10). If the flow of a cave stream increases with time, its passage widens. Small irregularities are overwhelmed and no longer affect the passage trend, but the larger humps and hollows continue to influence the trend. These wider passages still meander but with larger and more open bends.

Meanwhile, the overall trend of a vadose passage is in the direction of average dip, so the passages tend to be roughly parallel to each other, even though their smaller wiggles do not match. This is such a common occurrence that one can often identify local dip directions simply by the orientation of canyons. As a canyon cuts through one limestone bed into a lower one, each successive level of a canyon may follow a

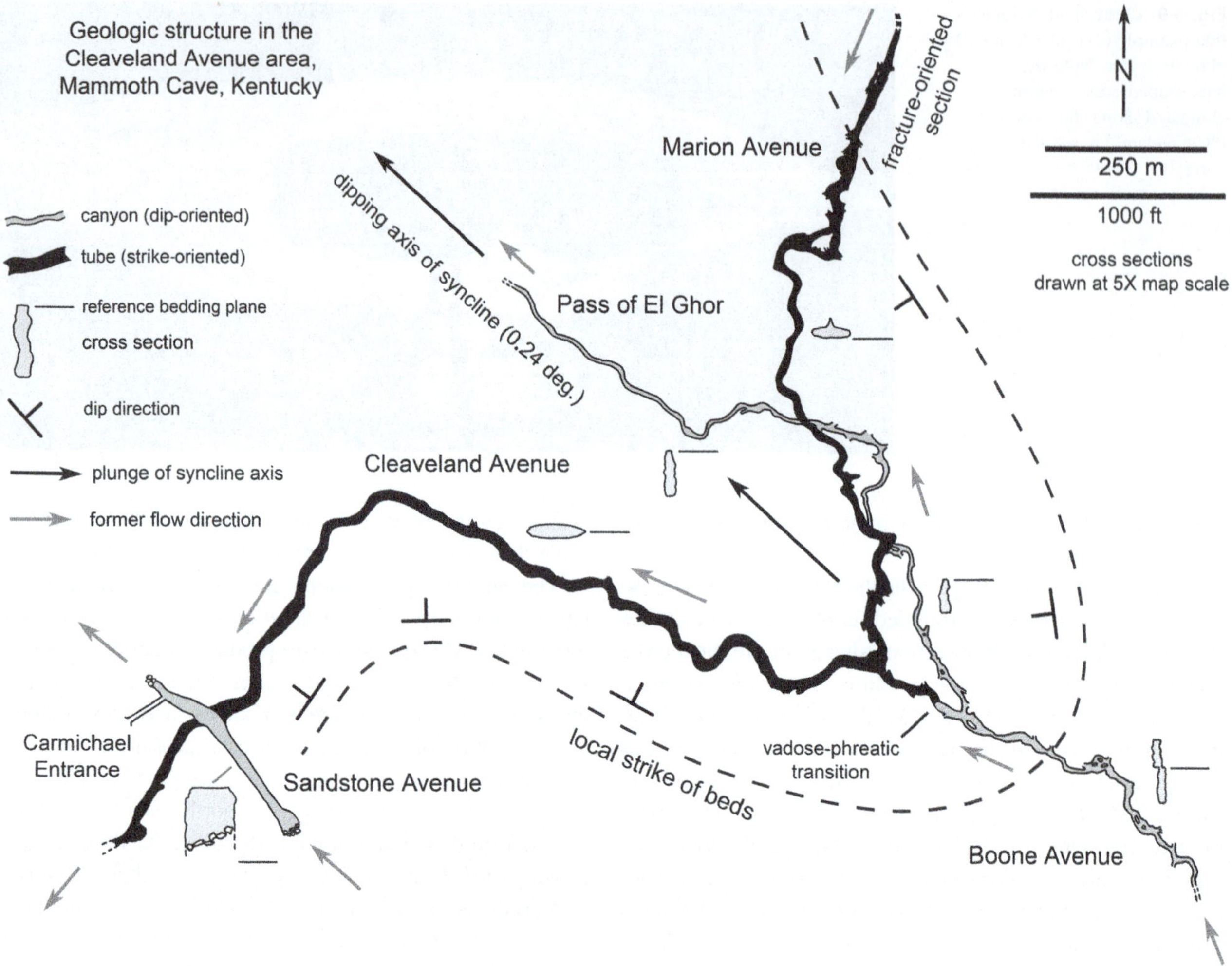

Fig. 6.10 Map of the Cleaveland Avenue area, showing the effect of irregularities in the dip of the strata. Cleaveland Avenue is wide enough that its trend was controlled only by broad structures in the beds. The narrower Boone Avenue and Pass of El Ghor have smaller and more closely spaced bends controlled by minor irregularities in the beds (compare with Fig. 6.8a)

slightly different path. These relationships account for much of the complexity of Mammoth Cave's pattern.

Phreatic passages follow the paths with the widest bedding-plane partings and/or fractures. Most phreatic tubes follow the strike of the beds (the direction perpendicular to the dip). This is because the openings tend to narrow with depth and are widest where they intersect the water table. The incoming vadose water is already following a favorable bedding plane, so there is little reason for the flow to select a different one below the water table. This strike orientation of phreatic tubes is not as strong as the down-dip tendency of vadose water, but it accounts for many of the seemingly odd passage trends in the cave (Fig. 6.10).

Surface rivers tend to deepen their valleys with time, and as a result, the water table also drops in elevation. Meanwhile, phreatic passages that were once at or near the water table gradually lose their water as it drains downward to form new passages at lower levels. If the escaping water first runs as a stream along the floor of the old phreatic passage for some distance, it may dissolve a narrow canyon to produce a passage cross section shaped like a keyhole—a tube at the top with a canyon in the floor. Where the water finally leaves the old passage, it usually drains down the dip of the rocks, a direction that is probably quite different from that of the original tube. It is very common to see vadose canyons escaping down the dip from the floors of older strike-oriented tubes.

A good example of the difference between vadose and phreatic passages can be seen on the Grand Avenue Tour. A descent from the entrance leads to a large dip-oriented canyon passage, a fragment of the Main Cave that has been isolated by breakdown (Fig. 6.10). Perpendicular to it is a long tubular passage, Cleaveland Avenue (Fig. 6.9). This follows several broad bends along the strike of the beds and is clearly a phreatic tube that formed along the water table. It is almost perfectly horizontal, with only about a foot (30 cm) of elevation difference over its length of nearly a mile (Palmer 1989). It leads to a high but narrow canyon passage (Boone Avenue) that has developed on several levels. The tour turns to the right into the canyon, up the dip and in the original upstream direction. This passage follows a trough-shaped warp in the limestone beds and was an ideal path for vadose water. In the opposite direction, not seen on the tour, the lower half of the canyon continues down the dip as the Pass of El Ghor.

When these passages were forming, the down-dip vadose flow in Boone Avenue reached the water table at the level of Cleaveland Avenue and turned sharply to follow the strike of the beds. It was the strike-oriented water at the former water table that formed the tube-shaped Cleaveland Avenue. Cleaveland is just a segment of an immensely long passage that once extended 4 miles (6.4 km), not counting bends, all the way to a spring far to the west, but it has been truncated into three segments by the deepening of surface valleys. The distance to the nearest point on the Green River was only half as much, but the longer path along the strike was more efficient.

When the water table dropped, Cleaveland Avenue was abandoned, and the water continued down the dip to form the Pass of El Ghor. This is a canyon that extends a great distance to another tubular passage (Silliman Avenue) 50 ft lower. This was the elevation at which the water table paused, just as it had during the origin of Cleaveland Avenue. But instead of following the strike along the water table, Silliman continued to descend below the water table to a depth of at least 70 ft (21 m). Farther downstream it rose again to a spring along the Green River. This looping segment is deep enough that it still contains a large stream, Echo River, at the present Green River elevation (Fig. 6.7b). Part of its downstream end can be seen at River Hall on the Historic Tour.

6.9 Solution Features and Deposits

Interpreting the history of water flow through Mammoth Cave is made easier by a variety of solution features that indicate the direction and speed of water flow in each passage. Also, deposits such as sediment, breakdown material, and chemical growths known as speleothems (or, more popularly, cave formations), all provide clues to past conditions.

Scallops are wave-shaped hollows dissolved in the limestone surfaces by flowing water (Fig. 6.11a). They are especially useful in determining former flow conditions in passages that are now dry. They range in length from about half an inch to a couple of feet, and they can be found in most canyons and tubes. The scallop shapes indicate the

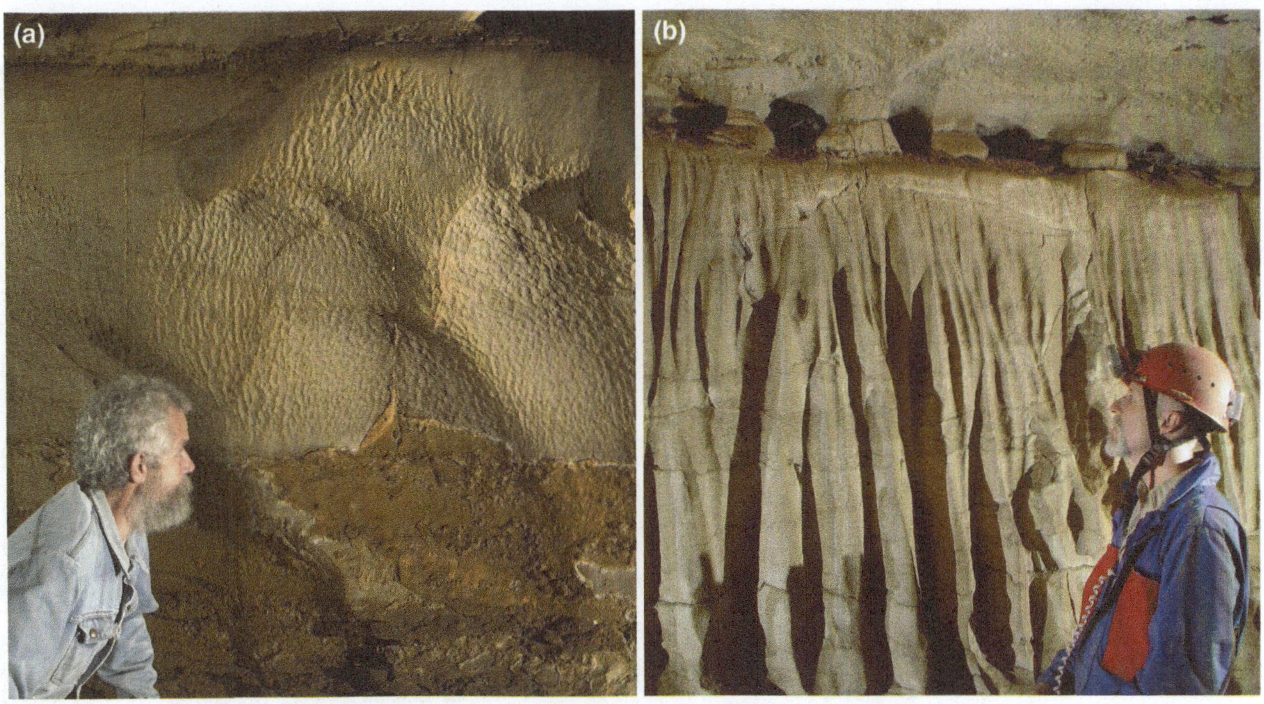

Fig. 6.11 **a** Solutional scallops in the walls of Grand Avenue (also known as Kentucky Avenue). The water was flowing from *right* to *left* at a slow velocity, about 2 in. (5 cm) per s. Later periodic rapid flow formed small scallops superposed on the larger ones with a flow rate of about 15 ft (4.5 m) per s. **b** Anastomoses along a bedding plane in the Ste. Genevieve Limestone, formed by periodic flooding when this passage was active. Flutes in the underlying wall were formed when water drained out of sediment in the anastomoses when the floodwaters subsided

direction of most recent water flow to enlarge the cave. The steep sides of the hollows are on the upstream side. Also, the scallop length decreases with faster flow (Curl 1974; Palmer 2007). Those about half an inch long (~1 cm) were formed by water moving about 10 ft/s (3 m/s), and those about one foot long (~30 cm) indicate flow rates of about 0.3 ft/s (~10 cm/s). Scallops are not formed in very slow water because the turbulent eddies are not stable.

Flutes are parallel vertical grooves dissolved by waterfalls or by water draining out of openings in the cave wall after periods of high water flow (Figs. 6.8b and 6.11b). Fluted walls are most common in shafts, and also in canyons formed by the upstream retreat of waterfalls. Thick-bedded rocks are most easily fluted.

Anastomoses are small sinuous tubes that interconnect in a braided, maze-like pattern along bedding planes (Fig. 6.11b). The tubes have roughly semicircular to circular cross sections and range in diameter from roughly half an inch to 2 ft (1–60 cm). They usually enlarge upward into the base of the overlying bed because their floors are armored by insoluble sediment. They normally form during floods, when water fills a cave passage under pressure and is injected into adjacent bedding-plane partings. As the floodwater subsides, the water drains back out and may leave flutes in the walls below. Today none of the tour routes in Mammoth Cave experience flooding, except in the vicinity of River Hall along the Historic Route, where the Green River rises and slowly backs up into the cave during high flow.

6.10 Cave Deposits

6.10.1 Sediment

Most cave passages contain a variety of sediments including clay, sand, and gravel. Their character and distribution can give information on where the water came from, how fast it was moving, and what the cave conditions were when the sediment was deposited. Most of this material consists of insoluble grains that are carried in by flowing water. Most caves contain sand and gravel that has been carried by streams. The flow velocity can be interpreted from the size of the sediment particles. Deposits of mud (clay) indicate slow-moving water and are most common where periodic floodwater ponds in the lowest levels. In cave streams, rock fragments break into smaller particles and become rounded in the downstream direction. Sandstone blocks litter the floors of many shafts, having been carried in by streams from the eroding edges of the plateaus.

Interpretation of cave sediment is fairly simple. For example, the deposit in Fig. 6.12 indicates several things: (1) There was an upstream source of coarse sand and gravel.

Fig. 6.12 Alternating beds of sand and gravel in an upper-level cave passage. The gravel indicates rapid flow, and also the original water source. The only local source is the conglomerate in the highest ridge tops, and it must have been carried into the cave by surface streams. Scale: the blue flash-bulb is 1.8 in. long (4.5 cm)

There is only one local source for the gravel—the highest ridge tops where Pennsylvanian conglomerate is present. (2) The velocity was great enough to carry this material far into the cave. (3) Slowing of the flow enough to deposit the sediment, perhaps by a change of passage slope, eddies sweeping the grains into slow-moving water on the insides of bends, or a diminishing of water flow (for example, during the last phases of a flood). (4) Fluctuating velocities to allow alternating sand and gravel deposits.

6.10.2 Breakdown

Breakdown is caused by the collapse of cave ceilings or walls. The term applies both to the process and to the fallen pieces that result from it. This process can greatly modify passage shape and perhaps make it difficult to determine the original solutional pattern. Breakdown material varies from small chips and flakes to large slabs and blocks (Fig. 6.5). Piles of breakdown in a cave give the impression that an entire ceiling has come down in a single catastrophic failure. In reality, breakdown usually subsides slowly, one block, or a few blocks, at a time. As the cave ceiling retreats upward by breakdown, it may stabilize into an arch. In wide sections of passage, or at intersections, the result is usually an oval or circular dome-shaped room. An example is the Rotunda on the Historic Route.

6.10.3 Speleothems

Water that infiltrates downward through the soil absorbs a large amount of carbon dioxide (CO_2) produced by organic activity. In the soil, the CO_2 reaches concentrations about 50–100 times greater than in the outside atmosphere. CO_2 concentrations in caves are somewhere between the two. As the water infiltrates through fissures in limestone or dolomite, it dissolves some of the rock. When the water enters a cave, much of the CO_2 is lost to the cave air, and limestone tends to be precipitated as speleothems, or cave formations composed of calcium carbonate—the same material as limestone, but in a purer and more crystalline form called calcite. These include the well-known stalactites and stalagmites. Dolomite precipitates slowly, usually only where evaporation is intense, and tends to form minor crusts. Where water infiltrates rapidly, or does not encounter much limestone, it tends to continue dissolving and produces shafts and canyon passages instead of speleothems.

Evaporation in dry parts of the cave can draw moisture from the bedrock walls and form other types of speleothems. These include flower-like deposits of gypsum ($CaSO_4 \cdot 2H_2O$) and other very soluble rocks that are rare at the Earth's surface. The various speleothems and the way they are deposited are described in detail in Chap. 10.

Here, however, it is sufficient to mention only that speleothem types are useful for determining the flow conditions and chemistry of cave water. For example, passages beneath the sandstone caprock tend to contain many evaporative minerals. These are mostly white and delicate, with curving bundles of mineral fibers or straight needle-shaped spears. Calcite speleothems tend to form beneath the edges of the ridges, where infiltrating water is more abundant. The transition from one speleothem type to the other can allow explorers to determine their location relative to the overlying geology and landscape.

6.11 Conclusion

This chapter has introduced what Mammoth Cave is like and how to interpret its origin. Visitors should easily be able to recognize and interpret the various cave features described here. The following chapter uses these concepts to reveal how the cave has changed with time.

References

Curl R (1974) Deducing flow velocity in cave conduits from scallops. Nat Speleol Soc Bull 36:1–5

Meiman J, Groves C (1997) Magnitude/frequency analysis of cave passage development in the Central Kentucky Karst. In: Proceedings, 6th science conference, Mammoth Cave National Park, KY, pp 11–13

Palmer AN (1981) A geological guide to Mammoth Cave National Park. Zephyrus Press, Teaneck, NJ

Palmer AN (1989) Stratigraphic and structural control of cave development and groundwater flow in the Mammoth Cave region. In: White WB, White EL (eds) Karst hydrology—concepts from the Mammoth Cave area. Van Nostrand Reinhold, New York, pp 293–316

Palmer AN (2007) Cave geology. Cave Books, Dayton, OH

Palmer AN, Palmer MV (eds) (2009) Caves and Karst of the USA. Nat Speleological Soc, Huntsville, AL, pp 114–115

Pohl ER (1970) Upper Mississippian deposits of south-central Kentucky. Ky Acad Sci Trans 31:1–15

Quinlan JF, Ray JA (1981) Groundwater basins in the Mammoth Cave region, Kentucky. Friends of Karst, Occasional Publication 2, map. Reprinted in: White WB, White EL (eds) (1989) Karst hydrology: concepts from the Mammoth Cave area. Van Nostrand Reinhold, New York, map in pocket

Swann DH (1964) Late Mississippian rhythmic sediments of the Mississippi Valley. Am Asso Petrol Geol Bull 48:637–658

White WB, White EL (eds) (1989) Karst hydrology: concepts from the Mammoth Cave area. Van Nostrand Reinhold, New York

Geologic History of Mammoth Cave

7

Arthur N. Palmer

Abstract

Mammoth Cave is an underground part of the vast Mississippi River drainage system. Although the cave occupies a relatively small area, it has preserved about 10 million years of the drainage history of the eastern USA. It is still actively growing today. As surface rivers vary in the shape and pattern of their valleys, the cave records these events as different levels and types of underground passages. The nature and timing of those changes are well preserved in the cave. Their ages can be determined by analyzing the sand and gravel carried into the cave while it was forming. The various cave levels record the long history of the Ohio River and the influence of continental-scale glaciers on surface drainage patterns. The many changes in the erosion level of the Green River help to account for the great complexity of the cave.

7.1 Introduction

The origin of a cave is closely related to the geologic history of the surrounding region. At the surface, much of the evidence for this history is destroyed by weathering and erosion, but it is well preserved in caves. This is especially true for Mammoth Cave. Reconstructing that history is of interest not only to geologists but also to cave explorers who want to predict how to find new passages. Throughout the evolution of the cave, the pattern of active passages has been roughly dendritic—i.e., branching, like a surface river system. But whenever a cave stream changes course to form a new passage, it leaves behind a dry passage as evidence of its old pattern. In Mammoth Cave, the result is an extremely complex tangle of passages, both wet and dry, which seem to make no sense when viewed all together on a map (see Fig. 7.1). However, detailed examination reveals a very clear history.

A.N. Palmer (✉)
State University of NY, Oneonta, NY, USA
e-mail: arthur.palmer@oneonta.edu

H.H. Hobbs III et al. (eds.), *Mammoth Cave*, Cave and Karst Systems of the World,
DOI 10.1007/978-3-319-53718-4_7

7.2 Stages of Cave Development

The origin of a cave passage in limestone involves two major steps. When underground water first begins to seep through narrow cracks in the rock, it moves very slowly and becomes almost entirely saturated with dissolved limestone after only a short distance. Many thousands of years are usually required before the crack widens enough to transmit water rapidly. Only then can a real cave passage begin to form. The time required for this beginning phase is called the "breakthrough time." That time is shortest if the original cracks are wide and the flow path is short. From then on, the passage enlarges at a much faster rate that can turn the narrow cracks into a humanly traversable cave within a few tens of thousands of years. After breakthrough, the enlargement rate depends mainly on the amount of water flow and its chemistry. A large passage can form only if it is fed for a long time by a large surface catchment area. When a cave passage is abandoned by its flow and becomes dry, it may continue to survive for a long time with no further enlargement. With its combination of many active and abandoned passages, Mammoth Cave contains a record of millions of years of geologic history.

Fig. 7.1 Aerial view of the Pennyroyal Plateau (foreground) and Chester Upland (background). Photograph courtesy of Gary Berdeaux

7.3 Relationship to the Land Surface

Even in passages that are now dry, the original positions of the water table can be determined from the distribution of vadose and phreatic passage types. The best examples are where canyons change downstream to become tubes. At those points, the passages also tend to change course from directly down the dip of the beds to a different path, typically along the strike of the beds (perpendicular to the dip). The passage slope also tends to decrease significantly at these points. Vadose passages descend continuously along their lengths, but phreatic passages tend to loop up and down along their lengths, perhaps almost imperceptibly.

As the Green River erodes its channel downward, the water table drops at roughly the same pace. Therefore, the vadose-phreatic transitions in the cave keep shifting downward. If the river erodes downward steadily without pause, phreatic passages may not be able to stabilize and grow large. But when the river pauses in its downward erosion for a long time, the water table tends to remain in the same position as well and the vadose-phreatic transition zones in caves become sharply defined. A consistent "level" of this type, where several large passages and their vadose-phreatic transitions cluster within a narrow elevation range, represents a major stage of cave development.

Each significant cave level correlates with equally important phases in surface landscape development, when rivers in the area stabilize at what is called their **base level**—the level below which rivers cannot erode without further uplift of the land or a drop in the level of their outlets. Widespread changes in base level relate to major shifts in continent shape, sea level, or climate.

There have been times when the base level of Green River rose, for example, when the continental surface was depressed by the weight of glaciers. During these times, the river partly filled its channel with sediment. Caves at the former river level became flooded and often partly filled with sediment, while higher-level passages could be reactivated and enlarged beyond their original size.

The relatively flat Pennyroyal Plateau surface (Fig. 7.1) correlates with major cave passages at elevations of around 620–600 ft (190–183 m). This erosional surface rises toward the southeast, away from the Green River. Since then, it has been modified in various ways, first by the accumulation of a thick sediment cover when the base level rose and later by the underground diversion of small surface streams as the major rivers deepened their channels. During the later event, the Pennyroyal developed many sinkholes and underlying caves. Some of these supply groundwater to the lowest levels of Mammoth Cave (Fig. 6.4). The various passage levels in the cave can help to clarify these surface events.

7.4 How to Interpret the Cave Levels

Although Mammoth Cave is complex, it is not difficult to understand the processes that formed it. A good first step is to assume that the uppermost passages are oldest and the lowest are youngest. This is not entirely true because vadose passages can form at any time, and some of the very highest are still active today. But the largest passages, which dominate the cave, did form in sequence from highest to lowest, with only a few episodes when water levels rose and modified them at later times. If we can understand some of the main passages, most of the remaining ones easily fall into place.

The Historic Tour route is a good place to start (Fig. 7.2). The entrance passage leads directly to some of the largest and oldest passages in the cave. These were long ago abandoned by their water and are now dry. At the first junction, the Rotunda, two large passages lead off into the darkness. Note the many rock layers in the walls. These are the lowest beds of the Girkin Formation (Fig. 6.2). Saltpeter miners in the early nineteenth century excavated a pit in the sandy floor, but the original bedrock floor lies at least 20–30 ft (6–9 m) lower. The passages in this part of the cave were once deep canyons that became partly filled with sediment by cave streams during a base-level rise. This is typical of the uppermost levels of the cave.

The passage that leads straight ahead is Broadway, and to the right is Audubon Avenue. Water originally flowed out of Broadway and continued into Audubon Avenue. This is known because Broadway and its tributaries extend a great distance up the dip to the southeast, away from the Green River and toward the recharge areas of the Pennyroyal Plateau. The right-angle bend into Audubon can be explained by the dip of the limestone beds. The dip can be detected only by careful surveys because its angle is only about one degree here. In this area, Broadway follows the strike of the beds and apparently formed at or near the water table. The dip makes a sharp bend at this point from northeast to northwest, and the passage runs perpendicular to those trends, first northwest and then southwest (see Chap. 6 for details on dip and strike control of passages). The entrance passage formed later, when the water escaped from

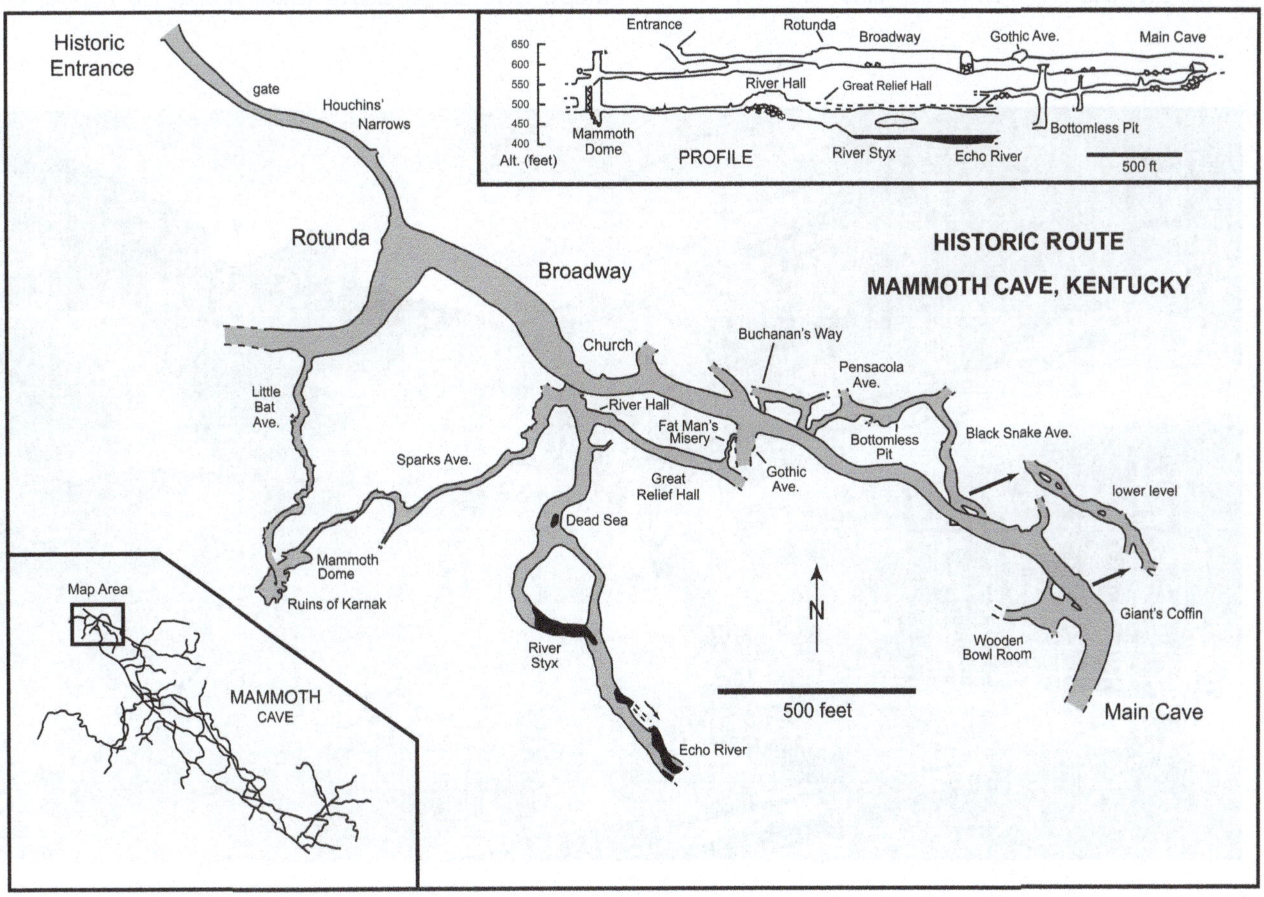

Fig. 7.2 Map and profile of the Historic Route (simplified from surveys by the Cave Research Foundation). Altitudes indicate feet above sea level. Refer to Fig. 6.1 for location relative to the entire Mammoth Cave System

Broadway along a more direct route to the Green River. It runs straight down the dip, which shows that it formed under vadose conditions as the Green River dropped.

The tour follows Broadway in the former upstream direction. After a short distance, the passage splits into two separate levels (Fig. 7.3). The upper level, Gothic Avenue, was the very first passage to form in this part of the cave. It is a wide tube with a small vadose canyon in the floor. The original flow was later diverted to form the lower level (Main Cave), which followed an independent route for most of its length but joined its earlier path, Gothic Avenue, at this point. Gothic Avenue is located in the bottom beds of the Girkin. Main Cave and most of the remaining routes on the tour are located in the Ste. Genevieve Limestone.

Main Cave is a wide canyon passage for most of its length, although sediment has filled much of it in this part of the cave. It is easy to see that Gothic crosses over Main Cave but curves around sharply to join Main Cave at the "Church" to form Broadway, which combines the heights of the two upstream passages (Fig. 6.5). Historic saltpeter vats are on display at the Gothic junction, and the Church contains remnants of hollow wooden pipes that carried freshwater into the cave and dissolved nitrates back out.

It is tempting to imagine that these passages were once entirely filled by huge underground rivers. This was probably never the case, except perhaps when the Green River was in flood and backed up into the cave. By observing still-active parts of the cave, it is clear that most of the time these passages were floored by shallow meandering streams bordered by low sediment banks. Most passage growth took place during periodic high flows when the streams spread across the entire passage widths.

A thousand feet ($\sim$300 m) beyond Gothic Avenue, a steep passage descends from Main Cave to join others at a lower level. This was one of the routes through which the original cave stream abandoned its upper level. The relatively small size of this passage shows the modest scale of the last stream to occupy this part of Main Cave.

Through this diversion passage, the tour descends into many complex galleries that meander through the limestone at various levels (Fig. 7.4). Many are tube-shaped with round or elliptical cross sections. In places, they are intersected by narrow canyons and vertical shafts. The shafts cluster together in certain areas and still contain active drips. The tour route has passed directly beneath the Main Cave and approached the northeastern flank of Mammoth Cave Ridge. Here the protective sandstone cap has been eroded

Fig. 7.3 Junction between Gothic Avenue (*above*) and Main Cave (*below*). The contact between the Girkin Formation and the Ste. Genevieve Limestone is the bedding plane at the base of Gothic Avenue

Fig. 7.4 Black Snake Avenue consists of several levels of tubes

away so that water can descend rapidly from the surface and produce the shafts. Note the vertical flutes in the shaft walls (compare with Fig. 6.11b). Heavy rain can turn the drips into showers.

Some of the passages contain conspicuous scallops, which give information on the direction and velocity of the water that formed them (Fig. 6.11a). Note the anastomoses along bedding-plane partings in Black Snake Avenue, with fluted walls beneath them. These show evidence for periodic flooding in the remote past, when water was forced into the partings under pressure and drained out as the flood subsided (Fig. 6.11b). This process is common in passages that are still enlarging near base level.

Bedding in the limestone has an obvious effect on passage shapes. Many ceilings are formed by nearly flat beds, even where collapse has taken place. Most tubular passages have elliptical cross sections elongated along the beds, with their widths greater than their heights. Very few passages make angular turns, as they would if they were following intersecting fractures. Instead, they have sinuous patterns like the meanders in a river. In places, such as the ceiling of Great Relief Hall, solutionally widened fractures are exposed in the ceiling, but they do not influence the passage pattern (Fig. 7.5). One exception is seen in Sparks Avenue, where greatly widened fractures have formed tent-like alcoves in the ceiling, and in one place the passage rises in the downstream direction along one of these widened fissures (Fig. 6.6).

River Hall, at the lowest point on the Historic Route, illustrates an important concept. As the tour enters from Great Relief Hall, the bedrock ceiling and floor drop steeply about 20 ft (6 m). River Hall is part of a major tributary that rises from a lower level on the left. Scallops in the ceiling show that the original water movement was upward into River Hall. The explanation for this pattern is described later.

The tour leads to Mammoth Dome, the largest shaft in this part of the cave, about 200 ft (60 m) high (Fig. 6.8b). The dark brown St. Louis Limestone is visible in the walls below the trail level. Nearby, some of the drips entering the cave are depositing limestone as stalactites and flowstone,

Fig. 7.5 Great Relief Hall, a tubular passage at one of the major cave levels (500 ft). Its ceiling shows several solutionally enlarged joints that cross the passage without affecting the passage trend. Also in the ceiling are resistant chert bodies (SiO_2). The pock-marked lower walls consist of a distinctive dolomite bed that can be traced through most of the cave in Mammoth Cave Ridge

rather than dissolving it. That water enters the ground through the soil directly into the limestone beds. The great amount of carbon dioxide supplied by the soil has allowed much limestone to dissolve from the walls of narrow cracks. When the water emerges into the cave, it loses most of its CO_2 and its ability to hold the dissolved limestone. These speleothems (cave formations) are not only decorative—they also indicate where we are in the cave relative to the land surface.

The major cave levels are arranged with the oldest at the highest elevation and with younger levels at successively lower elevations as the Green River valley was eroded downward. Meanwhile, vadose passages could form above the water table at any elevation, at any time during the cave's history. For example, the entrance passage to the Domes and Dripstones tour contains a series of vadose canyons and shafts at a higher elevation than the much older passages of the Historic Tour. The water that feeds the canyons and shafts is supplied by a small valley in the side of Mammoth Cave Ridge and has no relation to the Green River level.

7.5 Developmental History of Mammoth Cave

It is easy to reconstruct the developmental history of the cave at a local scale, but to understand the entire cave system requires a much broader view, which includes the geologic history of the major surface rivers throughout the eastern USA.

7.5.1 The Oldest Passages

Mammoth Cave, as we know it today, had many precursors farther up-dip to the southeast. Only a few isolated fragments are left from those times many millions of years ago. They are located in plateaus and knobs that stand high above the Pennyroyal Plateau along the Chester Escarpment.

Within the known limits of the Mammoth Cave System, the oldest passage by far is Collins Avenue in Crystal Cave (Fig. 7.6; Palmer 1981). It formed when the Green River

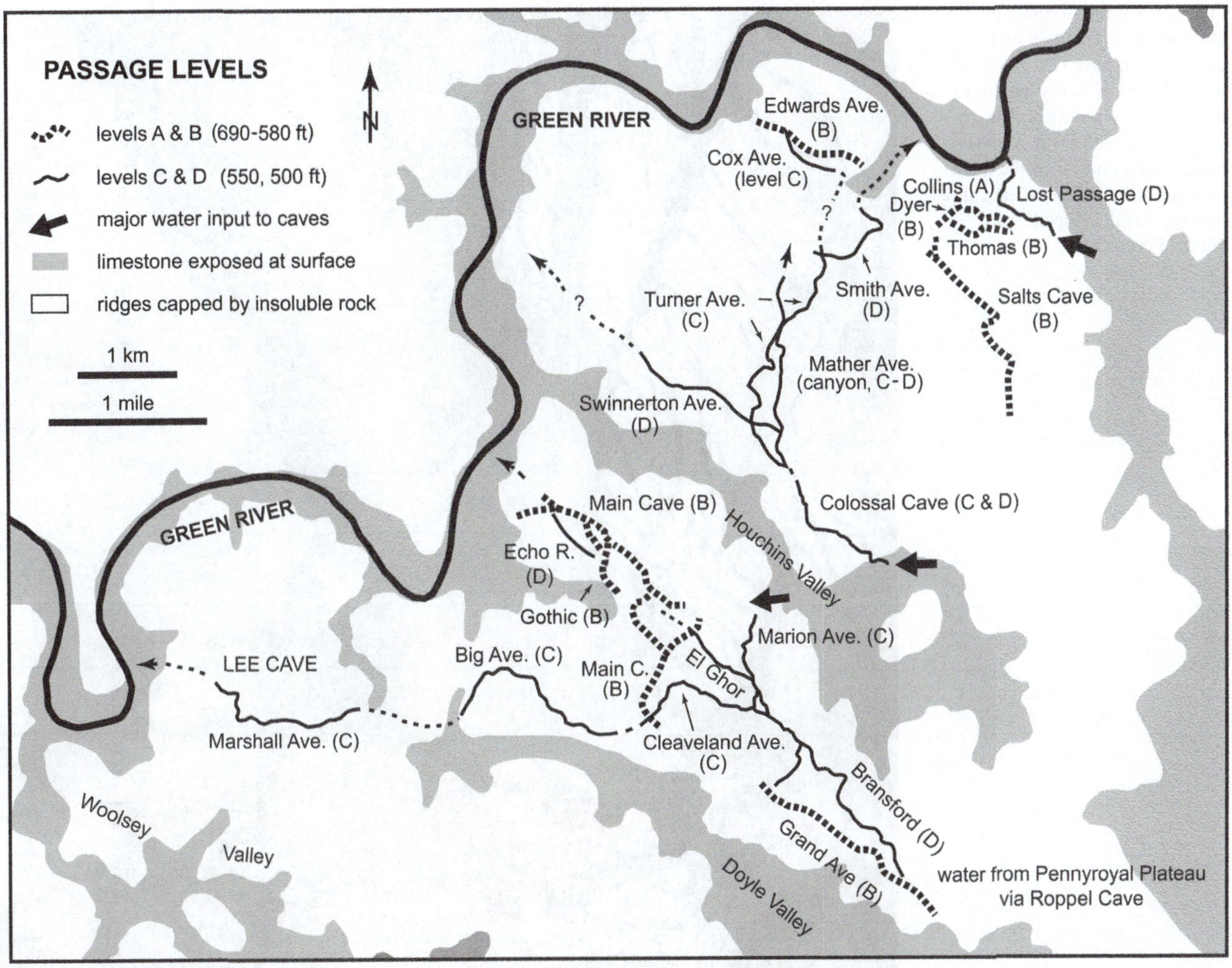

Fig. 7.6 Generalized map of the main upper-level passages in Flint Ridge and Mammoth Cave Ridge

first eroded through the sandstone cap to expose the underlying limestone. At the southern tip of a bend in the river, a spring developed that was fed by water from a higher stream valley to the east. The underground distance was rather short, about a mile (1.6 km). The overlying sandstone cap was thin and the cave was shallow—only 18 ft (5.5 m) below the top of the limestone. Joints and bedding-plane partings in the limestone were wide enough to allow water to pass through them easily. Rates of surface erosion at that time were very slow, so the flow of water through these beds would have persisted for a long time. All of these conditions made this the perfect location for the birth of Mammoth Cave.

As Collins Avenue developed, the region remained relatively stable, with very slow rates of river entrenchment. When the passage began to form, the Green River was the base level for the entire region, and no part of its drainage basin could have been lower in elevation. That includes the Pennyroyal Plateau. Today, the neighboring parts of the Pennyroyal lie about 60 ft (18 m) below the ceiling of Collins Avenue, which gives a hint of the lengthy surface erosion that must have taken place since the passage formed. So far, we have found no way to date the origin of Collins Avenue, but the great amount of surface erosion since then, coupled with known dates from lower passages, suggest that more than ten million years have elapsed since Collins Avenue began to form. This is a rough estimate, but it is unlikely to be an exaggeration.

As the surface streams eroded downward, the outlet for the Collins Avenue stream migrated downward as well. The cave passage deepened to form a large canyon passage extending through most of the Girkin Formation (Fig. 7.7). Only then did other passages begin to form in the Mammoth Cave System. These too were large, and they correlate roughly in elevation with the Pennyroyal Plateau surface of today. At that time, erosion of the land surface was much slower than it is today. Therefore, the Pennyroyal was not dissected by surface rivers, and there were few, if any, sinkholes to deliver water underground. Instead, its runoff formed large surface streams that supplied water to the

Fig. 7.7 Collins Avenue in Crystal Cave. This is a wide canyon 70 ft high (21 m), which is filled with sediment nearly to the ceiling by sediment. This is the junction with Dyer Avenue, a downstream portion of the main passage of Salts Cave. The entire passage has developed in the Girkin Formation

limestone at the base of the Chester Escarpment. This water drained underground through the limestone ridges of the Chester Upland to springs along the Green River and formed some of the largest cave passages in the region (Fig. 7.8).

The huge upper-level passage of Salts Cave was formed where water flowed through the eastern part of Flint Ridge (Fig. 7.6). This passage, in places more than 60 ft (18 m) high, formed in two stages within the lower Girkin Formation. Its earliest route was toward the northwest, down the dip, but where it reached the water table, it turned along the strike of the beds toward the northeast. Now severed from Salts Cave by a surface valley, this strike-oriented segment, known in Crystal Cave as Dyer Avenue, intersected the bottom of Collins Avenue and provided the only known connection between Collins and the rest of the cave system. With time, as the main Salts passage deepened, its lower half bypassed the diversion route to Dyer Avenue and continued down the dip to the northwest, to form Edwards Avenue, the main passage of Great Onyx Cave, near the base of the Girkin. Thus, the largest passages in both Crystal Cave and Great Onyx Cave are merely downstream segments of the huge Salts trunk passage.

Meanwhile, a stream in Houchins Valley, southwest of Flint Ridge, sank into the eastern flank of Mammoth Cave Ridge and drained underground to the Green River. This produced Gothic Avenue and its upstream segments

Fig. 7.8 Grand Avenue (also known as Kentucky Avenue) is the upstream end of Main Cave. This part of the passage is located in a thick-bedded section of the Ste. Genevieve Limestone about 30 ft (9 m) below the contact with the overlying Girkin

(Figs. 7.2, 7.3, and 7.6). They formed in the bottom part of the Girkin Formation, ignoring the entire upper two-thirds of that rock layer.

The next major event in Mammoth Cave Ridge was the origin of a tremendously long passage that extended the entire length of Mammoth Cave Ridge—a total of about 7 miles (11 km)—and probably farther in the upstream direction in parts of the ridge that have since been removed by erosion (Fig. 7.6). It is located in the upper Ste. Genevieve Limestone, slightly below the level of Gothic Avenue. These two levels converge to form the large passage called Broadway (Figs. 7.2 and 6.5).

In the vicinity of the Green River, which controlled the vertical position of these passages, the major levels that formed during this early phase are now located at about 620 and 600 ft (190 and 182 m) above present sea level. Collins Avenue is much higher at about 690 ft (210 m).

The next event was unusual. All of these upper-level passages were filled with sediment, to as much as 50–60 ft (15–18 m). Some passages were filled to the ceiling, but most were filled only part way. Thick sediment probably covered the Pennyroyal Plateau as well (Ray 1996). The sediment appears to have about the same age everywhere—roughly 2.6 million years (Granger et al. 2001). The fill may have accumulated in several phases, possibly interrupted by erosion. A rise in the Green River level is a likely trigger for events like this, and the fact that the sediments are thickest in passages near the Green River lends support to that idea.

7.5.2 Younger Passages

After these major upper-level passages developed, the water sources to Mammoth Cave began to fragment into smaller catchment areas. Passages became more numerous but smaller. Large passages still continued to form where many smaller ones happened to join, but they were surrounded by swarms of others that were fed by local catchment areas around the ridge flanks and between ridges of the Chester Upland.

Meanwhile, the Green River increased its entrenchment rate. The water table dropped rapidly and tributaries of the

Green River were diverted underground. The original stream valleys were converted into karst valleys studded with sinkholes. Each sinkhole was able to deliver its own local water flow to an independent vadose stream passage. The Pennyroyal also began to subdivide into small drainage basins as sinkholes diverted water underground and the number of inputs increased (Fig. 7.1).

As the Green River eroded its valley deeper, water in Mammoth Cave was diverted from the level of Main Cave and its contemporaries and established a new major level about 50 ft (15 m) lower. The cause of this abrupt drop is related to changes in the drainage pattern of the Ohio River (see Granger et al. 2001). Passages at this new level were mainly tubes with elliptical or lens-shaped cross sections. Cleaveland Avenue is a fine example (Figs. 6.9 and 6.10). It has almost no slope over its mile-long length, which suggests a lengthy period with an unusually stable water table. Turner Avenue in Flint Ridge is similar and is at the same elevation as Cleaveland Avenue (within a few feet). Because of the dip of the rocks, they are in entirely different beds, so there is no question of their having simply formed in a particularly favorable rock layer. The stability of the Green River at that time was the only feasible control.

A further drop in the Green River caused the water in Cleaveland Avenue and its contemporaries to descend another 50 ft (15 m). The water from Cleaveland formed a new canyon passage, the Pass of El Ghor, which extended straight down the dip of the limestone beds—a very abrupt change from the static water table that prevailed earlier (Fig. 7.6). This canyon fed a large tube at an elevation of 500 ft (152 m). Instead of following the water table by bending along the strike of the beds, as Cleaveland Avenue had, this phreatic passage dropped sharply downward across the beds below the water table, and wandered at various angles across the dip. While it was first forming, its lowest ceiling reached almost 70 ft (21 m) below the water table. Even today its lower parts still carry a large stream at the level of the Green River, called Echo River (Fig. 6.7b)—the passage that originally delivered water into River Hall, as described earlier in this chapter (Fig. 7.2). The Echo River passage is quite large in places because it has contained flowing water for a very long time.

The abandonment of Cleaveland Avenue was very abrupt, as shown by the uninterrupted canyon that descends all the way to the 500-foot level. It might seem that such a large drop in the water table would require a long period of gradual erosion by the Green River. Instead, the drop in river level must have been sudden, caused by a series of waterfalls and rapids progressing upstream—like smaller versions of Niagara Falls. As the rapids eroded their way upstream past Mammoth Cave, the local river level dropped rather suddenly, and so did the water table. Observations like this make it possible to learn a lot about the drainage history of the surrounding region.

Since then, further deepening of the Green River valley has allowed new passages to form at still lower elevations but gradually and with no major pauses. The lowest passages still contain active streams. These include an extremely long river passage, Logsdon River, which is fed by sinkholes in the Pennyroyal Plateau far to the east, emerges into Roppel Cave and joins Mammoth Cave and Proctor Cave (Fig. 6.1). The connections from Proctor to Mammoth, and later from Mammoth to Roppel, were made through this passage.

One of the last events to affect Mammoth Cave was filling of the Green River valley with about 50 ft (15 m) of sediment. The age and cause are still uncertain. It caused flooding of the lowest passages in the cave, although some passages at the lowest levels were already water-filled. The cause of the valley filling may be related to a sea-level rise during the melting of continental-scale glaciers around 14,000 years ago. In addition, a low dam on the Green River downstream from Mammoth Cave was built in 1906 to facilitate navigation. This raised the water levels in the cave about 6 ft (1.8 m). The lowest passages are now accessible only by diving, although this practice is not ordinarily permitted in the Park.

The geologic history of Mammoth Cave is described in more detail in Chap. 9.

7.6 Why Is Mammoth Cave the World's Longest?

With more than 405 miles (652 km) of mapped passages, Mammoth Cave is the longest known cave in the world, almost twice the length of the nearest contender. The second largest is Mexico's Ox Bel Ha, currently at 208.3 miles (335 km), in the Yucatan Peninsula.

Why is Mammoth Cave so long? The reasons are clear: (1) It has a huge drainage basin with plenty of rainfall, so that large amounts of water pass through the local limestone. (2) The gently dipping limestone is exposed over a large area. (3) The deep valley of the Green River has cut through almost 300 ft (90 m) of limestone over a distance of many tens of miles, and it serves as a convenient outlet for the underground water. (4) The Green River has eroded downward in small steps over many millions of years, giving time for many discrete levels of passages to form at different elevations. (5) The resistant caprock of sandstone protects many passages from being eroded away by surface streams. (6) Erosion along the flanks of ridges in the Chester Upland has produced many small valleys, each of them supplying water to feed cave passages. (7) Water from the Pennyroyal Plateau enters the ground through sinkholes and sinking streams over many square miles. This is a constant and abundant source of water that drains through the Chester Upland. (8) The limestone at Mammoth Cave consists of thin beds rather than a few massive ones. Hundreds of gently dipping bedding-plane partings are available to form discrete passages with no connection to those above or below. Rather than forming a few large passages, the water forms many smaller ones, each with its own unique path.

Nearby Fisher Ridge Cave, currently 125 miles (201 km) long, lies only a short distance to the east of the Mammoth Cave System (Quick 2009). Past experience suggests that a connection will eventually be found between the two, but at this time, there is no obvious route between them. Other large caves lie close to the perimeter of Mammoth Cave (Quinlan and Ray 1981). Searching for new passages is not a matter of guesswork. Surface karst features such as sinkholes, as well as the subsurface pattern of known caves, give strong hints as to the distribution of underground drainage. There are predictions that Mammoth Cave will someday be explored to more than double its present length. Although this is geologically possible, it will require much time and persistence.

References

Granger DE, Fabel D, Palmer AN (2001) Pliocene-pleistocene incision of the Green River, Kentucky, determined from radioactive decay of cosmogenic ^{26}Al and ^{10}Be in Mammoth Cave sediments. Geo Soc Am Bull 113(7):825–836

Palmer AN (1981) A geological guide to Mammoth Cave National Park. Zephyrus Press, Teaneck, NJ

Quick P (2009) The fisher ridge cave system. In: Palmer AN, Palmer MV (eds) Caves and Karst of the USA. National Speleological Society, Huntsville, AL, pp 114–115

Quinlan JF, Ray JA (1981) Groundwater basins in the Mammoth Cave region, Kentucky. Friends of Karst, Occasional Publication 2, map. Reprinted in: White WB, White EL (eds) (1989) Karst hydrology: concepts from the Mammoth Cave area. Van Nostrand Reinhold, New York, map in pocket

Ray JA (1996) Fluvial features of the karst-plain erosion surface in the Mammoth Cave region. In: Proceedings, 5th science conference, Mammoth Cave National Park, KY, pp 137–156

Hydrology and Hydrogeology of Mammoth Cave

8

William B. White and Elizabeth L. White

Abstract

Mammoth Cave and its associated drainage is the downstream portion of a much larger aquifer system which includes recharge from sinking streams at the southeastern edge of the Sinkhole Plain, internal runoff and diffuse infiltration from the Sinkhole Plain, and runoff from valley drains and the caprock of the Mammoth Cave Plateau. The aquifer drains through a sequence of large springs along Green River. Extensive tracing of flow paths with fluorescent dyes shows that each spring has a distinct drainage basin with some spillover depending on recharge. Basin area can be estimated from measured base flows of the karst springs. Green River flows in a narrow valley that produces high flood levels that backflow into the springs carrying muddy flood waters deep into the conduit system. There is a complex flux of clastic sediments, some from upstream and some back-flooded from the river, that moves through the conduit system in response to storm flow. Chemical analysis of spring water, cave stream water, and cave drip water allows the calculation of dissolved carbonate (hardness), chemical saturation state, and concentration of dissolved CO_2. Spring and shaft waters are undersaturated; drip waters are supersaturated. CO_2 concentrations exhibit a pronounced maximum during the growing season. Although the Mammoth Cave System contains more mapped passages than any cave in the world, the sad truth is that only a small fraction of the active drainage system is accessible to direct observation and survey.

8.1 Introduction

Ever since the discovery of Echo River in 1838, the rivers that flow under Mammoth Cave have been the subject of description and speculation. The Styx and Echo Rivers appear in all of the many descriptions and guidebooks that were written in the nineteenth and early twentieth centuries. What they all have in common is that the view is from inside the cave. What was not appreciated until much later is that an understanding of the behavior of water in Mammoth Cave requires an understanding of all sources and flow paths, an understanding that takes us outside the cave and indeed outside Mammoth Cave National Park.

The larger picture of Mammoth Cave hydrology was first recognized in the pioneering work of Park geologist Pohl (1936) although little of Pohl's work reached formal publication. Pohl recognized the significance of clastic sediments in the cave and first identified the importance of base-level back-flooding. The US Geological Survey worked in the park in the early 1960s, concerned with water supply for the park and with the problem of mud accumulation on the low-level tourist trails (Brown and Lambert 1963; Cushman et al. 1963). Hydrological investigations at Mammoth Cave picked up momentum with efforts by the Cave Research Foundation during the International Hydrologic Decade, 1964–1974, especially with the Ph.D. thesis of John W. Hess and the extensive stream tracing studies of Park geologist James F. Quinlan. Results from this period of research were summarized in a book (White and White 1989). More recent investigations have required instrumentation for continuous

W.B. White (✉) · E.L. White
Department of Geosciences, The Pennsylvania State University, University Park, USA
e-mail: wbw2@psu.edu

H.H. Hobbs III et al. (eds.), *Mammoth Cave*, Cave and Karst Systems of the World,
DOI 10.1007/978-3-319-53718-4_8

observation of flow behavior and chemistry of the hydrologic system. These have been carried out by Park hydrologist Joe Meiman and by Chris Groves and other investigators from the Center for Cave and Karst Studies at Western Kentucky University.

8.2 The Hydrogeologic Framework

8.2.1 The Green River Springs

Green River bisects the Mississippian limestone aquifer. North of Green River, small tributary surface valleys have cut through the overlying sandstones and shales into the underlying limestones, resulting in a sequence of small basins with underground drainage and small caves. The portion south of Green River is by far the most extensive with a drainage divide southeast of the Sinkhole Plain. The watersheds north and south of Green River drain to a series of springs along the river. Eighty-one of these springs have been recorded, but most are small springs draining hillsides or small side valleys (Fig. 8.1). For many details of the surface and underground drainage, see the water tracing maps of Ray and Currens (1998a, b).

Some of the larger springs south of the river, counting downstream from Munfordville, are Gorin Mill Spring, Hick's Spring, Lawler Blue Hole, Pike Spring, Styx River Spring, Echo River Spring, and Turnhole Spring. There are also springs on the north side of the river draining the various karst valleys. The Green River springs are alluviated. There is 5–10 m of silt under the Green River channel so that the spring flows, fed by conduits graded to the bedrock channel of the river, have to rise over the silt fill to reach the river. Most springs, therefore, are rise pools with water emerging from beneath rock ledges to flow out to the river through silt-filled channels of varying lengths. Pike (Fig. 8.2) and Turnhole Springs are rise pools immediately on the riverbank. Styx and Echo Rivers (Fig. 8.3) are connected to the Green River by short surface channels.

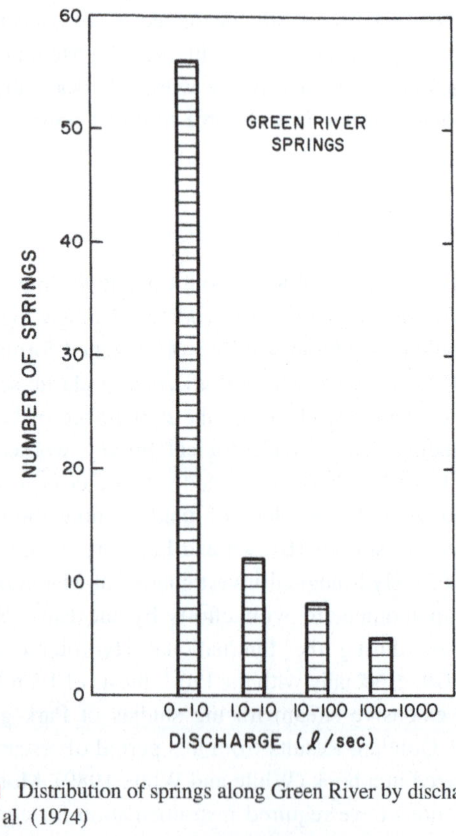

Fig. 8.1 Distribution of springs along Green River by discharge. From Hess et al. (1974)

8.2.2 Groundwater Basins

Like surface streams, large karst springs often have distinct drainage basins in which all recharge to the basin flows to the spring, while recharge beyond the basin boundary flows to a different spring. The drainage basins for the large Green River springs have well-defined boundaries, especially under low recharge regimes; however, under high recharge, water levels rise and there may be spillover into adjacent basins and basin divides may shift.

The catchments for the large springs in the Mammoth Cave area were mapped in elegant detail with more than 500 tracer tests and the construction of the piezometric surface by James F. Quinlan and his associates. The map, produced by Quinlan and Joseph A. Ray, was distributed as a single sheet as Occasional Paper No. 1 of the Friends of Karst in 1981. It was published with extensive discussion by Quinlan and Ewers (1989). The map is reproduced in outline form in Fig. 8.4. Three large basins account for most of the Mammoth Cave area. The Hidden River basin is the largest and extends from a drainage divide near Cave City eastward to a north–south divide closely paralleling the Little Barren River and discharges at Gorin Mill Spring with overflows to Hicks Cave Spring. The central portion of the area drains north to Turnhole Spring, the most downstream of the springs from Mammoth Cave emptying into the Green River. The springs between Gorin Mill Spring and Turnhole Spring each have relatively small catchments, especially Styx River and Echo River Springs, which share a small catchment encompassing only Mammoth Cave Ridge and bordering karst valleys. The southwestern-most basin sends its waters westward to Graham Spring on the Barren River. Two of the larger sinking creeks, Gardner and Little Sinking Creeks, drain to Turnhole Spring, but just west of these streams is a drainage divide so that Sinking Creek drains to Graham Springs near Bowling Green.

Within each of the spring drainage basins, there are multiple sources of recharge, each with somewhat different

Fig. 8.2 Pike Spring. Water rises on the edge of Green River from a 6-m-wide by 2-m-high conduit 6 m below river level. Flow is beginning to reverse as muddy water from the river begins to replace the clear water of the spring. Photograph by the authors

storm responses and chemical characteristics. These are sketched in flow sheet form in Fig. 8.5.

8.2.3 Sinking Stream Catchments

The southern drainage divide for all the large groundwater basins is formed by the collective southern divides of a very large number of generally small sinking creeks (Fig. 8.4). South of the divide the surface drainage is to Beaver Creek, a tributary of the Barren River. The many small creeks flow north or northwest on the mostly non-karstic Salem and Warsaw Formations and then sink into the lower units of the St. Louis Limestone. Typically, the larger streams cut blind valleys some distance into the limestone with a final swallet at the base of a limestone bluff. Debris washed downstream by storm flow piles up against the swallet, blocking any cave entrance, and also debris forms dams that may produce wet-weather lakes upstream in the blind valleys.

The sinking streams provide surface water recharge to the aquifer. Sinking stream water tends to be water of poor quality since it drains from pasture lands at the southern edge of the Sinkhole Plain and may be influenced by barnyard runoff, septic tanks, and other sources of contamination. Sinking streams also inject soil and other clastic sediment into the aquifer during storm flow (Fig. 8.5).

8.2.4 The Sinkhole Plain Catchment

There are no surface streams crossing the Sinkhole Plain. Instead, the Plain is a continuous tiling of closed depressions (Fig. 8.6), each of which acts as a small internally drained catchment. Most of the closed depressions (sinkholes) have substantial soil covers. Rainfall onto the Sinkhole Plain infiltrates through the soils into the epikarst and from there makes its way through fractures into the underlying cave passages. Storm flow that exceeds the infiltration capacity of the soils drains down the wall of the sinkhole. If there is an open throat at the bottom of the sinkhole, the storm flow continues downward into the aquifer as internal runoff. In some sinkholes the storm water collects as temporary ponds. In others, the clay-rich soils at the bottom of the sinkhole are sufficient to retain a perennial pond (Fig. 8.7). The soils at

Fig. 8.3 Echo River Spring. Water rises from a conduit about 6 m below the limestone ledge and feeds into 360 m of spring run to the Green River. Photograph by the authors

the bottoms of sinkholes often collapse by piping failure, thus providing an efficient pathway for internal runoff and also an injection of soil into the groundwater system.

Diffuse infiltration through the soils and internal runoff through open sinkhole throats are likely to have different chemical characteristics in addition to different rates of recharge. Much of the Sinkhole Plain is pastureland, and the perennial sinkhole ponds are used for livestock watering. The soil offers some filtration, but internal runoff will carry the same contaminants as the sinking streams.

8.2.5 Karst Valley Catchments

The Mammoth Cave Plateau (Chester Cuesta) is dissected by surface valley tributaries to Green River. These valleys occur on both sides of the river, and most are underdrained once the downcutting of the valley breaches the sandstone caprock into the underlying limestones. Underground drainage has completely disrupted many of the valley profiles so, for instance, Doyle Valley on the south side of Mammoth Cave Ridge is now separated from Green River by the high saddle of Sloans Crossing. Other valleys also have no direct surface pathway to the river although many have dry channels that are used during flood flow.

The karst valleys inside the Park are forested so that much of the rainfall infiltrates to the epikarst and from there into the underlying cave system. Intense rains may produce some overland flow, which reaches the valley bottom and ultimately flows into swallets. Because of the forest cover, water entering the aquifer from the karst valleys should be of good quality.

8.2.6 The Sandstone-Capped Plateau and the Haney Aquifer

Joppa and Mammoth Cave ridges are capped with the Big Clifty Sandstone. Flint Ridge, however, is capped with a thicker sequence of clastic rocks, which also contains the Haney Limestone. Rainfall on Flint Ridge, and also on the ridges north of Green River, infiltrates through the Hardinsburg Sandstone and other clastics to recharge the Haney Limestone aquifer as a perched groundwater body

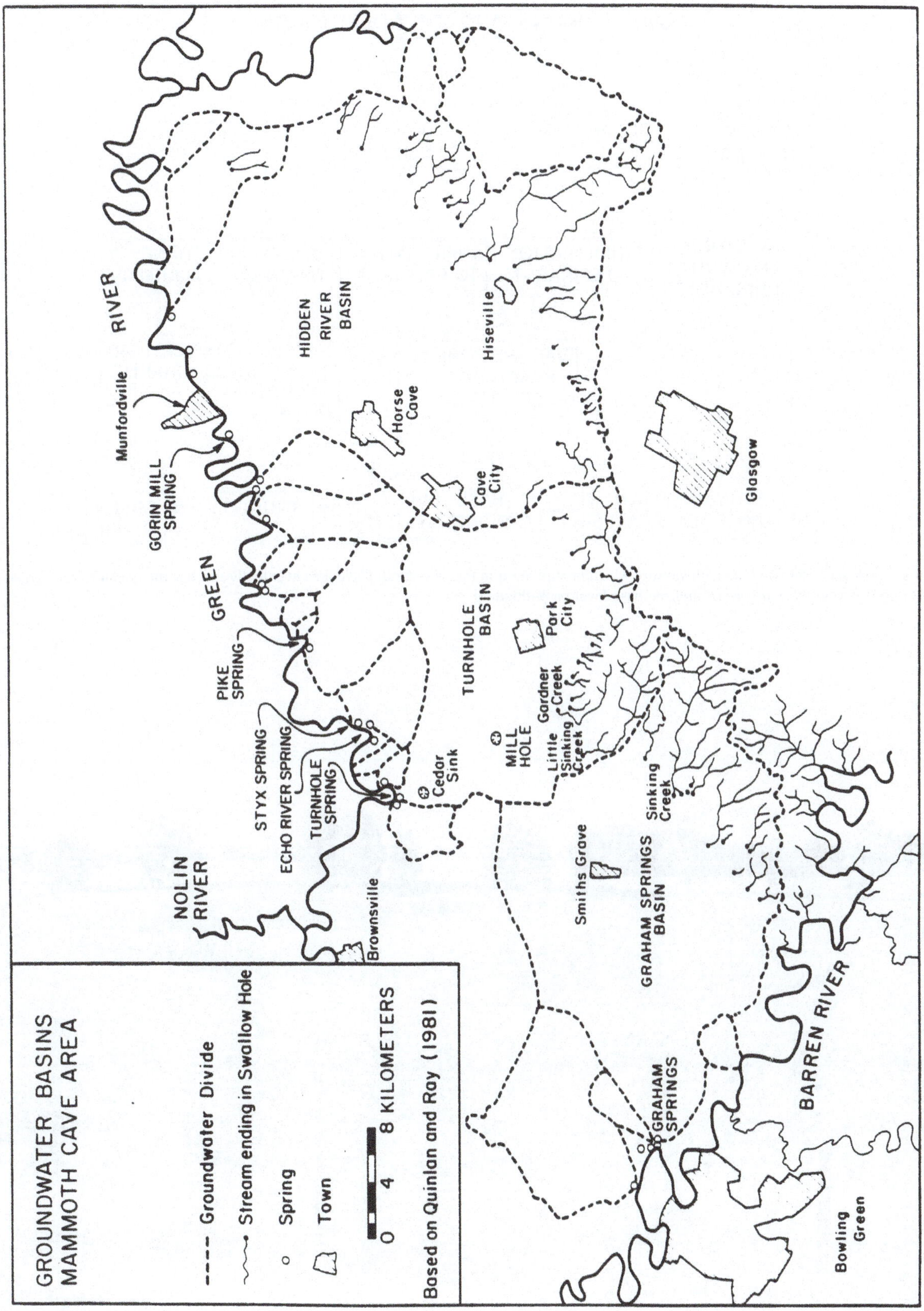

Fig. 8.4 Groundwater basins feeding the springs on the Green and Barren Rivers. This is an outline of the detailed colored map distributed by Quinlan and Ray (1981)

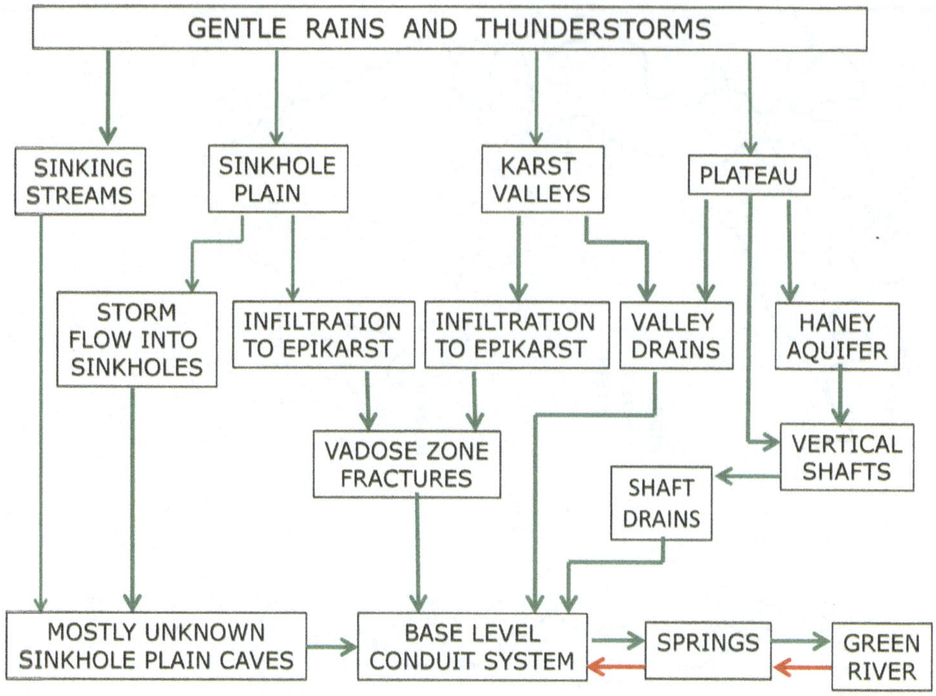

Fig. 8.5 Flow sheet showing the pathways by which precipitation in the Mammoth Cave area eventually reaches the Green River. *Green arrows* indicate flow directions; *red arrows* indicate base-level back-flooding

Fig. 8.6 A view across the Sinkhole Plain. Most of the sinkholes are relatively shallow and mantled with soil. All overland storm flow must drain into the sinkholes; there is no other pathway. Photograph by the authors

Fig. 8.7 A sinkhole pond. Farmers prize these ponds because they are convenient for livestock watering. However, the clay-rich soil plug that holds the pond can fail abruptly and the pond can disappear in minutes or hours. Photograph by the authors

above the Big Clifty Sandstone. The Haney Limestone aquifer is also a karst aquifer with the perched groundwater discharging to the surface through small springs along the perimeter of the ridge (Fig. 8.8). Some of the Haney springs have been capped and were used for a time as water supply by the National Park (Fig. 8.9).

Both water draining from the clastic caprock and also water from the Haney springs are highly unsaturated with respect to calcite. As a result, the caprock water enters the active drainage system through large cylindrical voids known as vertical shafts (Fig. 8.10). Shafts are dissolved by rapidly moving films of water streaming down the shaft walls. Observations and calculations suggest that the films on the shaft walls flow in a supercritical-laminar regime, a flow region very unusual in nature (Brucker et al. 1972). The shafts rarely span the entire thickness of limestone. Water enters at the top, usually through small canyons, reaches the floor of the shaft, and leaves through small drain passages, which eventually connect into the main conduit system. Why the volumes of the shafts are so large while both inlet canyons and exit drains are so small has not been satisfactorily explained. Vertical shafts are clustered along the Plateau margins and are the primary pathways for the movement of water from the plateau through the vadose zone to the underlying conduit system.

The rapid response to storm flow on the Plateau can be illustrated with a personal anecdote. In June of 2009, we took a class into Mammoth Cave through the Carmichael Entrance just as the black clouds of a thunderstorm were rolling across Mammoth Cave Ridge. The first drops of rain were falling as we entered the cave. It required an hour or a little more to hike through Cleaveland Avenue and begin our trek down Boone Avenue. In Boone Avenue we were greeted with a roar of rushing water. The Boone Avenue shafts, Thorpe's Pit, Edna's Dome, and Cathedral Dome were all howling waterfalls. The response time for the storm water to move through the vadose zone was less than one hour.

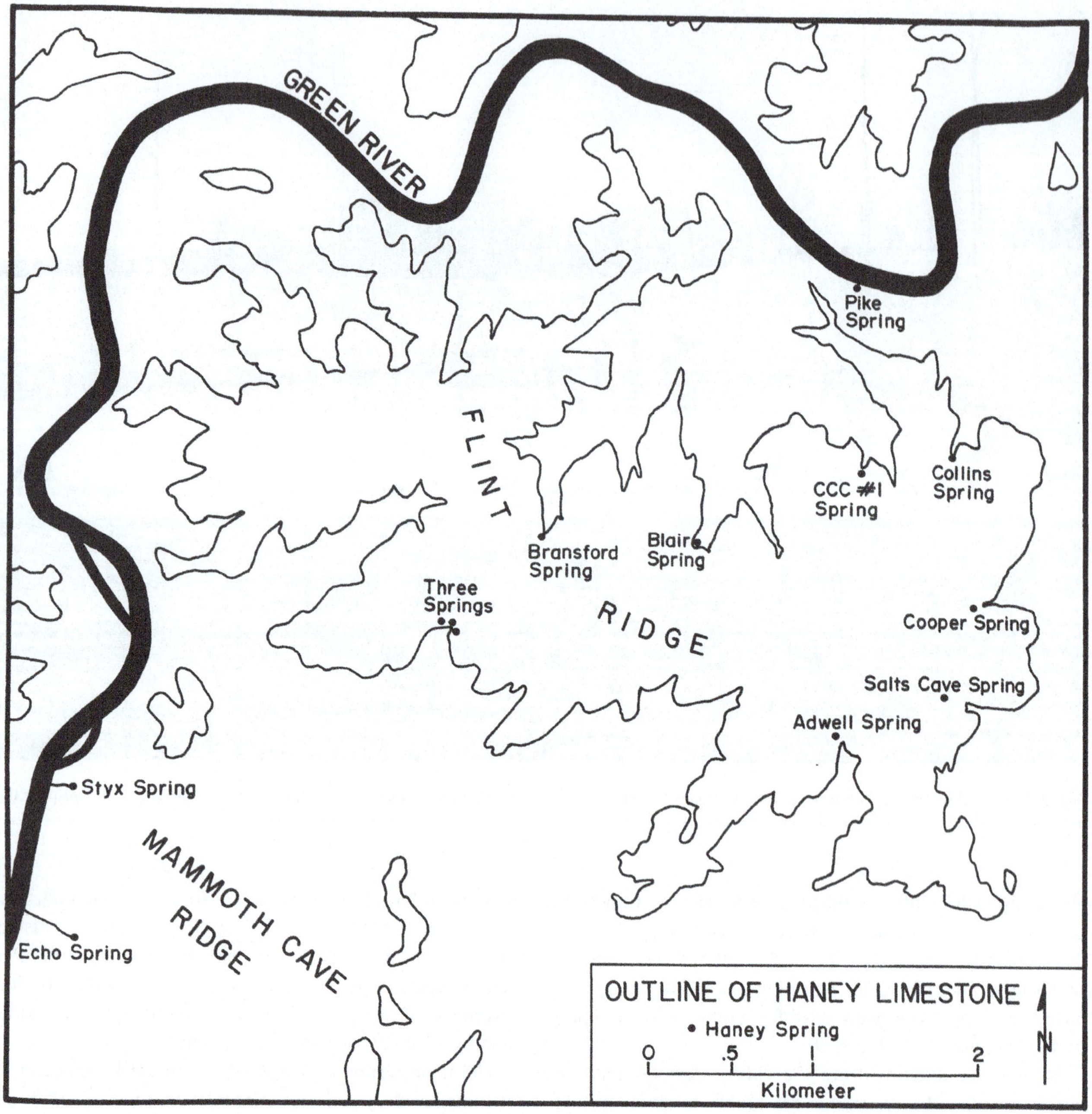

Fig. 8.8 The outcrop pattern of the Haney Limestone on Flint Ridge. The principal Haney springs are indicated

8.3 Physical Hydrology

The discussion of physical hydrology is drawn from the published scientific literature, from maps and records of the US Geological Survey, and from personal sources. The tables in this section were constructed from Hess and White (1989), but were recalculated to check for errors and miscalculations. The scientific literature presents data in the International System (SI) units. Topographic maps and river gauge records as well as many other data are in English units. At a certain risk of confusion, we use both and give both where possible.

8.3.1 Precipitation, Runoff, and Evapotranspiration

A rain gauge network of 27 stations was operated by John W. Hess in the Mammoth Cave area in 1971–1973. The data collected combined with data from eight National

Fig. 8.9 Collins Spring on Flint Ridge. Like other of the Haney springs, Collins Spring has been capped and connected to a collector system to provide water for the Park. Photograph by the authors

Oceanographic and Atmospheric Administration stations are shown in Fig. 8.11 and are displayed as deviation of monthly averages from the 30-year "normal." The potential evapotranspiration for the same years was calculated by the Thornthwaite method and is also shown in comparison with mean precipitation in Fig. 8.11. Evapotranspiration exceeds precipitation during the summer months. The long-term average annual precipitation, based on 34 years of record at Mammoth Cave National Park and nine years of record at Bowling Green, is 1264 mm (49.77 in.).

The US Geological Survey has maintained gauges on the Green River at Munfordville, at Brownsville, and on the Nolin River. These river records can be used to estimate the long-term hydrologic behavior of the Mammoth Cave area (Hess and White 1989). If measured basin areas and discharges at the Munfordville and Nolin River gauges are

Fig. 8.10 The bottom of Colossal Dome in the Colossal Cave section of Mammoth Cave. Note vertical walls and flat floor. The drain, known as Wretched River, is just large enough for exploration and continues for 1.5 km to a connection with larger conduits. Photograph by the authors

subtracted from the same data taken at Brownsville, the remainder can be attributed to the Mammoth Cave area. The Green River bisects the Mammoth Cave area into 246 km^2 (95 mi^2) north of the river and 744 km^2 (287 mi^2) south of the river. The data are summarized in Table 8.1, which gives the areas above the gauges and the mean annual discharge. The normalized discharge is the discharge divided by the basin area and is a useful characteristic of the river basin. N is the number of years of record—that is, the number of years that each of the gauges has been in operation.

The normalized mean annual discharge is about the same for the Brownsville and Munfordville gauges as might be expected since these measure large catchments with similar average characteristics. However, the Nolin River, with much of its catchment underlain by impermeable shales and siltstones, has a lower normalized annual discharge than the larger basins, and the Mammoth Cave area has a higher value. We propose that the difference is due to a higher evapotranspiration rate in the Nolin River Basin and a lower evapotranspiration rate in the karst of the Mammoth Cave area. This hypothesis can be checked if it is assumed that the average annual precipitation of 1264 mm is the same throughout the Green River Basin upstream from the Brownsville gauge. The evapotranspiration is the precipitation minus the runoff. These values are given in Table 8.2, recalculated as water depth. Runoff from the karst of the Mammoth Cave area is mainly underground through the conduit system where both evaporation and transpiration are low.

Calculations of mean minimum flow rates are more difficult. The minimum precipitation months in south-central Kentucky are September and October so that the long-term average stream flow during these months should be a measure of base flow. However, base flows at the Nolin River gauge at Kyrock and the base flows of the Green River are both impacted by impoundments. To avoid these influences six years of September and October data (1966–1971) from the Nolin River gauge at White Mills and the segment of Green River between Greensburg and Munfordville were used to compute base flow (Table 8.3).

Dry-season base flow is maintained in surface streams by groundwater discharging along the steam banks. The base flow from the Mammoth Cave area is much lower than that of the Green River Basin as a whole. The conduit system has

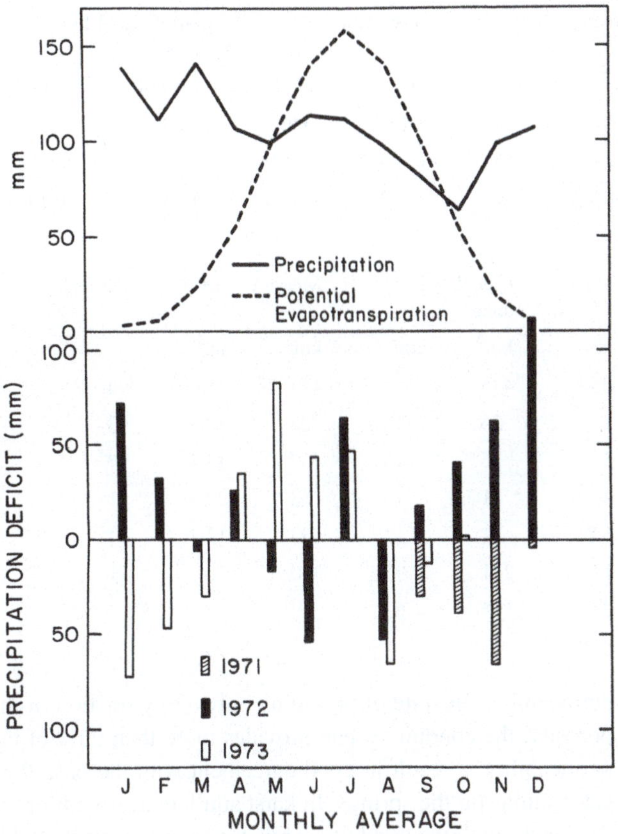

Fig. 8.11 *Upper frame* average monthly precipitation compared with the calculated potential evapotranspiration for 1971–1973. *Lower frame* monthly precipitation averaged over the observation stations expressed as deviation from normal

a very low hydraulic resistance and drains rapidly, leaving little water in storage to maintain base flow during dry periods. Indeed, the base flow from the Mammoth Cave area is so low that evaporation loss from the Green River itself, estimated to be 140 L/s or 5 cfs, becomes important. When this correction is taken into account, the normalized base flow for the Mammoth Cave area becomes 1.64 L/s/km^2.

8.3.2 Water Balance and Spring Catchment Areas

Although mean annual discharge from karst drainage basins depends on both precipitation and basin characteristics, base flow (also called sustained flow) is much more a characteristic of the groundwater system since it is groundwater discharge into stream channels that maintains their flow. In karst terrain, the stream channels are replaced with integrated conduit systems discharging through springs. Estimates for base flow of the Green River springs (Table 8.4) were obtained by flow measurements taken in September and October, but are essentially single measurements rather than long-term averages.

Base flow in karstic drainage basins has been found to be proportional to basin area (White 1977) with the normalized base flow as the proportionality constant. Using the base flow value, 1.64 L/s/km^2, obtained from river records, areas were calculated for each of the Green River springs (Table 8.4). These areas were then compared with the spring basin areas taken from the Quinlan and Ray map (shown in outline in Fig. 8.4). For four of the six spring basins, the area calculated from spring base flow is within 7% of the measured area. It should also be remembered that the basin boundaries sketched on the Quinlan and Ray map are estimated from dye tracing and potentiometric surfaces. Both sets of estimates are subject to considerable uncertainty.

Table 8.1 Mean annual discharge for gauges in the Green River Basin

Location	Area		Mean annual discharge		Normalized discharge		*N*
	km^2	mi^2	m^3/s	cfs	L/s/km^2	cms	
Brownsville Gauge	7154	2760	115.15	4086	16.1	1.47	40
Munfordville Gauge	4333	1672	72.64	2565	16.8	1.53	45
Nolin River at Kyrock	1831	706	23.22	785	12.7	1.11	23
Mammoth Cave area	989	382	19.29	681	19.5	1.78	

Primary data from the US Geological Survey

Table 8.2 Calculated runoff and evapotranspiration for gauges in the Green River Basin

Location	Runoff		Evapotranspiration	
	mm	in.	mm	in.
Brownsville Gauge	508	20	756	29.8
Munfordville Gauge	529	20.8	735	28.9
Nolin at Kyrock	400	15.75	864	34
Mammoth Cave area	615	24.1	649	25.6

Table 8.3 Estimation of base flow for the Mammoth Cave area

Location	Area		Base flow		Normalized base flow	
	km^2	mi^2	m^3/s	cfs	L/s/km^2	cms
Greensburg–Munfordville	2427	936	9.799	346	4.04	0.37
Nolin at White Mills	924	356	2.688	94.9	2.91	0.27
Mammoth Cave area	989	382	1.442	50.9	1.46	0.13

Table 8.4 Spring drainage areas calculated from base flow measurements compared with basin areas extracted from Quinlan map

Spring	Base flow		Calculated area		Measured area		Percent (%)
	L/s	cfs	km^2	mi^2	km^2	mi^2	
Gorin Mill	731	25.8	445	172	473	182	94
Garvin	42.5	1.5	26	10	28	10.8	93
Lawler Blue Hole	59.5	2.1	36	14	37	14.3	97
Pike	68	2.4	41	16	58	22.4	71
Echo River	57.5	1.8	31	12	33	12.7	94
Turnhole	396	14	241	93	340	131	71
Mill Hole	269	9.5	164	63			

For two springs, Turnhole and Pike, Table 8.4 shows about a 30% disagreement between basin area estimated from base flow and the basin area estimated by dye tracing. The Turnhole case is easy. Because Turnhole Spring is a rise pool on the riverbank, flow estimates were made where the feeder stream crosses the bottom of Owl Cave in Cedar Sink. Cedar Sink is a karst window formed by collapse of the master conduit. The Owl Cave stream reveals the portion of the Turnhole feeder that passes the west side of Cedar Sink. An additional quarter of the flow can be seen in Smith Valley Cave on the east side of Cedar Sink. When the divided flow path is taken into account, the two area estimates are similar. The explanation for the Pike Spring discrepancy remains unknown although an error in the drawing of the basin boundary seems the most likely possibility.

The excellent agreement between basin area calculated from base flow and basin area estimated directly allows another important conclusion. In maturely karsted aquifers such as the Mammoth Cave area, 90% or more of the groundwater discharge is through the conduit system to individual large springs. Diffuse flow from the aquifer to the river channel is less than 10%.

Worthington et al. (2000) conducted a comprehensive study of the permeability characteristics of the Turnhole Basin (Table 8.5). The estimated average porosity of the Mississippian limestone is 2.4%. The estimated conduit porosity of 0.06% may seem much too small considering the large volume of the master trunk passages. However, large as the cave volumes are, they are being expressed as a ratio of this volume to the volume of a block of limestone large enough to encompass the entire cave. Mature karst provides the most extreme contrast between porosity and permeability. In spite of providing only a tiny fraction of the porosity, the conduit system provides more than 98% of the permeability, a result in good agreement with the base flow calculations on the springs. In karst aquifers nearly all of the flow is through the conduits, while the matrix porosity holds nearly all of the storage. Given the extremely slow flow velocities through the matrix, the water in storage is essentially sequestered, while contemporary recharge flows to the springs through the conduits and, to a lesser extent, through the fractures.

8.4 Green River Floods

8.4.1 Flood Records and Recurrence Intervals

Green River flows in a deep gorge eroded into the Mammoth Cave Plateau. Because the valley is narrow with little or no floodplain, increases in the discharge of Green River result in a substantial rise in water level, giving considerably higher flood stages than is typical of rivers with wide floodplains. Because of the low gradient of the master conduit systems draining to the Green River springs, rises in the level of Green River drive water into the springs and thus contaminate the aquifer with river water. This backflow is common in all river systems and is known as bank storage, but in karst regions the effect is extreme.

The USGS discharge records for Green River at the Brownsville and Munfordville gauges can be used to construct the flood statistics for the Mammoth Cave area (White 1989). The water year is defined as the period from October 1 to October 1, corresponding from dry period to

Table 8.5 Hydraulic properties of the three components of karstic permeability in the Mammoth Cave area

	Porosity (%)	Fraction of storage (%)	Flow velocity	Permeability (%)
Matrix	2.4	95.6–97.3	0.1 mm/year	7×10^{-7}
Fractures	0.006–0.05	0.2–2	7–47 m/day	0.2–1
Conduits	0.06	2.4	0.004–0.2 m/s	98–99.8

Data from Worthington et al. (2000)

dry period. The highest discharge recorded during the water year is called the annual flood regardless of the date on which the high discharge occurs. The annual floods averaged over the years of record is called the mean annual flood. There may be multiple floods within a single year, and the floods vary considerably from year to year. The magnitudes of floods follow a log-normal distribution, and a number of models have been constructed to further tweak the statistics. For high-frequency, low-magnitude floods, the statistical distribution is easily described. It is the low-frequency, high-magnitude floods that are both the most important and the most difficult to fit to statistical models. Once a record of the floods in any particular drainage basin has been obtained, statistical fitting allows the calculation of the probability for the occurrence of a flood of any specified magnitude. Rather than probability, what is usually calculated is the return period that is the number of years that will pass before a flood of any specified magnitude will occur again. Thus, a 10-year flood is predicted to occur only once every ten years. But these are probabilities; there is nothing that forbids two 10-year floods in the same year. The mean annual flood has a return period of 2.33 years.

The relation between the discharge of Green River and the height of the water (the stage height) is shown in Fig. 8.12 for the Munfordville gauge. The mean annual flood would take the river level up to 460 ft (139 m) at Mammoth Cave, 40 ft (12 m) above the low flow pool stage at 420 ft (127 m), and would flood all cave passage up to that elevation. The 1962 flood at 485 ft (147 m) completely filled the Echo and Styx River passages as well as River Hall in Mammoth Cave. The high water mark is recorded on a brass plaque on a boulder near the top of the stairs leading from River Hall.

Green River floods for the Munfordville gauge (Fig. 8.13) and the Brownsville gauge (Fig. 8.14) are plotted as a function of return period. The fitted lines represent two commonly used models for flood statistics. Both work about equally well for the Brownsville data, but deviate considerably for the Munfordville data. The 1962 flood was plotted as a 100-year event at Munfordville but as only a 20-year event at Brownsville. This may be a manifestation of the effect of sinkholes and sinking streams on damping flood peaks as noted by White and Reich (1970) for floods in other karstic basins.

8.4.2 Flooding and Back-flooding in Mammoth Cave

The gradients of the base-level master conduits in Mammoth Cave are on the order 1 m/km or less. In principle, a river rise of 6 m would back-flood the cave to a distance of 6 km, the distance to the eastern boundary of the Park. The details of what happens depend on the distribution of rainfall. Thunderstorms on the Sinkhole Plain but not in the upper Green River Basin would flood the cave from the upstream end. Water piling up behind breakdown, sediment plugs, and other barriers would result in extensive water level rise upstream with resultant high hydraulic heads driving water through blockages in the conduits. If there are storms in the upper Green River basin but not on the Sinkhole Plain, the Green River will rise and back-flood into the spring orifices.

An illustration of effect of source area on flooding is the storm on July 22, 1973 (Fig. 8.15), that raised water levels along the master trunk feeding the Turnhole. At Mill Hole, a karst window upstream, water levels rose 11 m (36 ft) but at Owl Cave in Cedar Sink, downstream from Mill Hole, the water level rose only 4 m (13 ft), while the Green River displayed a more complex response but apparently rose only 2 m. Estes Well is located on the Sinkhole Plain about 500 m (1640 ft) west of the sink of Little Sinking Creek and is much higher (note scale) than the master conduit. The water level in the well rose less than a meter. Both Mill Hole and Owl Cave responded within 18 h of the storm, while Estes Well required more than two days. These hydrographs support the model that the low hydraulic resistance of the conduits causes them to drain rapidly and create a trough in the local water table. In a sense, the conduits can be thought of as horizontal wells with the water table trough as their drawdown.

More precise data are available from the instrumentation station set up in Logsdon River by Joe Meiman who took advantage of two observation wells drilled into the Logsdon River passage. By using a pressure transducer with a data logger on the surface, information can be obtained on Logsdon River even when the passage is completely flooded. Logsdon River, at low flow, is an air-filled stream passage that can be explored for 8 km upstream to its headwaters in the Roppel Cave section. Their data (Meiman and Ryan 1993) show that a rise of 3 m is sufficient to fill the passage to the roof. Increased recharge raises the pressure

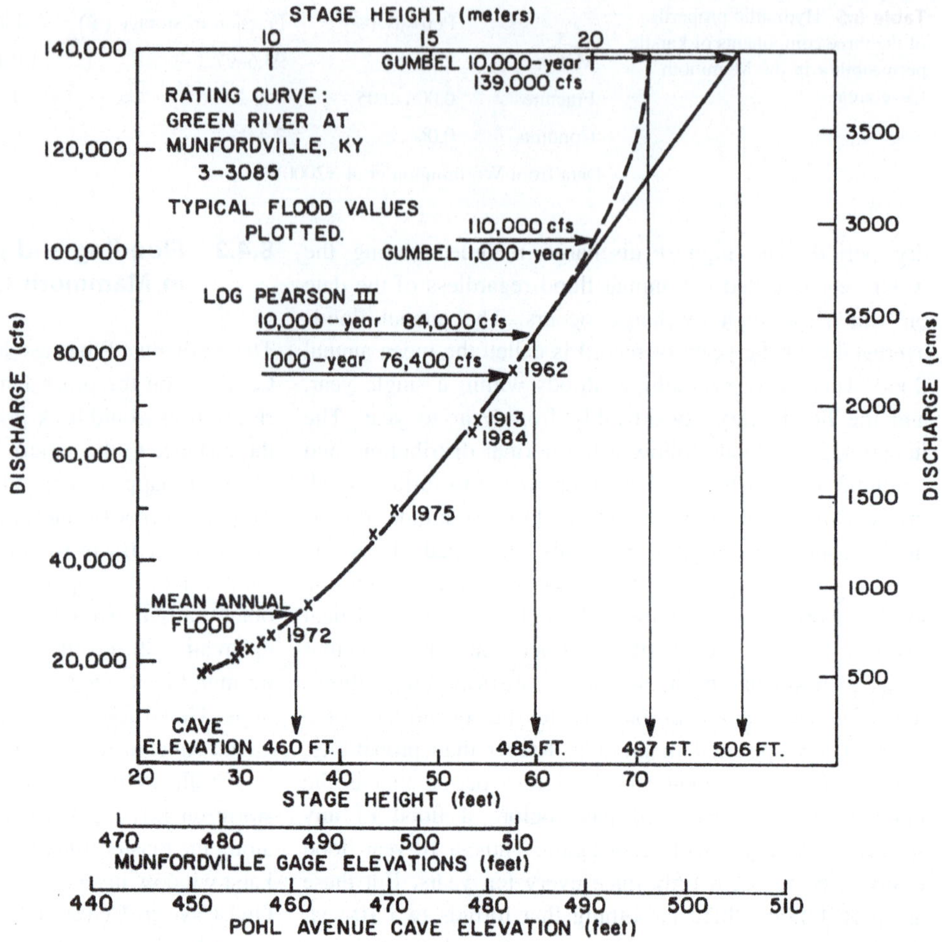

Fig. 8.12 Rating curve for the Green River at Munfordville. Three scales are shown: the stage height at Munfordville, the elevation of the water surface at Munfordville, and the water elevation corresponding to Pohl Avenue in the Flint Ridge section of Mammoth Cave. The dated points are significant floods

head in the water-filled passage with the well acting as a manometer. Water level rises in the well up to 23 m (75 ft) have been observed. Increased heads should produce high-velocity flow in the conduits and induce the transport of clastic sediment.

Back-flooding from Green River is a low-velocity flow with water driving up the spring feeder conduits by only a few feet of head as the river rises. Back-flooded water forms a vast underground lake occupying all cave passages, open fractures, and bedding plane partings up to the level of the final flood stage. As the flood recedes, the water drains back out through the springs. There is no mechanism for creating the high hydrostatic heads achieved by floods coming from the upstream direction. Because the muddy river water is ponded in the cave, there is sufficient time for the fine-grained sediment to settle out, creating the mud layers that were formally such an annoyance to Park management. However, natural back-flooding is an extremely important food transport mechanism for both aquatic and terrestrial cave life.

Flow reversal was observed directly at Pike Spring. The Green River was muddy and rising, but the spring was issuing green, slightly murky water. The observation was made in the autumn, and colored leaves were falling into the spring. At first, the leaves were floating outward and were swept away by the river. As the river continued to rise, the motion of the leaves stopped and then reversed. Tendrils of muddy water from the river moved toward the spring. The flow into the spring increased, leaves were swept into the spring, and the spring orifice was filled with muddy water flowing inward. The entire reversal process took about 15 min. Figure 8.2 shows the spring at the moment of pause between outflow and inflow.

8.4.3 Sedimentation

The USGS investigations of Mammoth Cave sedimentation in 1959–1962 (Collier and Flint 1964) determined from careful surveys that layers of fine-grained sediment from a

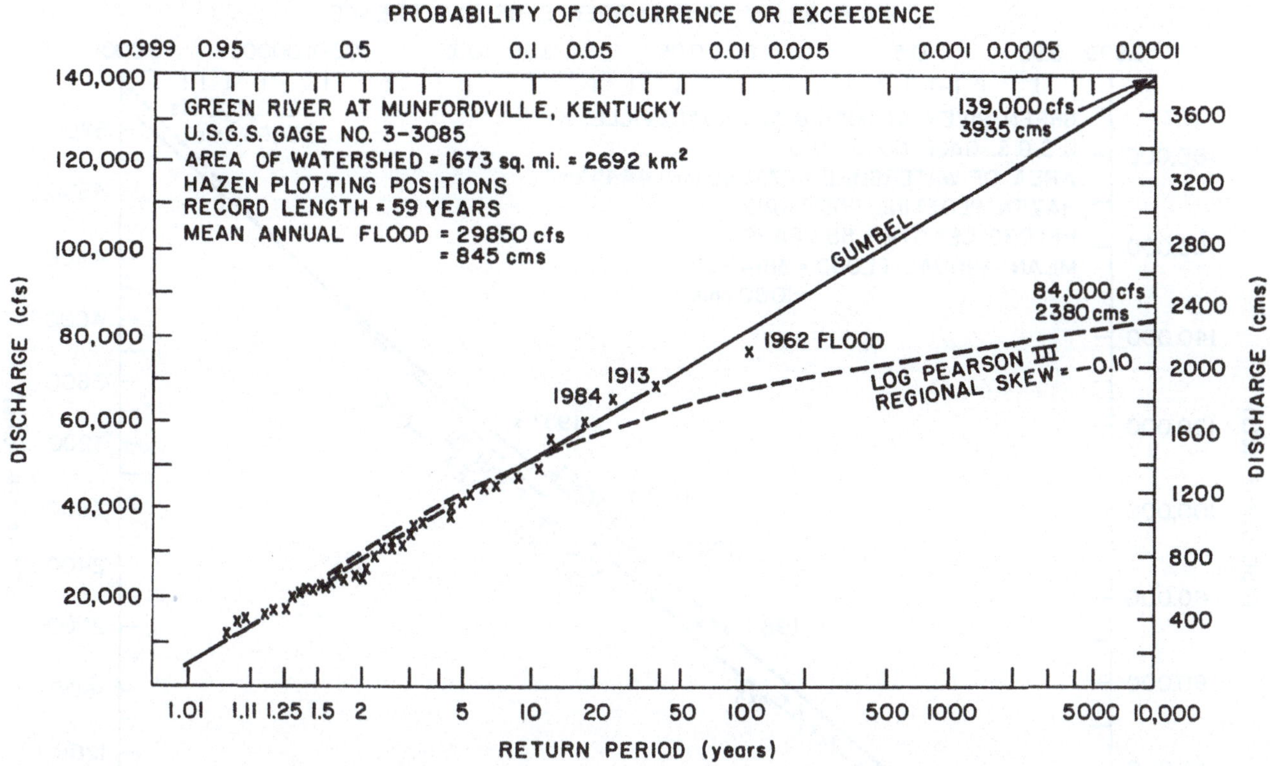

Fig. 8.13 Flood recurrence plot for the Green River at Munfordville. Discharge scales are given in cubic feet per second (cfs) and in cubic meters per second (cms)

few cms to 15 cm could accumulate on surfaces in the Echo River–Styx River passages. They concluded from an examination of the mineralogy that the silt was carried into the cave from Green River during back-flooding events. This conclusion was challenged by E.R. Pohl who argued that the bulk of the sediment was derived from inputs from sinking streams, from the Sinkhole Plain, and from caprock sediment carried into the cave through vertical shafts. As is frequently the case with geological arguments, both were right. Sediment comes from both upstream and downstream during flood events. The storm water from upstream meets the back-flooded water from the river, and the accumulated water simply fills up base-level passages as an underground lake with relatively little current. These are ideal conditions for fine-grained sediment to settle out on all upward facing surfaces.

In 1968 an attempt was made to establish a direction to base-level sedimentation as a senior thesis project by Roy Carwile and Edward Hawkinson from Ohio State University. The site selected was Columbian Avenue in the Unknown Cave section of the system. Columbian Avenue is an 800-m elliptical tube connecting Pohl Avenue 7 m above pool stage with Eyeless Fish Trail 3.7 m above pool stage. The 1- to 2-m-deep sediment fill was trenched at regular intervals and a stratigraphic column prepared for each trench. No evidence was found for a preferred direction in the sediment deposition along the passage (details in Bosch and White 2004).

Measurements near the junction of Logsdon and Hawkins rivers reveal the influence of source area (Fig. 8.16) (Bosch and White 2004). Sediment samples were collected from three locations in Logsdon River and two in Hawkins River. The distributions of particle sizes are plotted on the φ scale based on base-2 logarithms widely used in sediment studies. Converted to metric units, $\varphi = -3.322 \log(d/d_0)$ where d is grain size in millimeters and d_0 is the reference grain size, one mm in this case. Logsdon River receives sediment from the edge of the plateau and from the Cave City area. The sediment that reaches the stream confluence is almost entirely fine-grained material in the clay and silt range. Hawkins River which obtains its sediment from sinking streams and the Sinkhole Plain contains a much wider variety of particle sizes extending from clay and silt to coarse gravel.

8.5 Chemical Hydrology

Karst aquifers offer multiple sampling sites for the examination of chemical changes that take place as various sources of recharge move through the aquifer. Chemical analyses of the water provide raw data for the calculation of water hardness,

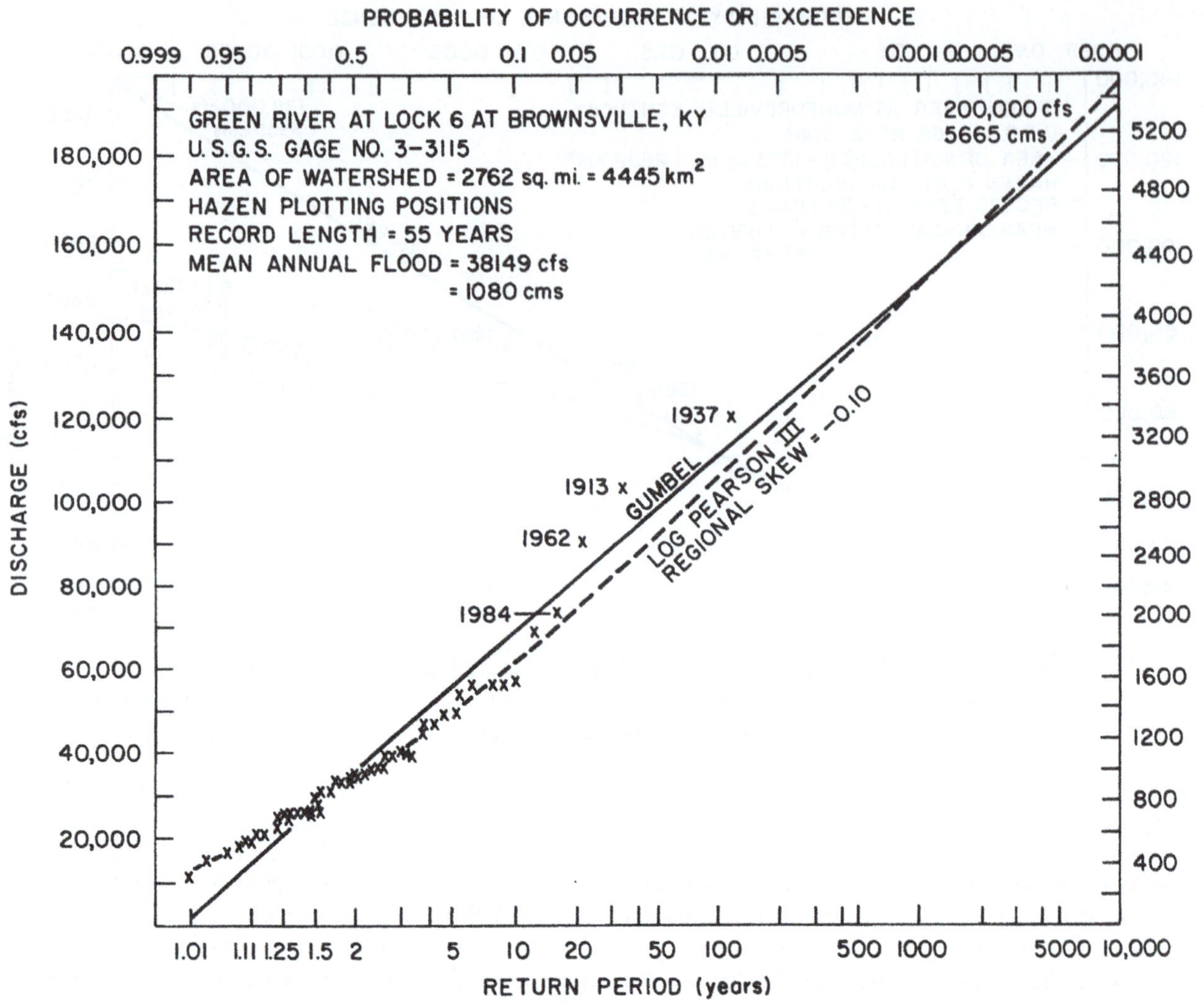

Fig. 8.14 Flood recurrence plot for the Green River at Brownsville

state of saturation with respect to calcite and dolomite, and the carbon dioxide partial pressure with which the water would be in equilibrium. These parameters, in turn, provide useful information on the hydrology of karst drainage basins. Specific conductance is a useful proxy for hardness and is the parameter most amenable to continuous monitoring, thus measuring chemical variations on short timescales.

Water chemistry in Mammoth Cave and the surrounding area was investigated by John W. Hess as part of his Ph.D. dissertation and more recently by Park hydrologist Joe Meiman and by students at Western Kentucky University. The sections that follow summarize some of the results.

8.5.1 Chemical Characteristics of the Karst Waters

Water analyses can be grouped according to source and averaged to give a sense of the chemical characteristics of the source (Table 8.6). Waters from the sinking creeks are near saturation with low carbon dioxide partial pressure because flowing streams allow dissolved CO_2 to be degassed to the atmosphere. Waters from the Haney springs are highly undersaturated. Shaft waters are undersaturated as they must be if the shafts are actively enlarging. Drip waters are supersaturated, again as they must be if the drip waters are depositing speleothems. Owl Cave is the furthest downstream sampling point for the master conduit feeding Turnhole Spring. In spite of the long flow path from the Sinkhole Plain, the water in the master conduit remains undersaturated. The other base-level springs, Pike, Styx River, and Echo River are also undersaturated. The averages conceal a great deal of variability due to seasonal changes and storm inputs (Hess and White 1993).

Carbon dioxide concentrations can be expressed as a volume percent, as a partial pressure (usually given as the logarithm of the pressure expressed in atmospheres), or as the ratio of the calculated partial pressure to the partial

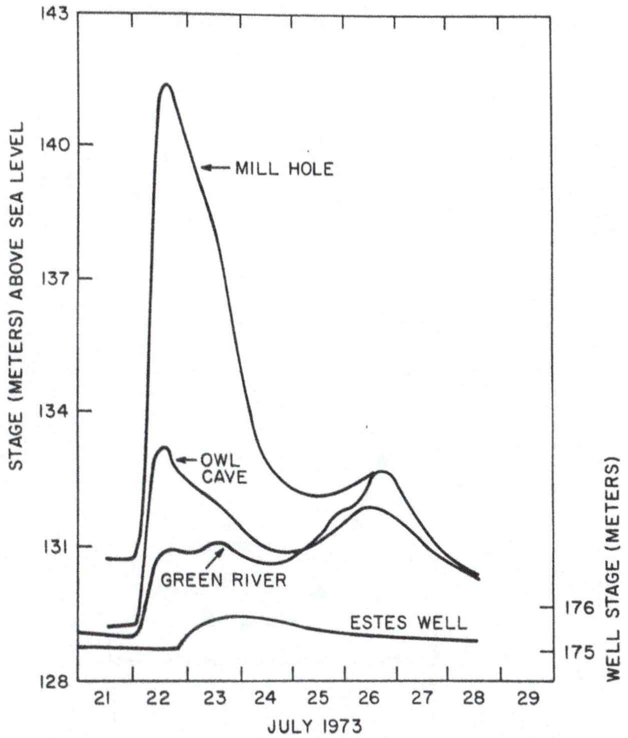

Fig. 8.15 Stage records for Mill Hole, Owl Cave (Cedar Sink), Green River, and the Estes Well for an intense storm on July 22, 1973

pressure of CO_2 in the surface atmosphere. The CO_2 concentrations in the various waters of the Mammoth Cave area are low compared to other karst areas with values only 13–22 times the atmospheric background. Carbon dioxide levels vary with season (Fig. 8.17) with a pronounced minimum during the winter months and a maximum in late summer and early fall. The seasonal curve tracks with the growing season, the highest CO_2 corresponding to high biological activity in the soil and the epikarst.

Logsdon River provides a site for the determination of water chemistry for a substantial distance along the flow path. A set of measurements in October 1996 produced a remarkable result (Fig. 8.18) (Anthony et al. 2003). Measured upstream from the Logsdon River observation wells, the CO_2 partial pressure decreases with distance upstream until it reaches a minimum at about 6000 m (20,000 ft) and then rises again to the final point in Roppel Cave. The saturation index tracks with the CO_2 pressure becoming supersaturated where the CO_2 pressure is a minimum. Chemical measurements in Logsdon River also allow an estimate of the rate of passage enlargement as a function of discharge (Groves and Meiman 2005). Much of the passage enlargement takes place during storms when Logsdon River is flowing pipe-full. Under low flow conditions, the water is flowing on a bed of clastic sediment and little enlargement takes place.

8.5.2 Chemical Responses to Storm Flow

It requires several days for CO_2-containing water to come into chemical equilibrium with limestone. In open conduits, water travels a long distance in several days. As a result, storm flow can move through the system quickly and dilute the water already present in the system. Continuous monitoring of specific conductance at Owl Cave produced the records shown in Fig. 8.19 (Hess and White 1988). Before the storms, the specific conductance—proportional to the concentration of dissolved carbonates—is high and constant.

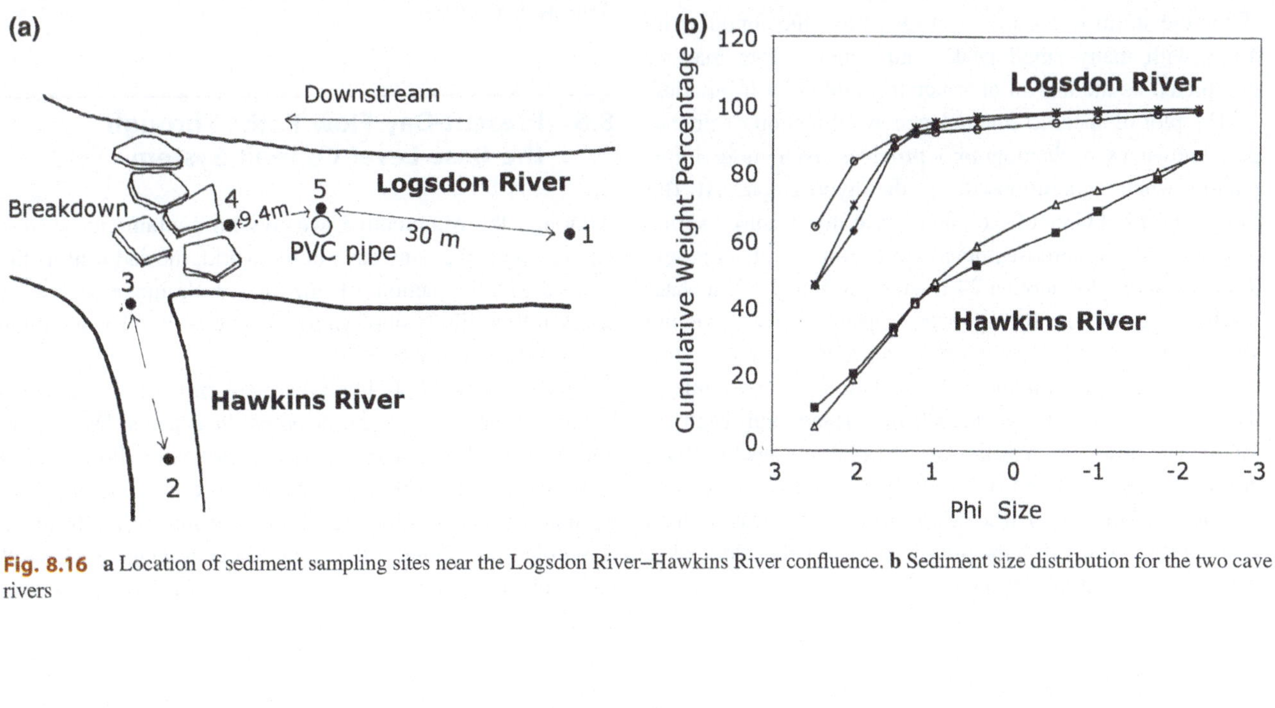

Fig. 8.16 **a** Location of sediment sampling sites near the Logsdon River–Hawkins River confluence. **b** Sediment size distribution for the two cave rivers

Table 8.6 Mean values for chemical parameters in the Mammoth Cave area

Source	SI_C	log P_{CO_2}	P_{CO_2}/P_{CO_2} (atm)	*n*
Sinking creeks	−0.09	−2.75	5	72
Haney springs	−1.384	−2.378	13	4
Shafts	−0.8295	−2.331	14	23
Drip water	+0.1828	−2.2877	16	16
Owl Cave	−0.21	−2.25	17	45
Base-level springs	−0.58278	−2.142	22	15

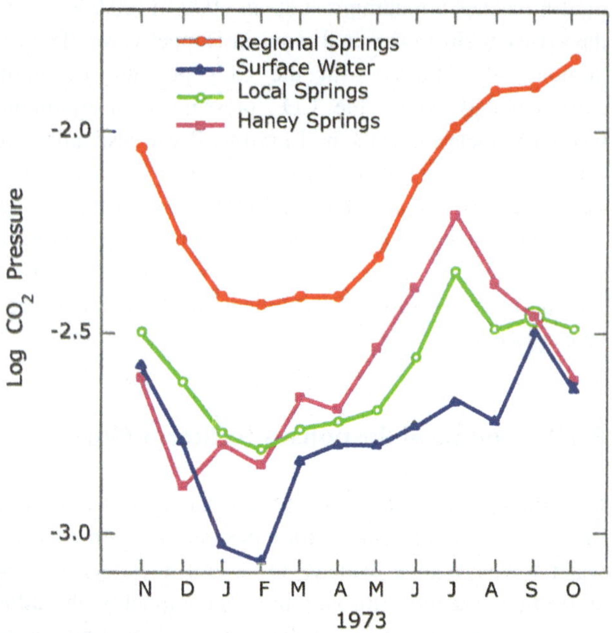

Fig. 8.17 Monthly averages of CO_2 pressure calculated from water analyses from sources indicated. Regional springs are averages of Owl Cave (Turnhole Spring) and Graham Spring. Local springs include Pike, Styx River, and Echo River springs. Data collected by John W. Hess for his Ph.D. thesis

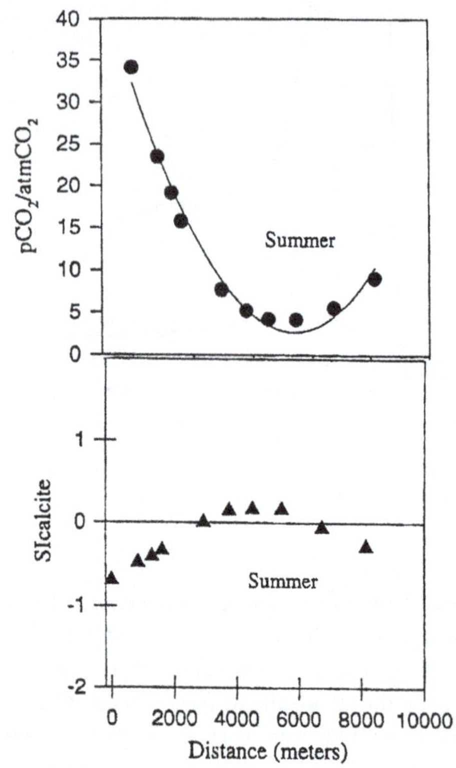

Fig. 8.18 Variation of CO_2 concentration and saturation index along Logsdon River measured upstream from the observation wells. From Anthony et al. (2003)

When the storm water arrives at Owl Cave the conductivity drops with many small peaks and valleys. This may be interpreted as the arrival of water from different tributaries.

The plot of specific conductance as a function of time (a plot known as a chemograph) provides even more information when superimposed on the hydrograph. At Big Spring, north of Green River, an intense storm caused spring discharge to rise almost immediately. But the chemograph did not respond for another 24 h when the actual storm water reached the spring. Rising heads upstream were forcing water out of flooded conduits. Comparison of the hydrograph and chemograph allowed calculation of the volume of the flooded conduits as 17,000 m^3 (Ryan and Meiman 1996). The observation station constructed in Logsdon River allowed a detailed comparison of chemographs and hydrographs as a function of whether or not the river was in open channel flow or was in pipe flow with the conduit flooded to the roof (Raeisi et al. 2007).

8.6 Present-Day Flow Paths Through the Base-Level Conduit System

Although the Mammoth Cave System contains more mapped passages than any cave in the world, the sad truth is that only a small fraction of the active drainage system is accessible to direct observation and survey. What we know is summarized in Fig. 8.20.

Under the Sinkhole Plain there are only a few fragments, mainly in Parker Cave, of the active flow paths. Dye tracing and water table mapping indicate a major flow line to Turnhole Spring. The Turnhole drainage line has two main tributaries. The southwestern-most collects water from the Sinkhole Plain south of Park City to the drainage divide with the Graham Springs basin to the west. Mill Hole provides a

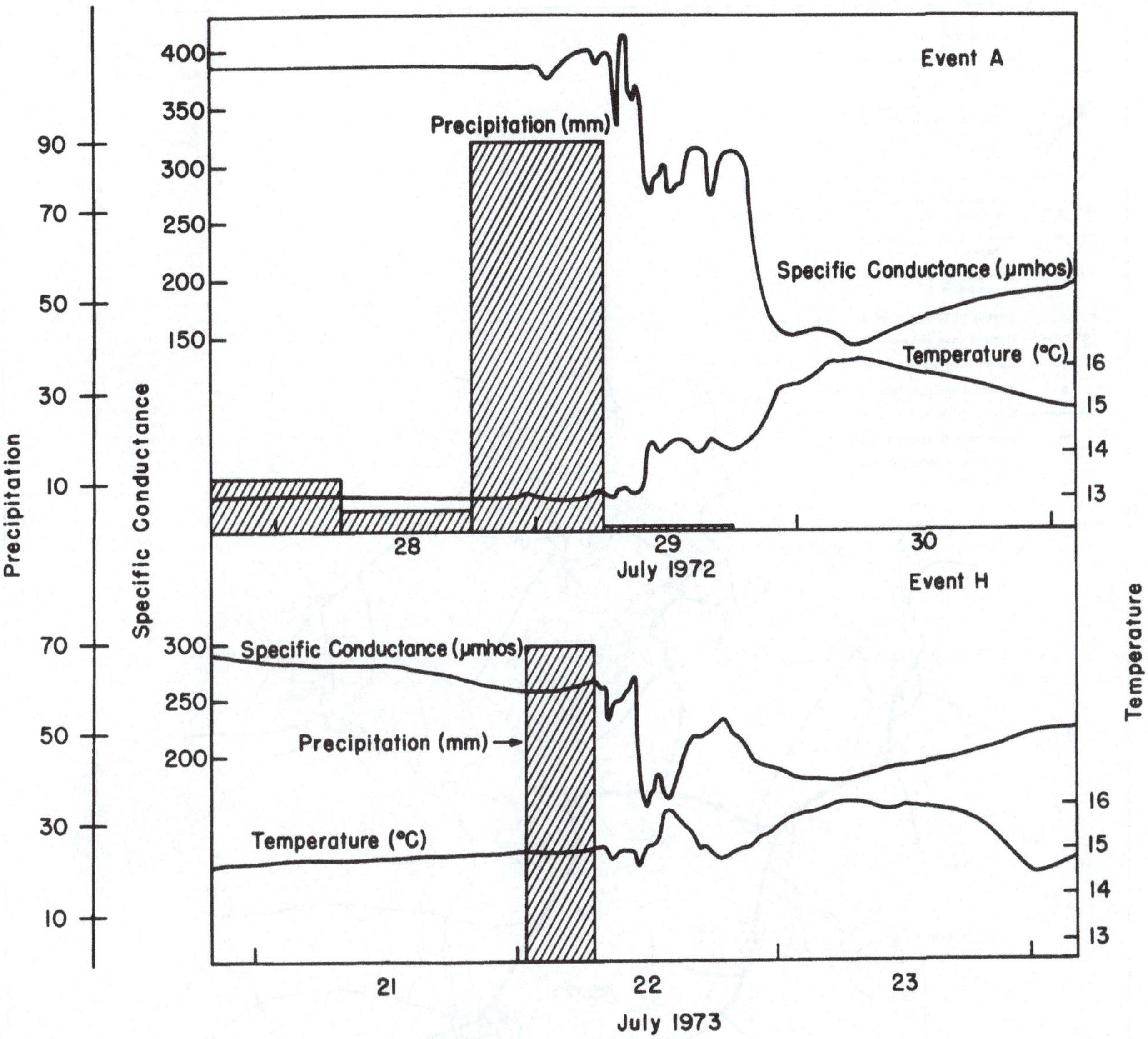

Fig. 8.19 Temperature and specific conductance responses to an intense storm. Measurements at Owl Cave on the Turnhole Spring master conduit

single access point to the Mill Hole sub-basin. The second tributary itself divides into two basins with a confluence accessible within the cave system at the Logsdon River–Hawkins River confluence. The Patoka Creek sub-basin drains the Sinkhole Plain east of Park City and appears in the cave from the upstream sump of Hawkins River. The Cave City sub-basin has its main collection area near Cave City. The upstream catchment includes parts of Roppel Cave, which forms the headwaters of Logsdon River, the only segment of the drainage system that can be traversed during low flow conditions. Logsdon River joins Hawkins River, and the combined stream sumps a short distance downstream. There is an observation point downstream from the Hawkins River Sump in Whigpistle Cave, but upstream from the confluence with the Mill Hole drainage line. There is a final observation point in Cedar Sink before the combined flow path reaches Turnhole Spring. Although the main trunk drainage seems likely to be accurate, there are multiple smaller tributaries feeding water from sinking streams, sinkholes, and shafts along the caprock that are almost entirely unknown.

What is known is that the present-day drainage from the Hawkins River Sump to Turnhole Spring is a relatively recent (geologically speaking) piracy route. The original pathway was through the large base-level conduits of Mammoth Cave to Echo River. It has been known that there is a large increase in flow of Echo River Spring in response to storms on the Sinkhole Plain. Quantitative measurements

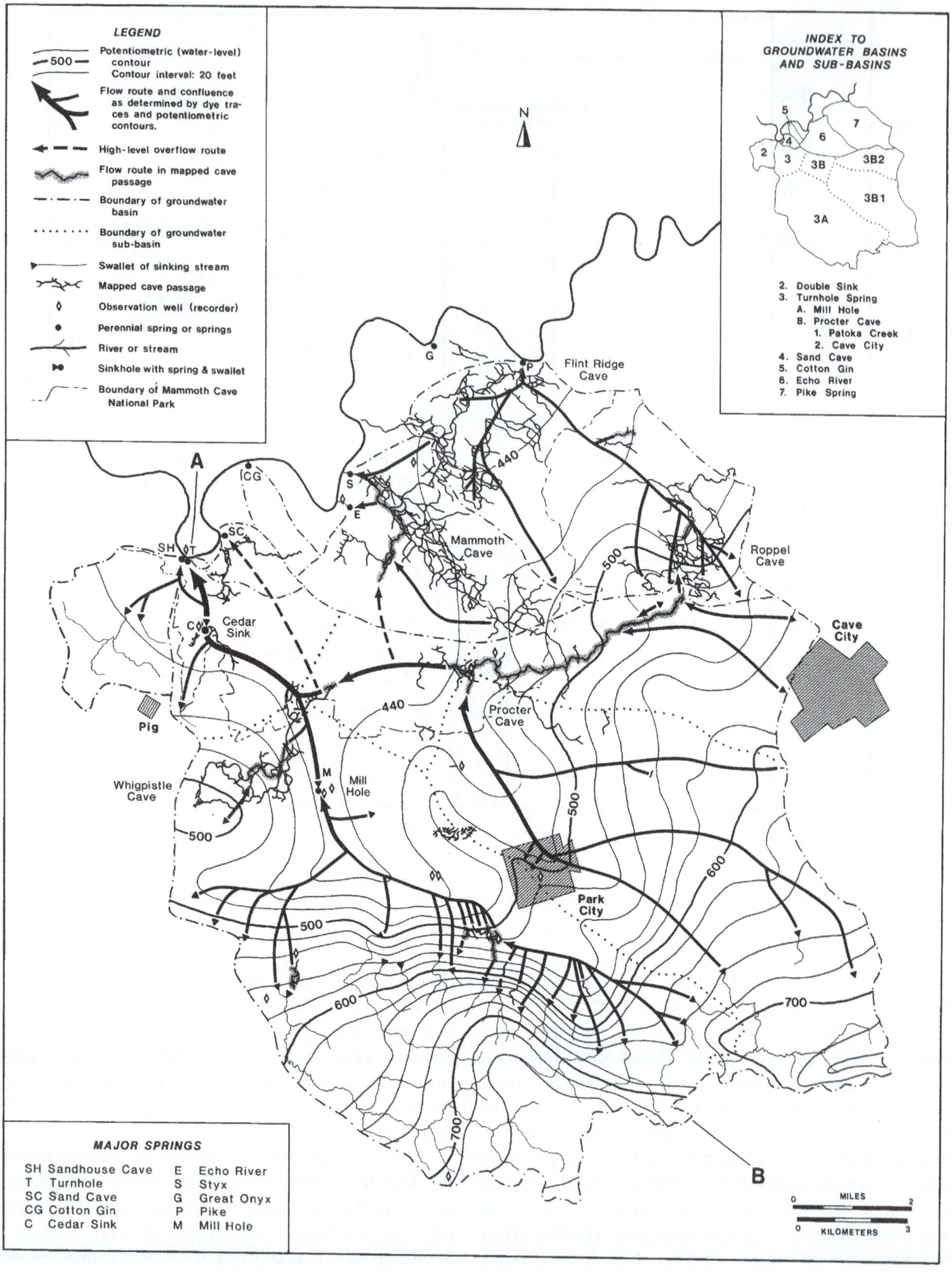

Fig. 8.20 The Turnhole Spring drainage map. From the work of James F. Quinlan

by Meiman and Ryan (1993) show that a 3-m rise at the Logsdon River instrument station will shunt water into the Echo River drainage.

Acknowledgements Much of the authors' own work in the Mammoth Cave area was under the auspices of the Cave Research Foundation. Quantitative physical and chemical hydrology, particularly the Ph.D. thesis work of John W. Hess, was supported by the U.S. Office of Water Resources Research under OWRR matching agreement number 14-31-0001-3638 to The Pennsylvania State University.

References

Anthony DM, Groves C, Meiman J (2003) Preliminary investigations of seasonal changes in the geochemical evolution of the Logsdon River, Mammoth Cave, Kentucky. Speleogenesis Evol Karst Aquifers 1(4):1–6

Bosch RF, White WB (2004) Lithofacies and transport of clastic sediments in karstic aquifers. In: Sasowsky ID, Mylroie J (eds) Studies of cave sediments. Kluwer Academic Publishers, New York, NY pp 1–22

Brown RF, Lambert TW (1963) Reconnaissance of ground-water resources in the Mississippian Plateau region, Kentucky. U.S. Geological Survey Water Supply Paper 1603, 58 p + 16 plates

Brucker RW, Hess JW, White WB (1972) Role of vertical shafts in movement of ground water in carbonate aquifers. Ground Water 10 (6):5–13

Collier CR, Flint RF (1964) Fluvial sedimentation in Mammoth Cave, Kentucky. U.S. Geological Survey Professional Paper 475-D, pp 141–143

Cushman RV, Krieger RA, McCabe JA (1963) Present and future water supply for Mammoth Cave National Park, Kentucky. U.S. Geological Survey Water Supply Paper 1475-Q, pp 601–647

Groves C, Meiman J (2005) Weathering, geomorphic work, and karst landscape evolution in the Cave City groundwater basin, Mammoth Cave, Kentucky. Geomorphology 67:115–126

Hess JW, White WB (1988) Storm response of the karstic carbonate aquifer of south central Kentucky. J Hydrol 99:235–252

Hess JW, White WB (1989) Water budget and physical hydrology (Chap. 4) In: White WB, White EL (eds) Karst hydrology: concepts from the Mammoth Cave area. Van Nostrand-Reinhold, New York, NY pp 105–126

Hess JW, White WB (1993) Groundwater geochemistry of the carbonate karst aquifer, south central Kentucky, USA. Appl Geochem 8:189–204

Hess JW, Wells SG, Brucker TA (1974) A survey of springs along the Green and Barren Rivers, central Kentucky karst. Nat Speleol Soc Bull 36(3):1–7

Meiman J, Ryan M (1993) The Echo River-Turnhole Bend overflow route. CRF Newsletter 21(1):16–18

Pohl ER (1936) Geologic investigations at Mammoth Cave, Kentucky. Trans Am Geophys Union 17:332–334

Quinlan JF, Ewers RO (1989) Surface drainage in the Mammoth Cave area (Chap. 3) In: White WB, White EL (eds) Karst hydrology: concepts from the Mammoth Cave Area. Van Nostrand-Reinhold, New York, NY, pp 65–103

Quinlan JF, Ray JA (1981) Groundwater basins in the Mammoth Cave region, Kentucky. Friends of Karst Occasional Publication No. 1, one sheet

Raeisi E, Groves C, Meiman J (2007) Effects of partial and full pipe flow on hydrochemographs of Logsdon River, Mammoth Cave Kentucky USA. J Hydrol 337:1–10

Ray JA, Currens JC (1998a) Mapped ground-water basins in the Beaver Dam 30 × 60 Minute Quadrangle. Kentucky Geological Survey, Frankfurt, KY, 1 sheet

Ray JA, Currens JC (1998b) Mapped ground-water basins in the Campbellsville 30 × 60 Minute Quadrangle. Kentucky Geological Survey, Frankfurt, KY, 1 sheet

Ryan M, Meiman J (1996) An examination of short-term variations in water quality at a karst spring in Kentucky. Ground Water 34:23–30

White EL (1977) Sustained flow in small Appalachian watersheds underlain by carbonate rocks. J Hydrol 32:71–86

White EL (1989) Flood hydrology (Chap. 5) In: White WB, White EL (eds) Karst hydrology: concepts from the Mammoth Cave area. Van Nostrand-Reinhold, New York, NY, pp 127–143

White EL, Reich BM (1970) Behavior of annual floods in limestone basins in Pennsylvania. J Hydrol 10:193–198

White WB, White EL (1989) Karst hydrology: concepts from the Mammoth Cave area. Van Nostrand-Reinhold, New York, NY

Worthington SRH, Ford DC, Davies GJ (2000) Matrix, fracture and channel components of storage and flow in a Paleozoic limestone aquifer. In: Sasowsky ID, Wicks CM (eds) Groundwater flow and contaminant transport in carbonate aquifers. A.A. Balkema, Rotterdam, pp 113–128

Mineralogy of Mammoth Cave

9

William B. White

Abstract

Mammoth Cave displays a wide variety of depositional environments that permit the deposition of some unusual minerals, although the cave overall is rather devoid of speleothems. Calcite in the form of stalactites, stalagmites, and flowstone is the common carbonate mineral and occurs where passages extend beyond the protective caprock. Aragonite also occurs but more rarely. Gypsum is the common sulfate minerals and occurs widely in dry passages beneath the caprock. Gypsum takes the form of crusts and curving crystals known as gypsum flowers. In the exceptionally dry portions of the cave are found other sulfate minerals including mirabilite, $Na_2SO_4{\cdot}10H_2O$, epsomite, $MgSO_4{\cdot}7H_2O$, hexahydrite, $MgSO_4{\cdot}6H_2O$, celestine, $SrSO_4$, and blodite, $Na_2SO_4{\cdot}MgSO_4{\cdot}4H_2O$. In addition, there is evidence for the rare sulfate minerals eugsterite, $2Na_2SO_4{\cdot}CaSO_4{\cdot}2H_2O$, and wattevilleite, $Na_2SO_4{\cdot}CaSO_4{\cdot}4H_2O$. The source of sulfate is likely the oxidation of pyrite, FeS_2 that occurs in the limestone and overlying sediments. Saltpetre of uncertain mineralogy occurs in the cave sediments and was mined in the nineteenth Century. Other minerals include black manganese oxides that occur on stream cobbles and on chert ledges.

9.1 Introduction

Mammoth Cave is not renowned for its speleothems. Visitors can trek for miles through bedrock tunnels without the smallest sight of the stalactites and stalagmites that are the pride of most show caves. From the viewpoint of the cave mineralogist, what Mammoth Cave has that many show caves do not is a very wide range of depositional environments in which unusual minerals can be formed.

The passages in Mammoth Cave cross beneath the sandstone-capped ridges but frequently terminate against the valley walls. The portions of the passages beneath the ridges are dry, but where they extend beneath the valley walls, runoff from the sandstone and water from the epikarst can penetrate and the passages are wet (Fig. 9.1). The dry passages are deposition sites for gypsum and other soluble sulfate minerals, while the wet ends of the passages provide the proper conditions for the deposition of carbonate minerals.

To get started, the list of the minerals that have (thus far) been identified in Mammoth Cave and associated caves is given in Table 9.1. This list may be compared with the complete description of minerals that have been found in caves (Hill and Forti 1997). The remainder of the chapter tells where the minerals are found, what they look like, and how they were deposited.

9.2 Carbonate Minerals

Calcite, $CaCO_3$, is the most common mineral found in caves, and Mammoth Cave is no exception. Aragonite, the high-pressure polymorph of calcite, also occurs, and hydromagnesite has been reported in Thomas Avenue in the Crystal Cave section (Palmer and Palmer 2003). No other carbonates or mixed carbonate–sulfate minerals have been reported.

W.B. White (✉)
Department of Geosciences, The Pennsylvania State University, University Park, USA
e-mail: wbw2@psu.edu

H.H. Hobbs III et al. (eds.), *Mammoth Cave*, Cave and Karst Systems of the World,
DOI 10.1007/978-3-319-53718-4_9

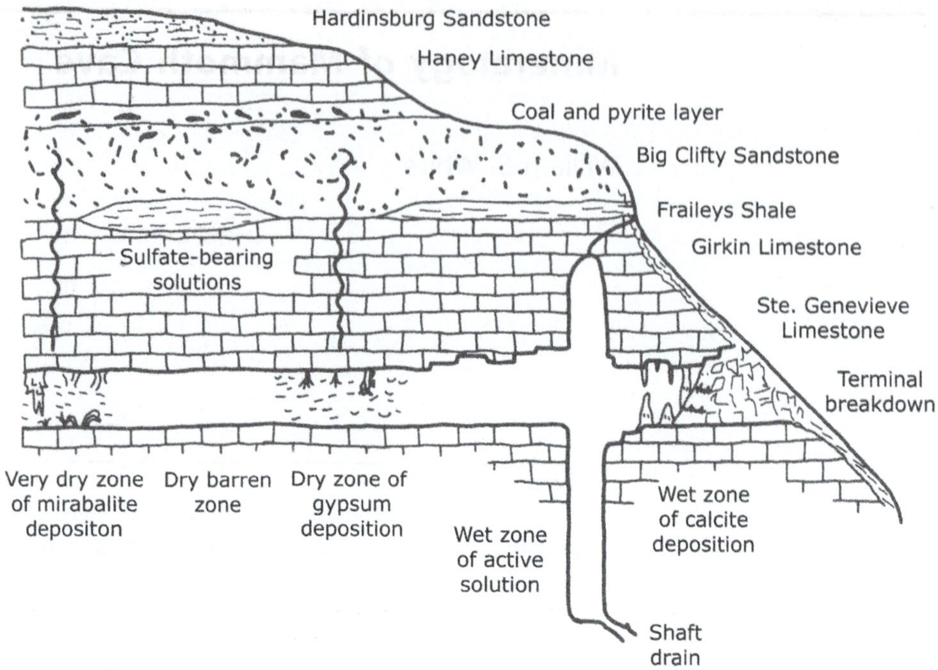

Fig. 9.1 Mineral depositional environments in Mammoth Cave

9.2.1 Dripstone and Flowstone in Mammoth Cave

Most of the calcite deposits, in the form of stalactites, stalagmites, and flowstone, are found in the wet areas where passages extend from beneath the caprock (Fig. 9.2). Notable speleothem locations are the Frozen Niagara and Violet City areas in traditional Mammoth Cave, the entrance area in Great Onyx Cave, Davis Hall near the Austin Entrance to the Crystal Cave section, and other locations where vadose water can reach the cave passage. A few of the flowstone deposits, including Frozen Niagara and the Amos Hawkins Formation, are quite massive. Massive flowstone also occurs in Boone Avenue near Thorpes Pit and along Kentucky Avenue. Now dry and sometimes partly re-dissolved, the origin of these speleothems remains a mystery.

Seventeen U/Th isotope dates were obtained in the 1970s by John W. Hess and Russell S. Harmon. At this time, isotope concentrations were obtained by α-particle counting which limited the age range that could be measured. As might have been expected, there is a great range in age from 2000 years for the youngest to six samples with ages that exceeded the limits of the measurements of 350,000 years. Because the speleothem deposition occurs after erosion has removed the protective caprock, there is no expected relationship between speleothem ages and the age of the passages in which they occur. Oxygen, carbon, and hydrogen isotope ratios were measured on a stalagmite from Davis Hall with dates spanning the interval from 230,000 to 100,000 years to provide a hint of changing climate conditions in the Mammoth Cave Area for this time period (Harmon et al. 1978).

9.2.2 The Occurrence of Aragonite

Aragonite occurs in the Fairy Grotto, a passage that branches from the Main Cave near the Cataracts, where it occurs as a bulk component of fragments of speleothem loose on the passage floor. Frederic Siegel (1965) found aragonite in speleothems in Great Onyx Cave. Few studies have been made of the carbonate minerals so that the distribution of aragonite through the cave system is not known.

The occurrence of aragonite in caves has always been something of a mystery. Aragonite is not thermodynamically stable under cave ambients. From the viewpoint of equilibrium thermodynamics, it should not have precipitated in the first place, and if it did precipitate, it should revert to the thermodynamically stable calcite. Cave aragonite does, in fact, eventually revert to calcite but does so on time scales of hundreds of thousands of years. What is needed is some mechanism either to enhance the metastable nucleation of aragonite or suppress the nucleation of calcite until supersaturation increases past the aragonite solubility curve (White 2012). In Mammoth Cave, speleothems that grow in the wet zone away from the caprock consist mainly of calcite. From mostly superficial observations, it appears that the speleothems in the transitional zone that contain both calcite and gypsum often also contain aragonite (Fig. 9.3).

Table 9.1 Minerals found in Mammoth and related caves. Crystallographic data from Mineral Data Base, 2013

Mineral	Composition	Symmetry	Space group	Unit cell parameters (Å)
Aragonite	$CaCO_3$	Orthorhombic	Pmcn	a = 4.959 b = 7.968 c = 5.741
Barite	$BaSO_4$	Orthorhombic	Pbnm	a = 8.878 b = 5.45 c = 7.152
Birnessite	$Na_{0.3}Ca_{0.1}K_{0.1}Mn^{4+}Mn^{3+}O_4 \cdot 1.5H_2O$	Monoclinic	C2/m	a = 5.174 b = 2.85 c = 7.336 β = 103.18°
Blodite = Blödite = Bloedite	$Na_2SO_4 \cdot MgSO_4 \cdot 4H_2O$	Monoclinic	$P2_1/a$	a = 11.126 b = 8.242 c = 5.539 β = 100.84°
Calcite	$CaCO_3$	Rhombohedral	R $\bar{3}$ c	a = 4.989 c = 17.062
Celestine = Celestite	$SrSO_4$	Orthorhombic	Pbnm	a = 8.359 b = 5.352 c = 6.866
Epsomite	$MgSO_4 \cdot 7H_2O$	Orthorhombic	$P2_12_12_1$	a = 11.86 b = 11.99 c = 6.858
Eugsterite	$2Na_2SO_4 \cdot CaSO_4 \cdot 2H_2O$	Monoclinic		
Hexahydrite	$MgSO_4 \cdot 6H_2O$	Monoclinic	A2/a	a = 24.442 b = 7.216 c = 10.119 β = 98.28°
Hydromagnesite	$4MgCO_3 \cdot Mg(OH)_2 \cdot 4H_2O$	Monoclinic	$P2_1/c$	a = 10.11 b = 8.94 c = 8.38 β = 114.58°
Mirabilite	$Na_2SO_4 \cdot 10H_2O$	Monoclinic	$P2_1/a$	a = 12.82 b = 10.35 c = 11.48 β = 107.66°
Nitrocalcite	$Ca(NO_3)_2 \cdot 4H_2O$	Monoclinic	$P2_1/n$	a = 14.477 b = 9.16 c = 6.285 β = 98.42°
Todorokite	$(Na, Ca, K)_2(Mn^{4+}, Mn^{3+})_6O_{12} \cdot 3\text{–}4.5H_2O$	Monoclinic	P2/m	a = 9.764 b = 2.842 c = 9.551 β = 94.14°
Wattevilleite	$NaSO_4 \cdot CaSO_4 \cdot 4H_2O$	Orthorhombic		

Note alternate names. First name given is preferred

9.2.3 Mechanisms for Carbonate Mineral Deposition

The standard model for the deposition of carbonate—mainly calcite—speleothems is sketched in Fig. 9.4. The Earth's atmosphere in 2013 contains about 400 ppm or 0.04 volume percent CO_2. The CO_2 dissolves in rainwater, which soaks into the limestone soils of the valley walls and the Sinkhole Plain. The CO_2 concentration in the soil is much higher, in the range of 0.5–4% in the Mammoth Cave area according to the measurements of Franz-Dieter Miotke (1974). As the rainwater infiltrates through the soil, it dissolves additional CO_2 and becomes highly acidic. This water dissolves the limestone at the soil/bedrock interface taking Ca^{2+} and

Fig. 9.2 Dripstone—stalactites and stalagmites—near the Violet City Entrance. Photo courtesy of David S. Carson

HCO_3^- ions into solution, thus creating the irregular bedrock surface of the epikarst. The water, now nearly saturated with calcite, continues its descent through fractures in the vadose zone with little additional chemical reaction until it emerges from the ceilings of underlying cave passages. In general, CO_2 concentrations in cave atmospheres are higher than the surface atmosphere, but in Mammoth Cave, the CO_2 concentration is higher only by about a factor of two. The drip waters degas CO_2, become supersaturated, and re-deposit the dissolved carbonate species as speleothem calcite.

9.3 Sulfate Minerals

Mammoth Cave contains an extensive suite of sulfate minerals. Gypsum was described in Cleaveland's Cabinet (Cleaveland Avenue) by Locke in 1842. Gypsum (often called alabaster in the early accounts) was apparently recognized soon after the cave beyond Echo River was discovered in 1838. Locke also noted the presence of "native sulfate of magnesia"—presumably epsomite. Gypsum is widely distributed throughout the upper level passages of the cave system. The other sulfate minerals are more localized and are described by location.

Modern investigations of sulfate mineralization began in the late 1950s partly because of discovery of Turner Avenue with its dramatic mineralization and partly because of the importance of gypsum and other sulfate minerals to archeological studies (Watson 1969, 1974). There have been many reports and abstracts but few formal publications. This chapter attempts to pull the various fragments of information together.

9.3.1 Gypsum Speleothems

Gypsum is only found in the dry passages of the cave system. A small amount of moisture on the passage walls is sufficient in most cases to dissolve and remove the gypsum, which has about ten times the solubility of calcite. Gypsum occurs as thin fibers (angel hair), as needle-like crystals, as granular and columnar crusts, and as "gypsum flowers" of various sizes and shapes (Fig. 9.5). Granular crusts vary in thickness from a few mm to several cm and are composed of

Fig. 9.3 "Old Granddad," a massive column of mixed calcite and gypsum in Turner Avenue. Photo by the author

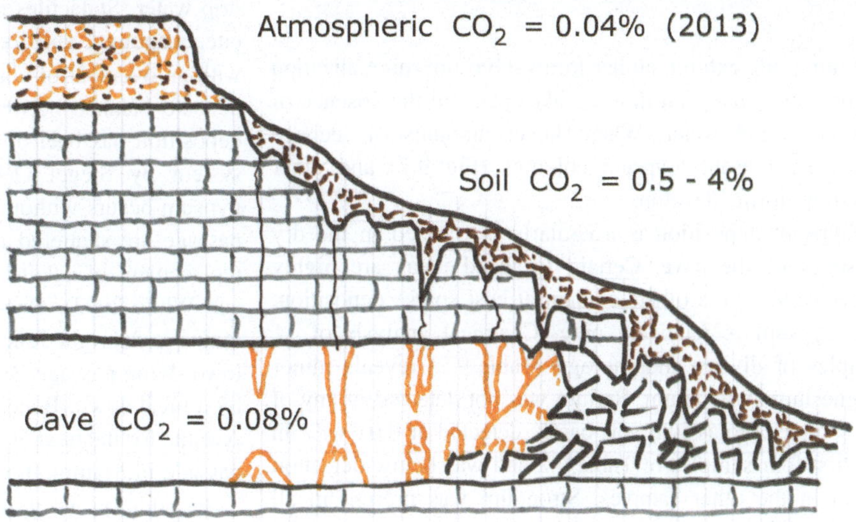

Fig. 9.4 Sketch showing the geochemical changes in rainfall as it eventually becomes drip water depositing speleothems. Specified CO_2 concentrations from Miotke 1974

equant crystals of gypsum typically in the mm size range. Columnar crusts are typically about one cm in thickness and are composed of tightly packed gypsum fibers with the long axis of the fibers perpendicular to the growth surface. Gypsum flowers (Fig. 9.6) are curved bundles of gypsum crystals that appear to grow from a common center. Examination of the flower "petals" under a polarized light microscope reveals bundles of fibers with curvatures accomplished by shifting the fibers. See the review by White (2012) for discussion of growth mechanisms. Both epsomite

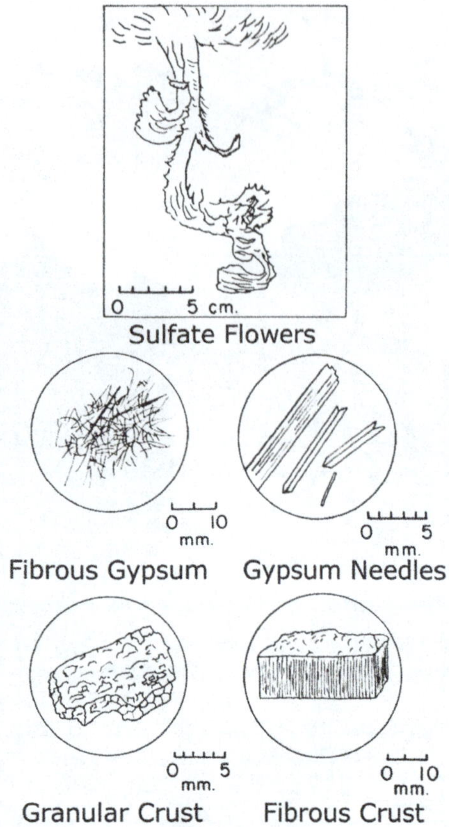

Fig. 9.5 Sketch of various gypsum morphologies. Drawings within *circles* represent views through a microscope

and mirabilite exhibit similar forms. Gypsum mineralization to form the crusts and flowers takes place in the absence of obvious liquid water. Where larger amounts of seepage water are present, gypsum columns (Fig. 9.7) and other dripstone forms develop.

Gypsum deposition is irregularly distributed in the dry passages of the cave. Certain localized areas are highly mineralized, while others contain at best sparse deposition. The gypsum is relatively pure. Chemical analysis of 11 samples of diverse morphology (Table 9.2) reveals minor magnesium and sodium. Barium was not detected in any of the samples. Potassium was barely detected at 0.01 wt% in the first four samples of Table 9.2 and was below detection limits in the other samples. Strontium was present in all samples. If the strontium concentration exceeds its solid solubility in gypsum, celestine will appear as a separate phase. Celestine occurs in Thomas Avenue just before Scotchmans Trap in the Flint Ridge section of the system as a light, sky-blue crust a few mm thick beneath a much thicker gypsum crust. According to observations by Carol Hill in 1981, the celestine wall area was 12 m long by 1 m wide.

9.3.2 Turner Avenue

Turner Avenue is a mid-level valley drain (Level C in Palmer's notation. See Granger et al. 2001) that extends roughly north–south for 3 km across Flint Ridge. The northern (downstream) end of the passage is against a valley wall near Green River where it ends in a massive breakdown. In the south (upstream), the passage also ends in breakdown against the wall of Houchins Valley, the enclosed valley that separates Flint Ridge from Mammoth Cave Ridge. Both ends of the passage are wet where they extend beyond the protective caprock. Most of the passage is beneath the caprock and is quite dry. A temperature and relative humidity profile measured along 2 km of passage beginning at the northern end (Fig. 9.8) shows that the relative humidity reaches a minimum of about 83% in the 1000–1500 m section of the passage where the most extensive mineralization occurs. However, calculation shows that the water vapor partial pressure is roughly constant at 10.8–11.0 Torr (1.44–1.47 kPa). The change in relative humidity and obvious increased dryness of the passage is due to a small temperature rise when the passage passes beneath the highest part of the ridge.

Sulfate deposition occurs in various amounts along the entire length of Turner Avenue but is very variable in both intensity of mineralization and in the minerals deposited. Near the northern end of the passage, there is some active drip water. Stalactites and columns occur consisting of calcite, aragonite, and gypsum. Mostly, the cores are calcite with an overcoat of coarsely crystalline gypsum. The rather delicate balance between carbonate deposition and sulfate deposition has been investigated in other parts of the cave system by Siegel (1965). South along Turner Avenue, gypsum occurs without carbonate. Walls and ceilings of the passage are shattered and broken into platy shards by the replacement of calcite in the limestone by gypsum (White and White 2003). One kilometer south of the northern terminus is Albright Junction where Turner Avenue crosses a lower-lying passage, Mather Avenue, and is connected to it by a breakdown. Beneath this breakdown and along Turner Avenue for the next 500 m is the most intensely mineralized passage in Mammoth Cave.

In addition to gypsum, mirabilite occurs extensively. Mirabilite takes the form of water-clear stalactites hanging from massive coarsely crystalline gypsum (Fig. 9.9) and as curved fibrous masses much resembling gypsum flowers (Fig. 9.10). Mirabilite, in the form of flowers and fibrous crystals, occurs along several hundred meters of the passage. The largest of these fiber bundles, more than a meter in length, is known as the "foxtail" (Fig. 9.11). Although mirabilite, $Na_2SO_4{\cdot}10H_2O$, is stable in the cave environ-

Fig. 9.6 Gypsum flowers in Turner Avenue. Photo by the author

ment, it dehydrates in a matter of seconds to minutes when exposed to surface ambients. The 10 waters of crystallization are lost in a single step, producing a white powder with the X-ray diffraction pattern of thenardite, Na_2SO_4. Samples for analysis were collected in tightly sealed bottles, which in turn were transported to the laboratory in an ice chest. All samples were stored under refrigeration. X-ray diffraction patterns were obtained by grinding the samples and preparing slides in an ice chest. Cold air saturated with water vapor was blown through the diffractometer, while the powder patterns were scanned. With these precautions, good agreement was obtained with the powder pattern for mirabilite given on card no. 11-0647 of the International Center for Diffraction Data.

The system Na_2SO_4–H_2O is a classic salt system, and its phase diagram is well known (e.g. Ricci 1951: 138). It is also a classic example of a system showing metastability. The metastable compound is $Na_2SO_4 \cdot 7H_2O$. Initially, there appeared to be some evidence for the occurrence of the metastable salt within the deposits. Because of the risk of hydration/dehydration during transport and sample handling, to determine whether the metastable salt might be present, nine samples were sealed in the cave into pre-weighted vials, which were then reweighed, opened, dried at 110 °C, and weighed again. The theoretical weight loss for the decahydrate is 55.9% and for the heptahydrate 47.0%. The mean weight loss of nine specimens was 55.6% with a standard deviation of 1.6%. The earlier indications of the metastable salt based on less stringent weight loss measurements were not supported.

Fred Benington (1959) dissolved some of the Turner Avenue mirabilite in a cold ethanol solvent and found a residue of acicular crystals which he identified as the "labile salt," $2Na_2SO_4 \cdot CaSO_4 \cdot 2H_2O$, found in laboratory experiments by Hill and Wills (1938). Benington's discovery was the first observation of the double salt in nature, but there were not enough follow-on measurements of crystallographic properties to allow the Mammoth Cave material to be established as a new mineral. Since Benington's report, the labile salt has been found in nature several times and has been formally described as the mineral eugsterite (Vergouwen 1981).

Fig. 9.7 Gypsum columns in Turner Avenue. CRF photograph

Table 9.2 Chemical analyses of gypsum from Mammoth Cave. Data are in wt%

Sample	Location	MgO	Na_2O	SrO
Clear selenite crystals	Turner avenue	0.09	0.01	0.24
Gypsum needles	Dismal valley	<0.005	0.02	0.03
Massive crystals	Turner avenue	0.08	0.02	0.11
Granular crust	Turner avenue	0.32	0.03	0.42
Gypsum flower	Turner avenue	0.01	0.01	0.06
Cave cotton	Turner avenue	0.06	0.02	0.21
Gypsum flower	Turner avenue	0.01	0.01	0.04
Columnar crust	Turner avenue	0.05	0.01	0.10
Granular gypsum flower	Cleaveland avenue	0.28	0.01	0.39
Columnar crust	Turner avenue	0.01	0.01	0.02
Fine gypsum crystals	Turner avenue	0.01	0.01	0.05

9.3.3 New Discovery

"New Discovery" is a section of Mammoth Cave discovered by exploration from within the cave system in 1938. An artificial entrance was constructed that gives access to Fossil Avenue and Big Avenue, the two main passages in New Discovery, but access has been severely limited by the Park Service and this section of the cave remains pristine.

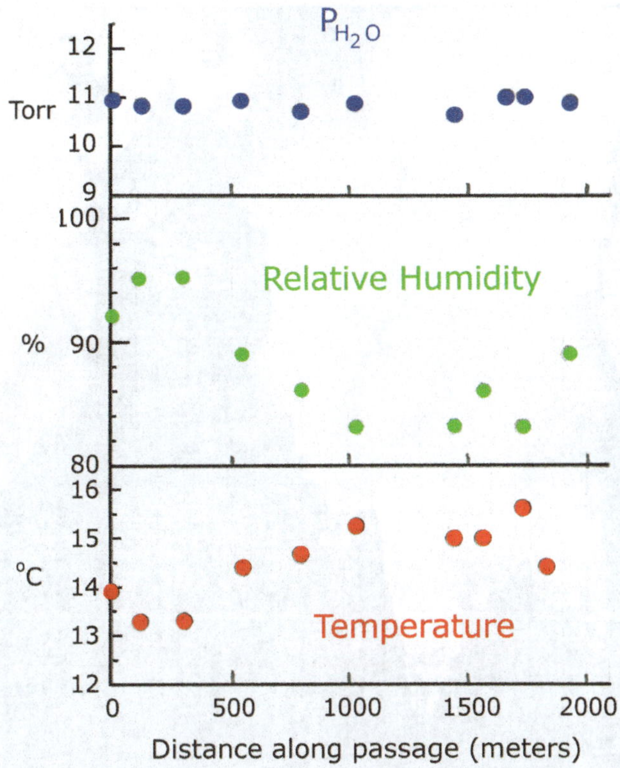

Fig. 9.8 Temperature, relative humidity, and water vapor partial pressure, measured along turner avenue southward from the top of Brucker Breakdown

Gypsum flowers with "petal" lengths of 30 cm or more occur in a canyon, Little Paradise, that cuts across Big Avenue (Fig. 9.12). Gypsum occurs, but rather sparsely, along both of the main passages.

A feature of great interest in New Discovery is found in a pile of irregular breakdown fragments along Fossil Avenue and has been called the "fuzzy rock" (Fig. 9.13). The fibrous crystals growing from the fuzzy rock have the X-ray pattern of epsomite, $MgSO_4{\cdot}7H_2O$. From a limited number of observations, the fuzz on the rock apparently appears only during the winter. During the spring and summer, the crystals collapse to a white powder only to grow again the following winter. An explanation is that temperature and water vapor partial pressure are exactly balanced on the epsomite—hexahydrite phase boundary so that slight lowering of temperature and increase in water vapor pressure during the winter are enough to stabilize epsomite, while slightly higher temperatures in the summer shift the conditions sufficiently to cause epsomite to dehydrate to hexahydrite. A somewhat similar phenomenon occurs in Cleaveland Avenue where a drill hole admits water and wets the sediment in an otherwise dry passage. Small hair-like crystals of mirabilite grow from the sediment in the spring and disappear during the summer.

9.3.4 Lee Cave

Lee Cave is a separate cave located southwest of Mammoth Cave proper (Freeman et al. 1973). The portion of interest is a km-long segment of large trunk passage known as Marshall Avenue. Marshall Avenue is a roughly east–west passage ending in breakdown against the walls of Deer Park Hollow to the east and Sand Cave Hollow to the west.

Sulfate minerals occur in Lee Cave as loose powdery material between shards of breakdown, as curved crystals shaped like gypsum flowers, as fine, hair-like crystals growing from bedrock surfaces, and as snow-like material which has drifted down over all upward-facing surfaces in the western and central parts of Marshall Avenue (Fig. 9.14). The mineralogy of these deposits is more complex than the deposits in Mammoth Cave.

Gypsum is moderately common and occurs as fragments of gypsum flowers and as fragments of gypsum crust. Epsomite occurs as hair-like crystals on many exposed surfaces. There is a massive epsomite "flower" 30 cm in length, and many fragments of epsomite crystals scattered on the passage floor. Epsomite also occurs in the loose crusts in the ceilings and somewhat more sparsely in the thick "snowdrift" crusts on the floor.

Fig. 9.9 Water-clear stalactites of mirabilite formed on masses of coarsely crystalline gypsum. Photo by the author

Some samples of the fibrous crystals contain up to 50% hexahydrite, $MgSO_4{\cdot}6H_2O$. To determine whether hexahydrite actually occurs in the cave or whether these observations were due to the dehydration of epsomite during transport and storage, samples were collected in sealed vials. The material appeared loose and powdery in the laboratory with no evidence of released water. X-ray powder patterns were clearly those of hexahydrite. Many of the large epsomite crystals appeared etched as if material had been dissolved and reprecipitated. The coexistence of epsomite and hexahydrite, along with the seasonal appearance and disappearance of epsomite on the "fuzzy rock" in New Discovery, suggests that the temperature and water vapor partial pressure in the cave lie very close to the pressure–temperature phase boundary that separates the stability regions of these two minerals.

The predominant mineral in the crusts is blodite, $Na_2SO_4{\cdot}MgSO_4{\cdot}4H_2O$. The mineral takes the form of clear white grains a few mm in diameter giving the floor crusts a rather sugary texture. X-ray diffraction analysis of the crusts shows that they are of variable mineralogy. In addition to blodite, the sulfate minerals epsomite, mirabilite, and wattevilleite also occur. Most unexpected was the appearance of wattevilleite, $Na_2SO_4{\cdot}CaSO_4{\cdot}4H_2O$, rather than eugsterite as the Na–Ca double salt. Wattevilleite was originally described in association with lignite deposits in 1879, but no crystallographic studies have been made. The mineral was identified only by its X-ray diffraction pattern and would be the first identification of this mineral in a cave.

The source of the magnesium for the magnesium sulfate minerals appears to be lenses of dolomite that occur within the limestone beds. Samples of the "snowdrift" materials were dissolved in water. X-ray diffraction patterns showed that the insoluble residue was mainly dolomite suggesting that the sulfates were formed by reactions within dolomite beds in the passage ceiling.

9.3.5 Other Sulfate Mineral Occurrences

The entrance to Great Salts Cave leads to a major trunk passage on the B level. Mirabilite was found as fibrous crystals growing on the wall at numerous locations along the passage. In a side passage known as "Dismal Valley," mirabilite occurs as hair-like crystals a few cm in length growing from the surface of a bank of sand and silt.

The entrance to Great Salts Cave, along with the Historic Entrance, is one of the most accessible natural entrances to

Fig. 9.10 "Flowers" of mirabilite in Turner Avenue. There is an open Brunton compass in the photograph for scale. Photo by the author

the cave system. The cave was explored by Late Archaic and Early Woodland peoples at least as early as 2000–3000 BP as determined by radiocarbon dating the soot from their torches (Benington et al. 1962). There was extensive traffic through the cave by these early explorers who were mining sulfate minerals from the cave walls (Watson 1969, 1974). Gypsum could be used as a pigment and the mirabilite used as a laxative. Cave walls and ceilings were blackened by soot from torches. In places, the gypsum crusts have peeled away exposing the limestone bedrock. Tufts of mirabilite crystals (Fig. 9.15) have grown on these exposed areas since the cessation of the mining, showing that the mineralization process remains active and gives a sense of the time scale required for the deposition of the mineral crystals.

9.3.6 Soluble Salts in Cave Sediments

Water-soluble sulfate minerals occur dispersed in the stream-transported clastic sediments that occupy most of the passage floors. Eighty-three sediment samples from all levels of Mammoth Cave proper and the Flint Ridge section were collected. Each sample was slurried in water and the coarse-grained insoluble particles allowed to settle. The water with its load of fine-grained suspended particles was decanted and the water allowed to evaporate at room temperature. X-ray diffractions patterns were obtained on the residues. Identification of water-soluble sulfate minerals in these residues does not identify the original mineral, but it does identify the dominant chemical composition.

Much of the suspended fraction in the slurries proved to be fine-grained quartz. Clays were surprisingly uncommon; kaolinite was identified in a few samples. About 30% of the sediment samples were found to contain water-soluble sulfates. Six of the samples were sodium-rich, nine were magnesium-rich, and eight had mixed sodium/magnesium salts. There was no obvious pattern to the distribution of the soluble minerals in the sediments. These results are consistent with the appearance of water-soluble sodium and magnesium sulfate minerals as speleothems on the cave walls but do not, of course, identify the actual minerals dispersed in the cave sediments.

Fig. 9.11 "Foxtail," a meter-long bundle of mirabilite fibers. Note hardhat for scale. Photo by the author

The most comprehensive study of clastic sediment mineralogy in Mammoth Cave is an unpublished document (Davies and Chao 1959). There are a variety of heavy minerals such as zircon, tourmaline, rutile, brookite, and anatase, which were clearly carried into the cave with the silt and sand that make up the bulk of the clastic sediment. Also identified were gypsum, celestine, and barite, which appear to be authigenic, that is minerals that formed in place in the cave environment. Celestine, in particular, appears as well-formed acicular crystals about one mm in length. It is very unlikely that such crystals would have survived transport with the clastic sediment. No mention was made in the report of any of the water-soluble minerals.

9.3.7 Depositional Mechanisms for Sulfate Minerals

To explain the occurrence of gypsum and other sulfate minerals, we must account for the source of sulfur, the essential common element in this family of minerals. Two hypotheses have been proposed with two variants on the second hypothesis:

1. The gypsum is derived from anhydrite layers that occur in the limestone. The anhydrite is dissolved and carried into the cave by ground water and re-deposited by evaporation.
2. The sulfur is derived from the oxidation of pyrite, FeS_2, which occurs in the limestone and overlaying rocks.
   - 2A. The source is a pyrite-rich layer at the top of the Big Clifty Sandstone. The *Pohl hypothesis* (Pohl and White 1965).
   - 2B. The source is pyrite distributed in the limestone near the cave passages. The *Palmer and Palmer hypothesis* (2003).

Evidence against the anhydrite source is provided by the few sulfur isotope measurements that have been made (Furman et al. 1999). Three samples of Mammoth Cave gypsum were analyzed giving $\delta^{34}S$ values of −5.06, −8.12, and −7.82, all in reasonable agreement with the values of −9 to −4.2 found for Mississippian pyrites but not in agreement

Fig. 9.12 Large gypsum flowers in New Discovery Section. Photo courtesy of Rick A. Olson

with the values of −19 to −14 found for St.Louis anhydrites. This bit of evidence supports the hypothesis that the ultimate source of the gypsum is the pyrite in the limestone or overlying formations, not the anhydrite that occurs interbedded in the limestones.

Distinguishing between hypotheses 2A and 2B is more difficult. A thick pyrite-rich layer does indeed exist at the top of the Big Clifty Sandstone. However, thin section measurements show a pyrite concentration on the order of 0.1% in the limestone. Both sources of pyrite are available. The shards of breakdown and the microscope evidence of replacement of calcite by gypsum in the wall rock are strong evidence that the chemical reactions that form gypsum and other minerals take place in the immediate vicinity of the cave passages. Gypsum is observed to be distributed along cave passages as regions with intense mineralization separated by relatively barren passages. In the Pohl hypothesis, this can be explained by the patchy distribution of the Fraileys Shale at the base of the Big Clifty. Intense gypsum mineralization would be found in passages where the shale was missing. In the Palmer and Palmer hypothesis, passages with intense sulfate mineralization correspond to zones of high pyrite concentration in the limestone.

In the Pohl hypothesis, pyrite in the layer above the Big Clifty is oxidized, the iron remains behind as $Fe(OH)_3$, and the sulfate ions and hydrogen ions migrate through the Big Clifty Sandstone and into the limestone in those areas where the Fraileys Shale is missing. Reaction with calcite in the limestone builds up CO_2 pressure that retards the reaction until the solutions reach a cave passage. The cave passage provides an escape for the CO_2 and the reaction proceeds with gypsum replacing calcite. Gypsum has a larger molar

Fig. 9.13 Fuzzy rock in Fossil Avenue, New Discovery Section. Photo by the author

volume than the calcite it replaces, and the pressure developed causes the observed spalling of plates and shards of breakdown. Seepage water rearranges the gypsum and other minerals to form the observed crystals, flowers, and crusts, including the crystals observed growing from silt banks and on breakdown blocks and chert nodules.

In the Palmer and Palmer hypothesis, the vadose seepage waters are moving in a closed system. Purely carbonate reactions between the water and the limestone consume all available carbon dioxide, driving the CO_2 pressures to values far below atmospheric values. These seepage waters also carry the oxygen necessary to oxidize pyrite in the limestone. Near the walls of cave passages, the seepage waters absorb CO_2 from the cave atmosphere so that the sulfate ions derived from pyrite can react and be precipitated as gypsum.

One model releases CO_2, the other absorbs CO_2. It would be easy to distinguish between them if we had a water sample. Unfortunately, the movement of seepage water and its chemical reactions and mineral depositions take place over hundreds to thousands of years. At any given moment, the volume of water is orders of magnitude too small to sample. As a final touch, it should be noted that these hypotheses are not mutually exclusive. Both mechanisms could be at work.

9.4 Manganese Oxide Coatings

Sandstone cobbles and chert layers in cave streams are frequently coated with black deposits often referred to as "manganese oxide." These deposits vary from a fraction of a mm to several cm in thickness. Some of the thickest coatings found in Mammoth Cave are on chert ledges in Bransford Avenue along the route to Cathedral Dome. The mineral grains are very small, disordered, and do not yield good X-ray diffraction patterns. From the marginal X-ray patterns

Fig. 9.14 Sulfate minerals drifted over silt bank in Marshall Avenue, Lee Cave. Photo by the author

and from infrared spectra, the Bransford Avenue sample was identified (tentatively) as birnessite, which contains both trivalent and tetravalent manganese as well as a variable mix of alkali and alkaline earth ions. Birnessite appears to be the most common manganese mineral in the stream cobble coatings. However, recent re-examination of the Bransford Avenue sample using synchrotron radiation (Florence Ling and Peter Heaney, private communication) suggests that the mineral is todorokite. The composition of todorokite (see Table 9.1) is quite similar to that of birnessite, but the crystal structure is different (Post and Bish 1988). Todorokite is the common manganese mineral in deep sea manganese nodules.

Chemical analyses are available for two samples from Mammoth Cave, one from Parker Cave on the Sinkhole Plain, and one from Priddy Cave, north of Munfordville (White et al. 2009) (Table 9.3). There are several points of interest. The black deposits are not pure manganese oxide. The infrared spectra indicate substantial amounts of carbonate and some silica. The manganese oxides are enriched in barium but depleted in strontium, exactly the reverse of the pattern found in gypsum and in carbonate speleothems. The manganese oxides contain amazing quantities of transition (iron group) metals—fractional percent concentrations of CoO, CuO, NiO, and ZnO. The manganese oxides act as

Fig. 9.15 Mirabilite crystals growing from a bare patch of cave ceiling in the Great Salts Cave Section. The mineral growth has taken place since the ceiling was scraped by aboriginal mineral miners 2000–3000 years ago. Photo by the author

scavengers for these and other metals, but the quantities are many orders of magnitude higher than the metal concentrations in cave streams.

As a final caveat, not all black deposits in Mammoth Cave are manganese oxides. Soot from torches and lanterns are the source of blackness in the historic portions of the cave. Carbon from decomposed organic matter and humic substances also occur (Hill 1982).

9.5 Saltpetre

Considering the importance of saltpetre mining to the history of Mammoth Cave, the mineralogy of saltpetre has been elusive. Clastic sediments—sand and silt—were collected into vats. The sediments in the vats were leached with water and the collected solutions piped to the surface where they were passed through wood ashes to exchange the Ca^{2+} in solution by K^+. The exchanged solution was then boiled to evaporate part of the water and cooled to crystallize KNO_3, niter, the desired ingredient for gunpowder (Eller 1981).

Compilation of 38 analyses from Mammoth Cave and Dixon Cave revealed nitrate concentrations from a few hundred ppm to 4% by weight (Hill 1981), but the actual nitrate-bearing mineral in the soil was not described. The assumed mineral is nitrocalcite, $Ca(NO_3)_2{\cdot}4H_2O$, but this has been difficult to confirm. Nitrocalcite can dissolve in its own water of crystallization, so it is expected to occur as films of fluid adsorbed to mineral grains. Crystalline nitrocalcite has not, apparently, actually been observed.

9.6 Final Word

Although this chapter summarizes much of what is known about the mineralogy of Mammoth Cave, it is not the end of the story. Much remains to be learned about the mineralogy and geochemistry of the unusual sulfate minerals, particularly in Turner Avenue and in Lee Cave. Unidentified lines

Table 9.3 Chemical analyses of manganese oxide coatings

Oxide	Mammoth Cave-A	Mammoth Cave-B	Parker cave	Priddy cave
Al_2O_3	6.62	7.56	5.76	2.77
B_2O_3	0.014	0.071	0.017	0.019
BaO	0.64	0.44	0.72	1.62
CaO	20.5	4.34	4.09	10.7
CoO	0.074	0.036	0.08	0.18
Cr_2O_3	0.038	0.047	0.006	0.1
CuO	0.56	<0.003	0.024	0.004
Fe_2O_3	4.09	6.58	5.86	2.59
K_2O	0.13	0.12	0.14	0.31
MgO	1.37	0.93	0.96	0.86
MnO_2	20.46	20.22	31.49	20.95
MoO_3	0.052	0.03	0.044	0.12
Na_2O	0.099	0.13	0.082	0.24
NiO	0.98	0.15	0.22	0.13
SiO_2	1.67	1.26	0.75	0.16
SrO	0.072	0.067	0.047	0.057
TiO_2	0.072	0.078	0.055	0.11
V_2O_3	0.037	0.071	0.035	0.24
ZnO	0.17	0.18	0.33	0.085
ZrO_2	0.17	<0.003	<0.003	0.22

Mammoth Cave-A Chert ledge in Hawkins Pass. *B* Chert ledge in Martel Avenue between Pinson's Pass and Nelson's Dome. Parker and Priddy cave samples are stream cobbles. All concentrations in weight percent

appear on some of the X-ray diffraction patterns implying other minerals to be described.

References

Benington F (1959) Preliminary identification of crystalline phases in a transparent stalactite. Science 129:1227

Benington F, Melton C, Watson PJ (1962) Carbon dating prehistoric soot from salts cave, Kentucky. Am Antiq 28:238–241

Davies WE, Chao ECT (1959) Report of sediments in mammoth cave, Kentucky. Administrative Report for U.S. National Park Service by U.S, Geological Survey

Eller PG (1981) Chemical aspects of the conversion of cave nitrates to saltpeter. Bul Nat Speleol Soc 43:106–109

Freeman JP, Smith GL, Poulson TL, Watson PJ, White WB (1973) Lee cave, mammoth cave national park, Kentucky. Bul Nat Speleol Soc 35:109–125

Furman FC, Gregg JM, Palmer AN, Shelton KL (1999) Sulfur isotopes of gypsum speleothems in the central Kentucky karst system: indications of pyrite and anhydrite sulfur sources, and sulfuric acid karstification. MSM Spelunker 42(1):20–24

Granger DE, Fabel D, Palmer AN (2001) Pliocene-pleistocene incision of the Green River, Kentucky, determined from radioactive decay of cosmogenic ^{26}Al and ^{10}Be in Mammoth Cave sediments. Geo Soc Am Bul 113:825–836

Harmon RS, Schwarcz HP, Ford DC (1978) Stable isotope geochemistry of speleothems and cave waters from the flint ridge-mammoth cave system, Kentucky: implications for terrestrial climate change during the period 230,000 to 100,000 years B.P. J Geo 86:373–384

Hill AE, Wills JH (1938) Ternary systems. XXIV. Calcium sulfate, sodium sulfate and water. J Am Chem Soc 60:1647–1655

Hill CA (1981) Origin of cave saltpeter. Bul Nat Speleol Soc 43:110–126

Hill CA (1982) Origin of black deposits in caves. Bul Nat Speleol Soc 44:15–19

Hill C, Forti P (1997) Cave minerals of the world, 2nd edn. National Speleological Society, Huntsville, AL

Locke J (1842) Alabaster in the mammoth cave of Kentucky. Amer J Sci 42:206–207

Miotke FD (1974) Carbon dioxide and the soil atmosphere. Abhandlungen zur Karst-und-Höhlenkunde RA, Heft, p 9

Palmer AN, Palmer MV (2003) Geochemistry of capillary seepage in mammoth cave. Speleogenesis Evol Karst Aquifers 1(4):1–8

Pohl ER, White WB (1965) Sulfate minerals: Their origin in the central Kentucky karst. Am Mineral 50:1461–1465

Post JE, Bish DL (1988) Rietveld refinement of the todorokite structure. Am Mineral 73:861–869

Ricci JE (1951) The phase rule and heterogeneous equilibrium. D. Van Nostrand, New York, NY

Siegel FR (1965) Aspects of calcium carbonate deposition in great onyx cave, Kentucky. Sedimentology 4:285–299

Vergouwen L (1981) Eugsterite, a new salt mineral. Am Mineral 66:632–636

Watson PJ (1969) The prehistory of salts cave, Kentucky. Illinois State Museum reports of investigations no. 16, USA
Watson PJ (1974) Archeology of the mammoth cave area. Academic Press, New York, NY
White WB (2012) Speleothem microstructure/speleothem ontogeny: a review of Western contributions. Int J Speleol 41(2):331–360
White WB, Vito C, Scheetz BE (2009) The mineralogy and trace element chemistry of black manganese oxide deposits from caves. J Cave Karst Stud 71:136–143
White WB, White EL (2003) Gypsum wedging and cavern breakdown: studies in the mammoth cave system, Kentucky. J Cave Karst Stud 65:43–52

Mammoth Cave Meteorology

10

Rickard A. Olson

Abstract

Cave atmospheric conditions are vitally important to conservation of a broad range of cave resources. These include hibernating bats, cave invertebrates and amphibians, the park's most significant historic structures, thousands of organic prehistoric artifacts, and evaporite minerals such as gypsum snowballs. Sophisticated computer modeling of cave atmospheric conditions to predict the effects of alternative management actions has begun, and this powerful tool can help park managers avoid costly mistakes. For example, artificially enhanced influx of cold winter air in the Historic Entrance resulted in an increased rock fall rate, which is both a safety and conservation issue. A prodigious amount of data on cave atmospheric conditions has been acquired, and these data need to be considered in a comprehensive way.

10.1 Introduction

In the hot summer months, visitors to Mammoth Cave's Historic Entrance are treated to a current of refreshing cool air flowing out. At night and early in the morning, it is possible to see a layer of fog at the boundary layer between the cool cave air below and the warm surface air above (Fig. 10.1). The difference in temperature has noticeable effects on the kinds of plants we see growing in the entrance zone. Fragile ferns (*Cystopteris protrusa*), Jack-in-the pulpit (*Arisaema triphyllum*), and wild germaniums (*Geranium maculatum*) are especially abundant in this cool zone and are not found otherwise in the immediate area.

10.2 Drivers of Cave Airflow

It did not take long for people to notice the seasonality of airflow in Mammoth Cave. "It is this difference of temperature that causes the air to rush into the narrows with such violence; but Mr. Miller informed me, that during the hot summer months, it rushes with equal violence in the contrary direction." (Blane 1824). During a visit to Mammoth Cave in 1850, Benjamin Silliman of Yale University (Silliman 1851) related guide Stephen Bishop's observation that the cave air "…flowed out when the external air was above 60° and inward when this was below 60°."

Mammoth Cave and most other caves in the area have multiple openings where air exchanges with the surface. Because these openings are usually at different elevations, air currents are created due to density differences between cave and surface air. So in the summer, the colder, denser cave air flows out of lower entrances and surface air is drawn in upper openings, most of which are too small for people to use. In winter, the relatively warmer cave air rises out of upper entrances, such as Salts Cave entrance, and cold air from the surface is drawn in via the lower entrances. It works like a chimney (Atkinson 1985), hence the name "chimney effect airflow." Barr and Kuehne (1971) presented theoretical plots showing the number of days that Mammoth Cave entrances would have air flow in, flow out, and also how many days there would be airflow reversals, neglecting any possible barometric effects. In extensive caves out west with single entrances, barometric pressure changes drive cave airflow.

In the Mammoth Cave System, barometric pressure change has little effect on cave airflow. Researcher Bruce

R.A. Olson (✉)
Cave Research Foundation, Cave City, KY, USA
e-mail: rickardolson16@gmail.com

H.H. Hobbs III et al. (eds.), *Mammoth Cave*, Cave and Karst Systems of the World,
DOI 10.1007/978-3-319-53718-4_10

Fig. 10.1 Humidity in warm surface air condenses into a thin cloud where it contacts cool cave air flowing out of the Historic Entrance in summer. Note the lush ferns below the fog layer growing in the cool moist air. *Photograph* by Paula Cormany

Zerr braved chilly temperatures on a fall day and night for 26.5 continuous hours during October 10–11, 1992, and measured airflow at Doyel Valley entrance every 12 min [1]. Analysis of these data, collected in this feat of endurance, showed that airflow was 95% determined by temperature differences between surface and cave with little influence from changes in barometric pressure. Severe barometric pressure changes, such as during thunder storms, can have temporary effects, but airflow patterns quickly return to normal.

Single entrance caves in the region, such as Dixon Cave, often function as cold traps where airflow is driven by cold air sinking into the entrance and displacing warmer air above, which goes out the entrance at ceiling level. These cold trap caves can be important bat hibernation sites.

Collapse may have played a role in at least one extreme airflow event in the Flint Ridge portion of Mammoth Cave. In Turner Avenue, there are small dunes known as the Aeolian Deposits. Sand and fine gravel were blown up through small connections between Dead Bat Avenue into Turner Avenue. These wind-blown deposits and the many bat mummies in Dead Bat Avenue have led Roger Brucker to speculate that there may have been an entrance at the far end of Dead Bat Avenue and that a significant passage collapse may have simultaneously closed this entrance and created the strong air flow required to create the Aeolian Deposits (Roger Brucker personal communication 2016).

10.3 Significance of Airflow

Exchange with surface air is important for many reasons. For example, maintenance of a fairly normal atmospheric composition is achieved via exchange with the surface. Even with the typically high humidity in caves, airflow accelerates evaporation of water. Some mineral deposits in caves are dependent on evaporation of water, and for this reason are called evaporites. Examples of evaporite minerals in the Mammoth Cave area include gypsum in the form of "flowers" and "snowballs" and calcite in the form of "popcorn" (see Chap. 9 this volume for more on this). Sites like the Snowball Room are popular because of these beautiful mineral deposits, and their popularity has put them at risk. Restrooms at this location, needed for the large crowds of people, block airflow into the Snowball Room and reduce natural evaporation. Falls of gypsum off the ceiling have become common.

Cave airflow can have some interesting effects on cave minerals. Directional popcorn, caused by airflow-linked evaporation, is common in Mammoth Cave and many other caves. Sometimes condensation and evaporation work in tandem with unanticipated effects. In Vanderbilt Hall, and in Big Avenue, atmospheric water condensed in small channels, dissolved some of the bedrock, and redeposited that as rims (not rimstone) when the moisture evaporated

[1]Unpublished manuscript on file at Mammoth Cave National Park by Zerr, B., Hardin, W., & W. Lewis. titled "Air flow in Doyel Valley Entrance to Mammoth Cave, Kentucky: An Analysis," accession number: MACA-0754).

Fig. 10.2 A rim in Big Avenue within Mammoth Cave's New Discovery section. These secondary mineral deposits are common in some western caves such as Jewel, Lechuguilla, and Carlsbad Caverns, but are extremely rare in the Mammoth Cave System

(Fig. 10.2). Such rims are common in western caves such as Carlsbad Caverns, Lechuguilla, and Jewel Cave. Art Palmer (2007) explains this process very well. Bare spots in otherwise heavy gypsum crusts in Dan's Avenue may be related to airflow because they all face the same direction (Fig. 10.3). Some rillenkarren features (vertical grooves) in Fossil Avenue and also in Big Avenue within the New Discovery portion of Mammoth Cave may be due to condensation corrosion. These include enigmatic rills in ceiling channels, a topic needing more research.

10.4 General Patterns of Airflow in Mammoth Cave

In winter at the Historic Entrance, air flows into the Rotunda and splits between Audubon Avenue and Broadway. Along both forks, much air leaves these upper level trunk passages as cold dense air sinks to lower levels. In Audubon Avenue, much goes down Little Bat Avenue to Mammoth Dome, and some returns to Rotunda along the ceiling as a counter current. In Broadway, the flow of air is reminiscent of a "losing stream" in that some air sinks to lower levels via the Vanderbilt Hall breakdown, some via the Corkscrew, some sinks through breakdown at Methodist Church to a crawl off Pensacola Avenue, and some filters through breakdown now blocking a former connection between the Church and Harveys Avenue. From Harveys Avenue, air partly sinks down to Inas Hall through a former stream piracy route, and then down to Great Relief Hall where it is carried along with moist air heading through Fat Mans Misery to Black Snake Avenue, and ultimately, back up to the Main Cave level at Cyclops Gateway and Giants Coffin. This area has had airflow routes mapped, but most of Mammoth Cave has not had airflow mapped.

In most caves in the region, chimney effect airflow causes the direction of flow to reverse seasonally. At least in the Historic Section of Mammoth Cave, there are some passages that do not seasonally reverse direction of airflow (Olson 2002). Gothic Avenue, Black Snake Avenue, and Watson Trace are three prominent examples of passages where airflow is always out toward the entrance regardless of season. Dr. Warren Campbell of Western Kentucky University has suggested a reason for this. Airflow in Mammoth is based on density differences in air masses. The air emerging from these passages is always at cave temperature and high in humidity. To us, humid air feels more dense than dry air (recall phrases like "so humid you could cut it with a knife"), but in fact humid air is less dense, so it rises up through less humid air even if the two air masses are at the same temperature (Warren Campbell, personal communication).

From 1976 to 1984, Bobby Carson (personal communication) made many airflow observations in Mammoth Cave using ventilation smoke tubes procured from Mine Safety Appliance Co. and later switched to children's soap bubbles, achieving similar results through 1989. One point of these efforts was to detect and characterize a ceiling level air current that runs counter to the normal winter airflow directions. This was important because condensation occurs at various points along Broadway and these observations helped us understand why this happens.

For the same reasons, and without knowledge of Carson's earlier observations, in February 2003, I used a small helium balloon to trace ceiling level air currents in Broadway and Audubon Avenue (Fig. 10.4). The balloon was made to be neutrally buoyant, and traveled from Gothic Avenue down into Broadway, and out to the Rotunda where it stalled. I next took the balloon to the far end of Audubon Avenue, and it flew at ceiling level all the way back to the Rotunda where it again stalled. Apparently these ceiling level counter currents are forced down where they meet at the Rotunda and mix with the inflowing cooler air below.

Fig. 10.3 Matt Keller points to bare spots in otherwise heavy gypsum crusts in Dan's Avenue. The bare spots all face the same direction, but a cause for this is not clear. Seasonal airflow may be a factor

Fig. 10.4 Helium balloon flying between Booths Amphitheater and Rotunda. A neutrally buoyant balloon is an excellent way to trace subtle air currents in caves

10.5 Early Airflow Measurements

On November 30, 1902, in Little Bat Avenue, Biologist Carl Eigenmann (1909) measured airflow at 8640 ft/h with an unspecified anemometer. This equates to 2.4 feet per second (fps) or 0.73 meters per second (mps). Weather on that day, from records in Bowling Green, KY, indicates a high of 54F (12.2C) and a low of 34F (1.1C) with a small amount of precipitation in the form of snow, sleet, and hail. He did not give a time for this measurement, but at the Historic Entrance gate on the same day at 7:00 AM, he measured the airflow at 4.3 mps. At 9:40 AM, he again measured at the gate and got 4.7 mps, so it is probable that the Little Bat Avenue measurement was taken in between these two readings at the gate. Modern airflow measurements in Houchins Narrows cannot be compared due to the different cross-sectional areas where these historical and modern measurements were made. Little Bat Avenue is fairly consistent in cross section, so comparison with modern measurements can be made. Searching modern records for Bowling Green, weather very similar to that on November 30, 1902, occurred on January 11, 2014 and January 14, 2014. Average airflow in Little Bat Avenue between 8:15 and 8:30 AM on these dates was 0.27 and 0.12 mps, respectively. Using the 0.27 mps value, this is only about a third the velocity of 0.73 mps measured by Eigenmann in 1902 and is very surprising because the old iron gate (Fig. 10.5) at the entrance would have restricted airflow considerably more than the modern gate (Fig. 10.6). Even

Fig. 10.5 Old iron gate in 1902 when Carl Eigenmann made his airflow measurements in Little Bat Avenue

with the smaller opening at the Historic Entrance in Eigenmann's time, he described seeing ice stalagmites at the end of Little Bat Avenue, which we never see today. Also, during cold snaps, such as on January 29, 2014, when surface temperature dipped to 0.9F (−17.3C), we saw a maximum airflow velocity in Little Bat Avenue of 0.5 mps. In the 44 years that I have been active at Mammoth Cave, I have never seen ice stalagmites in Little Bat Avenue.

With airflow at the Historic Entrance restricted by the old iron gate in 1902, then another input of cold air would be needed to cool Little Bat Avenue and provide the much greater flow measured. The only known input of air other than the Historic Entrance in modern times is Olives Bower. We know that air formerly entered Mammoth Cave by that route due to observations in 1897, and from geological evidence (Olson and Toomey 2005). The only other place with close proximity to the surface and signs of frost shattering at some time in the past is Lookout Mountain. The species of bat remains in the Lookout Mountain area are also consistent with a former entrance at that location. However, we have no historical accounts of such an entrance.

From October 23, 1936, to November 6, 1937, airflow observations and other atmospheric parameters were made in Mammoth Cave between "Sand Break" (aka Bunker Hill) and the Mushroom Beds. These field notebooks (rescued from a trash can by Bobby Carson) are stored in the Curatorial Facility in the park (MACA Accession # 327) and could be valuable for future cave atmospheric research.

10.6 Modern Measurements

Barr and Kuehne (1971) reported airflow plus temperature and relative humidity along a 2.5 km course from Austin Entrance through Turner Avenue on November 5, 1961. Airflow velocity ranged from 4 to 27 meters per minute. This equates to 0.22–1.48 fps or 0.066–0.45 mps.

Fig. 10.6 Modern gate at Historic Entrance with two openings much larger than in 1902. Rick Olson and Colleen O'Connor Olson are standing in front of the Plexiglas panels which they installed twenty years ago. These adjust airflow to approximate predevelopment rates

A wind vane anemometer was installed in Houchins Narrows in January 1996 and collected data until 1999 (Johnathan Jernigan personal communication). Ultrasonic anemometers were installed in Houchins Narrows and Gothic Avenue in December 2012, and data from these are displayed in the Visitor Center. These data are also stored in a server to be used for mathematical modeling along with data from stations in Broadway, Corkscrew, Audubon Avenue, Little Bat Avenue, Booths Amphitheater, Standing Rocks, Rafinesque Hall, and Vespertilio Hall (Toomey and Olson 2012).

10.7 Management of Airflow

With the discovery of radon in park caves in 1975, a study on the effect of airflow restriction at the Historic Entrance was carried out from October 10 through November 20, 1977 (Carson 1981). It was common practice to cover the gate with sheet metal panels in winter to protect visitors from a strong draft of cold air that would otherwise enter the cave, but the effect on radon levels in the cave had not been determined. Radon measurements were taken with the sheet metal panels on the gate and off the gate. The impact on radon levels caused by the panels was significant. The result was a 54% increase in radon levels that people were exposed to. An alternative strategy to mitigate radon levels and provide for visitor comfort was adopted in early 1979: One panel on the gate was hinged and could be opened when the last tour left the cave, and closed when the first tour of the day entered.

Intensive environmental restoration work in the Historic Section of Mammoth Cave has been ongoing since March 1, 1996, when I finally obtained permission to install Plexiglas panels on the open-grid gate in place at that time. This was over two years after unrestrained −16° F (−27 °C) air entering Rotunda caused a 40 ton slab to fall from the ceiling onto the War of 1812 saltpeter mining works. The Plexiglas was installed to restore the influx of cold winter air to approximately what it would have been before alterations to Houchins Narrows began in the early 1800s (Olson 1996). Even with installation of open-grid stairs and removal of a concrete plug in Crevice Pit above Mammoth Dome in 2004, restoration of winter temperatures in Little Bat Avenue has been limited by a condensation problem at Booths Amphitheater, which causes water to drip on the second set of saltpeter vats, also from the War if 1812 (Fig. 10.7). To achieve a simultaneous solution (i.e., conservation of vats and bats), an extensive network of environmental sensors has been deployed to support sophisticated computer modeling of the cave environment (Toomey and Olson 2012). Alternative solutions to prevent dripping on the saltpeter vats

Fig. 10.7 Condensation drips onto War of 1812 artifacts at Booth's Amphitheater in winter. This is caused warm moist air out of Gothic Avenue meeting cold air from Historic Entrance. Interaction between the two air masses is exacerbated by the artificial hill of mining spoil piles that reduces Main Cave's cross section by about half

will be explored via computer modeling. Novel options to remedy the situation include dehumidifier units placed up in Gothic Avenue just before the air reaches Booths Amphitheater and trenching through the artificial fill at Booths Amphitheater as an archaeological dig to create a channel for cold air to pass as it would have before mining spoil piles filled up half the passage. This channel would be covered by an insulated boardwalk at the current trail level with open stairs at each end to allow cold air to pass without interacting with warmer moist air coming out of Gothic Avenue. This latter option seems extreme, but if computer-modeling supports creating a cold air pass to protect the saltpeter vats, then Archaeologist Dr. George Crothers would not oppose this idea.

The feasibility of airflow restoration in Vespertilio Hall with the goal of reestablishing bat hibernation conditions was recently supported computer modeling (Bird et al. 2016). Historically, there was a major Indiana bat hibernation site cooled by an opening to the surface at the end of Olives Bower. Hovey and Call (1897: 21) wrote: "Thousands of bats in the winter season, suspended in great clumps, may be seen. A single catch one night gave the writer six hundred and seventy individuals, most of which went to the United States National Museum." Evidence including this historical account and geological aspects of this passage were summarized in a paper written for park staff with a recommendation to reopen this connection to the surface (Olson and Toomey 2005).

Fig. 10.8 This artificial wall of wood and plastic sheeting was installed in Gothic Avenue to block the flow of moist air toward Booth's Avenue. Condensation was greatly curtailed, but the wall was not put back up after this experiment due to concern for wildlife being trapped behind it. *Photograph* by Art Palmer

In 2002, there was an unfortunate episode in cave airflow management. I was directed to seal off Gothic Avenue (Fig. 10.8) to prevent moist cave air from reaching Booths Amphitheater as an experiment to see whether that would prevent condensation and dripping on the War of 1812 saltpeter vats (Olson and Jernigan 2002). This was widely criticized by Science and Resources Management staff, and rightly so because the airflow distortion caused by the spoil piles at Booths Amphitheater was being mitigated by distorting natural airflow in Gothic Avenue (Fig. 10.8). Nobody knows how many bats were trapped behind the wood and plastic wall installed just beyond the Blacksmiths Shop on January 30, 2002. Dripping on the saltpeter vats was significantly reduced, but the experiment was not continued in following winters due to the high potential for negative impact on bats.

10.8 Temperature, Relative Humidity, Evaporation, and Condensation

The primary controlling factors governing the temperature of a given cave passage are latitude, elevation, proximity to entrances, and water influx. Geothermal influences are also possible, but that has not been a factor in the Mammoth Cave area since the Cretaceous Period, which predates the cave as we know it. In winter, cold water can pour in sinkholes or back-flood through springs and cool otherwise remote cave passages. In summer, the same process can warm passages in surprising ways. For example, when warm water from Green River back-floods into River Styx Spring, River Hall becomes foggy. The water droplets are carried by normal summer airflow up the Corkscrew, collect on the ceiling of Broadway, and drip onto the floor below.

So if you see water dripping in Broadway near the Corkscrew, then you know that River Styx has back-flooded. This has been researched in an exemplary citizen science project led by Shannon Trimboli of the Mammoth Cave International Center for Science and Learning (Trimboli et al. 2016).

In 1971, Biologists Thomas Barr and Robert Kuehne (1971) did a significant amount cave environmental work in Mammoth Cave and the Flint Ridge Cave System, which had not yet been connected with Mammoth. In the Constant Temperature Zone of Mammoth Cave, they report a range of 13.2–14.0 °C with a mean of 13.6 °C. Even in dusty dry passages, they reported that relative humidity rarely falls below 80% and that areas occupied by cave-adapted species are typically in the range of 94% to saturation. These same researchers employed Livingston atmometers to gage evaporation in the Historic Entrance and Frozen Niagara entrance areas during winter and spring of 1956–1957. They showed that in winter, the evaporation rate was 15.5 times higher in Audubon Avenue than in the Radio Room near the Frozen Niagara entrance.

Condensation and evaporation instruments were deployed at 11 locations in Mammoth Cave beginning in March 1994 as part of a Cave Atmospheric Monitoring (CAM) program (Fry and Meiman 1995; Jernigan and Fry 2007). Not surprisingly, rates of evaporation are highly variable in different locations. For example, in River Hall during February 1995 evaporation occurred at a rate of 22 ml per day, and at the same time, water was condensing in Frozen Niagara at a rate of 3 ml per day. Airlock construction to restore natural airflow at Frozen Niagara took place between October 1995 and February 1996, so this data point is pre-airlock (Fry 1996).

Physical Scientist Johnathan Jernigan has done a lot of cave atmospheric work beginning in 1997, which has included mathematical modeling. This modeling is based upon convection as the underlying driving force of airflow, and he constructed empirical models using atmospheric data from Mammoth Cave. In particular, data from Houchins Narrows were used to predict temperatures at other locations in the cave (Jernigan and Swift 2001). His model was successful in being able to predict temperatures in other parts of the cave based upon data from Houchins Narrows.

Bat Conservation International deployed temperature and relative humidity dataloggers in park caves with hibernating bats from the summer of 2001 to the summer of 2005. This was part of a nationwide study to determine the environmental parameters important in bat hibernation sites (Tuttle and Kennedy 2002). There is a large amount of data that need to be analyzed in context with all the other data from different sources.

10.9 Importance of Cave Atmospheric Conditions

There are four ecological zones in the terrestrial cave ecosystem, and two of them are defined by temperature range. These are the Variable Temperature Zone and the Constant Temperature Zone (Eigenmann 1909). The Entrance Zone and the Twilight Zone can also be affected by cave temperatures, depending upon the direction of airflow. We have already seen that in summer, exiting cave air is pleasantly cool. In winter, air exiting upper entrances such as Salts Cave may exhibit a vapor plume visible from hundreds of feet away and make overhanging tree branches wiggle in the breeze

The nine species of bats known to hibernate in Mammoth Cave and nearby caves have different temperature and relative humidity (RH) requirements, so proper management of entrances is crucial to maintain hibernation sites. Cave-adapted invertebrates are more vulnerable to desiccation than their surface relatives, and in some cases, a difference of 1% in RH can be vitally important (Toomey 2009). Evaporation rate, important in the stability of some cave minerals, is extremely sensitive to RH (Buecher 1999).

10.10 Management of Temperature and Relative Humidity

In the early 1990s, there were many cave gating projects to protect bat hibernation sites from illegal intrusion and also to replace gate structures that distorted natural airflow. At the same time, artificial entrances to Mammoth and other caves were being fitted with airlocks to prevent infiltration of surface air into parts of caves where naturally there would have been little or none (Fry 1996). In many cases, terrestrial cave communities had become established where artificial entrances were blasted in, and so it was argued that these diverse assemblages of cave life should not be completely cut off from surface access. To that end, airlocks were fitted with small openings where cave crickets, bats, and wood rats could commute between cave and surface. Biological monitoring pre- and post-airlock construction showed manifold benefits to this management strategy for artificial entrances (Poulson et al. 1996).

10.11 Atmospheric Components and Potential Utility of Cave Air

The basic constituents of the atmosphere in Mammoth Cave are typical of surface air. However, perhaps because of the cool temperature and the lack of irritants such as pollen, people in the 1800s thought that cave air might have special properties. For instance, a gentleman (Anonymous 1810: 109) said this about the cave atmosphere: "Generally speaking, the cave is very dry, and the air salubrious; we judge of the latter from the facility with which combustion went on, our tapers burnt much more brilliant than when in the common air, owing we suppose to the presence of a greater proportion of the oxygen gas." Robert Bird (1838) wrote: "Its purity, judging from its effects upon the lungs, and from other circumstances, is remarkable, though in what its purity consists I do not know." He continues: "I recommend that all broken-hearted lovers and dyspeptic dandies to carry their complaints to the Mammoth Cave, where they will undoubtedly find themselves "translated" into very buxom and happy persons before they are aware of it."

10.12 Speleotherapy

Little did Bird know that in four years John Croghan would attempt to cure consumption (tuberculosis) patients with Mammoth Cave air (see Chap. 3 this volume for more on this topic). This plan failed of course because cave air cannot cure an infection, and the smoke from fires needed to keep the patients warm fouled the air.

Interest in speleotherapy for treatment of consumption did not end with the Croghan experiment. In nearby Long Cave, there was a move toward building a hotel and sanitarium that would be cooled by air from the cave. The major difference from the experiment in Mammoth Cave was that the cave air would be unadulterated by smoke from fires to keep the patients warm. They could stay in nice hotel rooms cooled by the cave air, and yet enjoy sunshine. To test this idea, a boring five inches in diameter and 225 ft deep was drilled into Long Cave, which can still be seen today at the junction of Grand Avenue and Echo Passage. A small experimental building was placed over the well, and by use of a "Sturtevant exhaust fan" cave air was brought into the building and cooled it nicely (Crump 1890). It was estimated that the cave could cool "…40,000 rooms 16 × 18 × 10 ft," which was probably too optimistic but such reported capacity would attract investors. The other aspects in favor of the site were "…its situation in the midst of a virgin forest of oak and hickory, with a sandy soil (resulting from the wear of the Chester sandstone) and splendid drainage—indeed everything seems to conspire to make this a favored spot for sanitary purposes."

Speleotherapy might seem a foreign concept, but in many Eastern European countries children exercise underground in specially qualified caves and salt mines, and derive documented clinical benefit. Dr. Stanley Sides attended an international speleotherapy conference in Aggtelek National Park in Hungary in 2001 and discussed the limited American experience with speleotherapy. His participation in the conference was sponsored by the National Park Service as part of the international relations program. He attended presentations outlining the science behind the selection of caves and mines for potential therapeutic benefit including those with increased levels of radon. He concluded, however, that it was unlikely western medicine would allow utilization of ionizing radiation linked with radon in the air of caves and mines as a treatment option for children. (Stanley D. Sides, M.D. personal communication, May 10, 2014).

10.13 Natural Air Conditioning

In 1959–1960, the park service constructed a ventilation shaft and tunnel between Houchins Narrows and the Visitor Center (VC) to cool the VC and administrative offices with cave air. The system worked well until it was discovered the cave air being supplied included radon and radon decay products. In July 1976, the ventilation shaft was sealed to prevent radon decay products from entering surface facilities. According to guide Joe Duvall, the air handlers that drew air up into the facilities were so noisy that it made talking at Rotunda difficult (Personal communication 2014).

10.14 Radon

Radon, a noble gas that is produced by the radioactive decay of uranium, radium, and thorium (Yarborough 1980) was found to exist inside park caves in 1975. Long-term exposure to radon decay products can pose a risk of lung cancer (US Surgeon General Report 1985). Mammoth Cave National Park began documenting employee exposure to radon decay products in the summer of 1976. However, among cave explorers and guides at Mammoth Cave, there has not been an obvious elevated lung cancer rate compared with the general population, which may be due to an extremely low sample population size.

10.15 Carbon Dioxide

Mammoth Cave air, at different times and locations, has elevated levels of carbon dioxide (CO_2) when compared with surface air. CO_2 concentrations vary in complex ways, but generally levels are highest in the summer growing season. This is because microbial activity in the soil peaks in summer, and infiltrating water picks up more carbon dioxide as it percolates through the soil. Today, surface CO_2 levels are at about 400 parts per million (ppm). Levels in the cave sometimes exceed 1600 ppm or about four times higher than surface air (Johnathan Jernigan personal communication). This is due to natural out-gassing from groundwater as it enters the cave, and these concentrations are not nearly high enough to pose a health risk. The threshold for symptoms from CO_2 exposure is at about 2% of the atmosphere (Palmer 2007), and 1600 ppm is equal to 0.16%. Most caves have CO_2 levels around 10 times higher than surface air (Palmer 2007), and one reason for the lower concentrations observed in Mammoth Cave may be capillary seepage entering the cave, which is depleted in CO2 and therefore absorbs it from cave air (Palmer and Palmer 1995).

10.16 Methane

Methane concentrations lower than in surface air were detected in three caves in Kentucky, including Mammoth Cave, during the summer of 2012. The largest methane concentrations in the cave atmosphere were observed at cave entrances, and decreasing methane concentrations were noted further from entrances (Webster et al. 2012). It is thought that methane oxidizing bacteria on cave walls may be responsible for the lower concentrations at deep cave sites.

10.17 Concluding Thoughts

A tremendous amount of atmospheric data has been taken in the Mammoth Cave System and other area caves. Complete coverage of even the published work is beyond what is possible in this chapter due to space limits. Much of the data have never been analyzed because the people involved were overwhelmed with work load. So there is significant opportunity for cave atmospheric research to be done without taking a single temperature measurement. Further work is needed of course because there will always be new questions.

References

Anonymous (1810) The subterranean voyage or mammoth cave partially explored. The Enquirer, Richmond Va 6(109)

Atkinson T (1985) Chimney effect winds in caves. Can Caver 17:10–15

Barr TC Jr, Kuehne RA (1971) Ecological studies in the mammoth cave system of Kentucky II. The Ecosystem: Annales de Speleoogie 26(1):47–96

Bird R (1838) Peter pilgrim: or a Rambler's recollections. Lea and Blanchard, Philadelphia, pp 109–110

Bird A, Olson R, Toomey R, Addison A, Bosch R (2016) Use of multiphysics simulation to model environmental conditions associated with hibernacula including preliminary indication of impacts on saltpeter vats in Mammoth Cave. In: 11th research symposium at Mammoth Cave National Park, p 162

Blane W (1824) Mammoth cave, winter of 1822–1823. In: An excursion through the United States and Canada. Baldwin, Craddock, and Joy, London

Buecher R (1999) Microclimate study of kartchner caverns, Arizona. J Cave Karst Stud 61:108–120

Carson B (1981) Summary and findings of the radon daughter monitoring program at mammoth cave national park, Kentucky. Radiation hazards in mining: control, measurement, and medical aspects, pp 654–658

Crump M (1890) Experiments with cave air. Science 15(379):290–291

Eigenmann C (1909) Cave vertebrates of America. Carnegie Institution, Washington, DC, p 7

Fry J (1996) Eighteen cave gates and airlocks: conclusion of a three year project to restore entrance dynamics at mammoth cave national park. In: Proceedings of the fifth annual mammoth cave national park science conference, pp 69–83

Fry J, Meiman J (1995) Cave atmospheric monitoring project in mammoth cave national park. In: Proceedings of mammoth cave national park's fourth science conference, pp 205–206

Hovey H, Call R (1897) The mammoth cave of Kentucky, an illustrated manual. John P. Morton and Company, Louisville, Kentucky

Jernigan J, Fry J (2007) CAM: monitoring atmospheric conditions in mammoth cave. Unpublished report on file at Division of Science and Resources Management, Mammoth Cave National Park

Jernigan J, Swift R (2001) A mathematical model of air temperature in mammoth cave, Kentucky. J Cave Karst Stud 63(1):3–8

Olson R (1996) This old cave: the ecological restoration of the historic entrance ecotone of mammoth cave, and mitigation of visitor impact. In: Proceedings of the fifth annual mammoth cave national park science conference, pp 87–95

Olson R (2002) Unpublished airflow map based upon field observations (MACA accession # 0754). Div of Sci and Resources Management, Mammoth Cave National Park, KY

Olson R, Jernigan J (2002) A pilot study on condensation prevention at Booth's Amphitheater in the Historic Section of Mammoth Cave. Unpublished report on file at Mammoth Cave National Park

Olson R, Toomey R (2005) Olive's Bower: evidence for a recent air vent to the surface critical to Indiana bat hibernation habitat in the historic section of mammoth cave. Unpublished paper for park staff on file at Mammoth Cave National Park

Palmer A (2007) Cave geology. Cave books, 4700 Amberwood Drive, Dayton, OH 45425

Palmer A, Palmer M (1995) Geochemistry of capillary seepage in mammoth cave. In: Proceedings of mammoth cave national park's fourth science conference, pp 119–133

Poulson TL, Lavoie KH, Helf KL (1996) Effects of natural changes and entrance retrofitting on numbers and sizes of cave crickets. In: Proceedings of the fifth annual mammoth cave national park science conference, pp 15–24

Silliman B (1851) On the mammoth cave of Kentucky. Am J Sci Arts 11:333–339

Toomey R (2009) Geological monitoring of caves and associated landscapes. In: Young R, Norby L (eds) Geological monitoring. Boulder, CO, Geo Soc of America, pp 27–46

Toomey R, Olson R (2012) Environmental modeling and management for protection of endangered bats in Mammoth Cave. NPS Project Management Information System (#102319)

Trimboli S, Weber R, Ryan S, Toomey RS (2016) An overview of the reverse flow patterns of river styx in mammoth cave, Kentucky: 2009–2012. In: 11th research symposium at mammoth cave national park, pp 128–129

Tuttle M, Kennedy J (2002) Thermal requirements during hibernation. In: Kurta A, Kennedy J (eds) The Indiana Bat: biology and management of an endangered species. Bat Conservation International, Austin, TX

US Surgeon General (1985) The health consequences of smoking: Cancer and chronic lung disease in the workplace: a report of the Surgeon General, Chapter 11, ionizing radiation and lung cancer, pp 442–450. http://profiles.nlm.nih.gov/NN/B/C/B/N/

Webster KD, Etiope G, Drobniak A, Schimmelmann A, Pratt L (2012) Measurement of terrestrial methane concentrations comparable to proposed methane concentrations on Mars. In: International workshop on instrumentation for planetary missions, Abstract #1009, Greenbelt, Maryland, 10–12 Oct. http://www.lpi.usra.edu/meetings/ipm2012/pdf/1009.pdf

Yarborough KA (1980) Radon- and thoron-produced radiation in national park service caves. Nat Radiat Environ III 2:1371–1395

Under Foot: The Paleontology of Mammoth Cave

11

Mona L. Colburn

Abstract

This chapter focuses on the fossils of vertebrates (animals with backbones) found in the caves of Mammoth Cave National Park. Bat remains are the most ubiquitous kind of fossil, but several extinct animals have been identified. The Pleistocene is well-represented. The oldest radiocarbon-dated remains are a deposit from the Sangamon Interglacial Episode. In order to orient the casual reader to the paleontology of Mammoth Cave, the following background is provided:

- brief overview of some basic principles of geology,
- common types of fossils located in the limestone in which the cave is formed,
- vertebrate fossils found outside the boundaries of the park, and
- methods used by cave paleontologists.

The vertebrate fossils are summarized from the first scientific discovery in 1959 to the most recent studies in 2015. The bulk of the evidence is from multi-year paleontological studies of Mammoth Cave conducted from the mid-1990s to the early 2000s. First, the vertebrate fossils are presented by sections of the cave that a visitor might tour. Then, the significant discoveries are presented chronologically in a discussion of Pleistocene environmental change in the central Kentucky region.

Paleontologists' dream
entering
the egress
the paleosearchers
crawl through the twilight
crawling as daylight turns to night
light turns to dark and darkness to light
searching searching
crawling onward
quietly under slumbering bats
searching the dark void
for signs of the past
hoping to discover enlightenment
before exiting the entrance to daylight

M.L. Colburn (✉)
Illinois State Museum, Springfield, IL, USA
e-mail: monacolburn@gmail.com

H.H. Hobbs III et al. (eds.), *Mammoth Cave*, Cave and Karst Systems of the World,
DOI 10.1007/978-3-319-53718-4_11

11.1 Introduction

Caves. Fossils. These words capture the imagination. They conjure up adventure, exploration, maybe a little danger, and the dark mystery of hidden places and unknown pasts. These are the intrigues of studying the paleontological remains in Kentucky's Mammoth Cave National Park (MCNP). Visitors walking the trails are struck by the lush beauty of the woods and the rugged topography. However, this terrain is like the proverbial tip of the iceberg—most of Kentucky's geological story lies hidden in layers of rock hundreds of meters below our feet. For millions of years, geological events have created layer upon layer of sediment and rock. Think of it as a layer cake with the oldest strata (layers) at the bottom.

Geologists are scientists who study Earth's physical characteristics and its formation processes. They have been trained to interpret these strata to tease out the details of Earth's history. The strata may contain evidence of past life that is preserved as fossils (mineralized physical remains), sub-fossils (non-mineralized remains), and trace remains (tracks, burrows, impressions, scat, chemical signatures) of plants, animals, and other life forms (algae, bacteria, fungi). The fossils are studied by specialized geologists called paleontologists. The geologists and paleontologists collaborate with other scientists to understand what the evidence tells us about Earth's ancient geological activity, environments, climates, and the evolution and diversification of life forms.

11.2 Fossils in Kentucky's Rock Layers

Geologists divide Earth's history into a number of major geological eras and numerous subdivisions, which are related by stratigraphy, time, and fossil content. The fossils of Mammoth Cave are from two vastly distant geological periods: the Mississippian Period, which occurred more than 323 million years ago (abbreviated as "Ma"), and the Quaternary Period, which began about 2.6 Ma. The Quaternary is divided into two epochs, the Pleistocene (2.6 Ma to 11,700 years ago) and the Holocene (11,700 years ago to present). We are living in the Holocene. Fossils from other geological time periods do not occur at Mammoth Cave. Dates for geological time periods in this chapter are those in the International Chronostratigraphic Chart version (2015/01) published by the International Commission on Stratigraphy.

Throughout geologic time, Earth's continental plates have tectonically joined and separated, forming and reforming various landmasses. During the Mississippian Period (359–323 Ma), the area of Kentucky was part of the landmass called Euramerica (North America, Greenland, and northern Europe). Kentucky was south of the equator, and it was covered by warm, shallow salt water that was teeming with life (Fig. 11.1). The deep ocean was about 250 miles (402 km) south, and the nearest dry land was far to the east in what is now Virginia. During episodes of land erosion, rivers deposited sand, clay, silt, mud, and gravel into the sea. Calcium carbonate from the bodies of marine animals that fell to the sea bottom accumulated, compacted, and became layers of limestone and dolomite all over the world. Millions of fossil brachiopods, bryozoans, blastoids, corals, sponges, crinoids, and other animals are exposed in the limestone walls, floors, and ceilings of Mammoth Cave. Based on the habitat preferences of animals in the fossil record, we know that the shallow sea was less than 50 feet (15 m) deep.

Many of these fossils are fragments of animals. Crinoids fragmented because their thin stalks and arms were composed of numerous flat cylindrical segments known as columnals (sometimes called Indian beads). Though rare, whole crinoids occur in the cave. If you look closely and long enough in some passages, you might see shark fossils. Usually, sharks are represented by isolated teeth and fin spines. The partial shark skeleton (which is fossilized cartilage) in Mammoth Cave is a rare find. As you read this chapter, passages are forming in lower levels of the cave. Who knows what treasures will be revealed when the water level drops.

Over much time Euramerica became part of the supercontinent Pangaea, which later separated into Gondwana and Laurasia. During the Cretaceous (145–66 Ma), Laurasia was separating into the familiar continents of North America and Eurasia (Europe and Asia, which included China but not India). Earth's tectonic plates continued to move during ensuing geological periods; and, today geologists can find the same fossil species in a particular rock layer separated by thousands of miles. Here are two examples based on conodonts (tooth-shaped fossils from a primitive eel-like chordate). Fossils of *Gnathodus bilineatus* are found in Mississippian Period limestone formations in the Midwestern USA (Willman et al. 1975) and in South China (Qi and Wang 2005). Another species, *Gnathodus texanus*, occurs in the Harz Mountains of Germany (Gradstein et al. 2004), and it occurs in the Paoli limestone of the Visean Stage of the American Midwest (Rexroad and Liebe 1962). In the midcontinent of North America, the *G. texanus* zone ranges from 335–346.7 Ma (Ogg et al. 2016: 108). Thus, we know the approximate age of the Paoli limestone, which is in level B of Mammoth Cave.

Big Bone Lick in northern Kentucky was the first famous Pleistocene (aka "Ice Age") site in the USA. It has been known since the early 1700s. The soft boggy, spring-fed ground preserved the bones of large mammals such as ancient bison (*Bison antiquus*), helmeted muskox (*Bootherium bombifrons*), muskox (*Ovibos*), stag-moose (*Cervalces scotti*), caribou (*Rangifer*), American mastodon

Fig. 11.1 Museum exhibit of the shallow, warm waters of the Mississippian Sea (*Photo* by Doug Carr, Illinois State Museum) and common fossils found in the Mississippian Limestone of Mammoth Cave: *bottom left*, colonial coral at wall and ceiling junction in Long Cave (*Photo* by Mona Colburn); *center*, tricolored bat on a rugose horn coral in Colossal Cave (*Photo* by Rick Olson); *bottom right*, fossil hash containing crinoid and brachiopod fossils exposed in Long Cave ceiling (*Photo* by Rick Olson)

(*Mammut americanum*), woolly mammoth (*Mammuthus primigenius*), and perhaps complex-toothed horse (*Equus* cf. *complicatus*) (Schultz et al. 1963). Cold-adapted megafauna became extinct by the end of the Pleistocene when climatic conditions warmed, continental ice sheets melted, and the environment changed. Some species adapted to the new conditions; others moved north to cooler environments where they could survive. Many of the animal species survived the changing climate and exist today, e.g., modern bison (*Bison bison*), elk, deer, squirrels, small rodents, waterfowl, and turkey.

The first record of fossil bones in a Kentucky cave may be Samuel Brown's 1804/1805 discovery of the skull of a large pig (later identified as the extinct flat-headed peccary) and the bones of a "Great Bos" (cow) in Great Saltpeter Cave. Brown thought that his "Bos" bones matched drawings of bones from a Virginia cave described by President Thomas Jefferson (Wilson 1985). Later, the bones were identified as a species of giant ground sloth that had been named *Megalonyx jeffersonii* to honor President Jefferson. Bones of extinct mammals have been found in more than twenty Kentucky caves located outside of MCNP (Table 11.1).

11.3 Studying Vertebrate Fossils Found in Caves

The stable, low energy, constant temperature conditions of caves are great for preserving biological remains. Because decades of human activities in Mammoth Cave have destroyed cave biota, archaeological artifacts, and paleobiological remains, paleontologists search protected areas that may have escaped trampling and mining. They look for isolated bones, partial and whole skeletons, partial and whole mummified animals, in situ deposits of bones, and trace fossils (e.g., animal scratch marks, footprints, trails, roost staining, and scat). Remains have been found in alcoves, breakdown cavities, sediments, on ledges, and in ancient undisturbed passage fill. Bones have even been found in sediment adhering to walls, ceilings, and the underside of large slabs of breakdown.

Visitors often ask two questions about a fossil: "What is it?" and "How old is it?" These are the very things that cave paleontologists want to know! After finding a fossil or sub-fossil, we record its location on a cave map, take measurements, take a photograph, estimate its developmental

Table 11.1 Extinct mammals found in Kentucky caves located outside of Mammoth Cave National Park (based on FAUNMAP Working Group 1994; Wilson 1980, 1981, 1985)

Common name	Scientific name	County location
Beautiful armadillo	*Dasypus bellus*	Bullitt
Jefferson's giant ground sloth	*Megalonyx jeffersonii*	Franklin, Barren
Dire wolf	*Canis dirus*	Woodford
Giant short-faced bear	*Arctodus simus*	Franklin
Jaguar	*Panthera onca*	Bullitt
Giant beaver	*Castoroides ohioensis*	Trigg
Complex-toothed horse	*Equus complicatus*	Mercer
Horse	*Equus* sp.	Franklin, Woodford
Hay's tapir	*Tapirus haysii*	Scott
Vero tapir	*Tapirus veroensis*	Rockcastle
Long-nosed peccary	*Mylohyus nasutus*	Bell, Logan
Leidy's peccary	*Platygonus vetus*	Caldwell
Flat-headed peccary	*Platygonus compressus*	Boyle, Bullitt, Fulton, Hart, Logan, Rockcastle, Wayne, Woodford
American mastodon	*Mammut americanum*	Barren, Bullitt, Fayette
Mammoth	*Mammuthus* sp.	Woodford

age (juvenile, adult, or old), identify the species, and write a description. Taxonomic identifications are based on morphological characters, measurements, published information, and sometimes by comparing to museum specimens. The degree to which a specimen can be identified varies depending on the kind of material and condition. Some remains and traces cannot be identified to a particular species. For example, bat guano, staining, and many bones may only be identifiable to the bat order, Chiroptera. Although uncommon, a mummified bat can be ideal for identifying species because distinguishing characters such as ears, fur color, forearm length, and feet may be visible. The skull is the best skeletal element for identification; we evaluate size, shape, tooth count, and measurements. Interesting, puzzling, or unique specimens are collected for study and for curation in a museum where they will be available for future research. Information about fossils is computerized, analyzed, and reported.

Paleontologists determine the age of fossils using relative dating methods and chronometric (absolute) dating methods. Rock stratigraphic layers, faunal associations, and contextual clues can help determine the age of one fossil relative to another. This method is often used for fossils found in the limestone. In the best-case scenario, the researcher has an "index" species—one that is characteristic of, or restricted to, a particular limestone formation. For example, rugose corals of the genus *Lithostrotionella* are characteristic of the St. Louis Limestone, and *Pugnoides ottumwa* brachiopods and *Platycrinus penicillus* crinoids are common in Ste. Genevieve Limestone, but rarely occur in St. Louis Limestone (Willman et al. 1975). The coral *Caninia veryi* is restricted to the Girkin Formation (Johnson 2002; Pohl 1970).

Some absolute dating techniques use the known rates of decay of radioactive isotopes to daughter products and calculate their relative amounts in a fossil or rock. Radiocarbon dating is a well-known radiometric dating method that measures the radioactive decay of ^{14}C to ^{12}C and calculates a date based on the ratio found in a specimen. Other radiometric dating techniques are potassium-argon, uranium-thorium, and aluminum-beryllium. In the earlier example of a conodont species found oceans apart, the age of the *Gnathodus texanus* zone was determined with uranium-lead dating. Specialists can analyze, date, and interpret climate records from speleothems (e.g., stalactites, stalagmites, and flowstone). Speleothems contain isotopic carbon, oxygen, and uranium and can provide proxy data for ancient climate and vegetation. Speleothems can be dated by calculating the proportion of uranium to thorium. If fossils are sandwiched between dated layers of flowstone, then those dates bracket the age of the fossils. Some researchers analyze the paleomagnetism of small blocks of sediment. They want to know whether the magnetic grains (which align with magnetic north) in a sediment block show normal polarity (meaning that the sample's magnetic north is the same as it is today) or whether it is reversed (meaning that magnetic north was near the South Pole). The dates of Earth's magnetic reversals have been calculated. The reversal that occurred 780,000 years ago in Mammoth Cave sediments (Schmidt 1982) has the potential to help date a deposit if fossils were found associated with those particular sediments.

Determining the age of remains in caves can be problematic because remains rarely are found in stratified contexts. Often, non-contemporaneous remains occur near each other. A major problem with bones in caves is that of time averaging—this occurs when very little sediment has accumulated, but much time may have passed since an animal or its remains entered the cave. The following hypothetical scenario is an example of the confusion that can be caused by low deposition rates. A free-tailed bat bone associated with guano dated to >48,000 years old might be lying on the cave floor next to an Indiana bat mummy that is 2000 years old. Nearby might be a mouse mummy that was collected five years ago by a woodrat, which in turn might be close to a modern odiferous gray bat that is being scavenged by crickets. Mixing has been caused by foot traffic and moving sediment from one area of the cave to another for trail repairs. Even crickets cause bioturbation of cave sediments. Another problem is that the jumble of limestone breakdown on the cave floor provides many cavities into which fossils can fall. A piece of bone might fall into a cavity and end up resting at a lower level than an older bone that may have fallen in hundreds of years before. All these possibilities mean that researchers must pay close attention to context in order to minimize errors.

Based on aluminum-beryllium dating of quartz crystals in gravels, we know that the final flooding of passages at the Main Cave level took place about 2.3 Ma (Granger et al. 2001). Animal remains lying on the surface are younger, but we want to narrow down the age of remains more precisely than 2.3 million years. Radiocarbon dating can be performed on organic remains (e.g., bone, mummified tissue, bat guano, and raccoon scat); however, the method is not capable of determining age for materials older than about 50,000 years.

Because external sources of carbon can skew results, contaminants must be removed from samples prior to dating. Examples of contaminants are small pieces of charcoal from fires and prehistoric cane torch fragments, soot from prehistoric torches, historic period oil lamps, and thrown cloth torches (a practice that continued on a limited basis into modern times). Thus, materials without a soot coating might postdate the early 20th century; however, old bones covered by sediment can also be soot-free. In addition, because the ratio of ^{14}C to ^{12}C in the atmosphere has not been constant, radiocarbon dates need to be converted (calibrated) to calendar years. For this chapter, uncalibrated radiocarbon dates were converted to calendar years before present (cal BP where BP is the year 1950) at 2 sigma using the INTCAL13 calibration curve program accessed online at Calib.org (Stuiver et al. 2014). Dates in this chapter are given as YBP (years before present) or as years ago. Because it is not possible to radiocarbon date every specimen, there are many unanswered "How old is it?" questions.

11.4 History of Vertebrate Paleontological Research at MCNP

Early scientific interests at Mammoth Cave centered on exploration, biology, and geology; but little attention was given to old bones. In the 1800s, the only fossils noted were those of invertebrate marine animals found in the limestone. Recent research focused on millions of bat remains and fossils of larger animals that were discovered in the cave system. About 1928, Andy Collins (brother of Floyd, the famous cave explorer) found mastodon bones near the entrance to Collins' Onyx Cave (Thompson and Thompson 2003). Around the same time, former cave owner George Morrison made the first written report of animal bones found inside Mammoth Cave when he wrote that Fox Avenue was given its name because bones of three foxes were found near the end of the passage. Forty-six years later, a Cave Research Foundation caver found only a single skeleton of a raccoon. A diligent search in the 1990s located a raccoon skull and the skull and partial skeleton of a second raccoon.

The first scientific investigation was the discovery by Davies and Chao (1959) that the Chief City sediments they were studying were actually part of a massive pile of bat guano. Radiocarbon dating showed that the guano was more than 38,000 years old, which was the limit of radiocarbon dating in the 1950s. The first scientific papers on vertebrate fossils in Mammoth Cave were made as recently as the early 1960s (Jegla 1963; Jegla and Hall 1962). Jegla and Hall identified bones associated with the Chief City guano as Mexican free-tailed bat (*Tadarida brasiliensis*), a species that does not live in Kentucky today. Jegla also reported the accumulations of *Myotis* bat bones found in Bat Cave and the New Discovery Section of MACA. In the late 1970s–early 1980s, Ron Wilson (Carnegie Museum of Natural History in Pittsburgh) and Cave Research Foundation members found isolated bones of extinct Pleistocene mammals in remote areas of the Mammoth Cave system. In the 1990s, the National Park Service and scientists from the Illinois State Museum intensively searched for vertebrate fossils in MCNP caves. We searched major passages in the Mammoth Cave system, re-examined previously known fossil localities, and investigated several small caves and pits (Colburn 2005, 2007, n.d.; Colburn et al. 2000, 2015a b; Toomey et al. 1995, 1998; Widga and Colburn 2015).

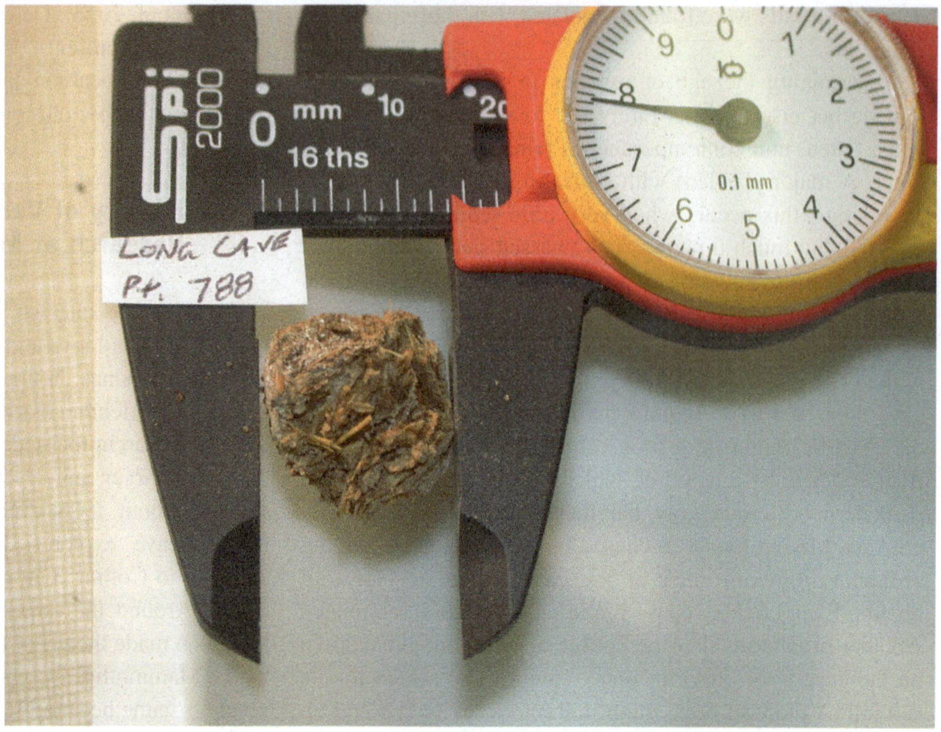

Fig. 11.2 Cross section of raccoon scat full of broken bat bones found in Long Cave (*Photo* by Mona Colburn)

11.5 Paleo Secrets of Mammoth Cave National Park

We continue to make discoveries in these caves that have been visited by millions of people over the last several thousand years. More than sixty different kinds of animals have been identified. Remains of bats are most abundant, followed by those of raccoons (*Procyon lotor*) and woodrats (*Neotoma*). Isolated bones of medium and large-sized extinct Pleistocene mammals have been discovered in a handful of locations.

Bat evidence includes roost stain (left by the oil and dirt on bat feet), bones, guano, mummified remains, and bones in raccoon scat. Based on accumulations of these remains, several long-term or intensively used bat roosts have been identified at specific areas in the cave. Most bat species in the paleontological record still inhabit the region, although their populations were greater in the past.

Raccoon bones are rare finds in the caves, but their feces are common. Raccoon scats inform us about bats because they are full of bat bones (Fig. 11.2). Paleontologists found areas with numerous pieces of scat and fragmented bones from disintegrated scat. We think these accumulations are communal raccoon latrines that were located near hibernacula used by colonial bats of the genus *Myotis*, which hibernate in tight clusters composed sometimes of a few hundred individuals. The colonial species include the following: southeastern bat (*M. austroriparius*), northern long-eared bat (*M. septentrionalis*), gray bat (*M. grisescens*), little brown bat (*M. lucifugus*), and Indiana bat (*M. sodalis*). The last three taxa form the largest clusters, but clusters often contain multiple species. During hibernation, when hundreds of torpid bats were hanging together, they were easily plucked from the walls by hungry raccoons who reached them from ledges and the floor. Raccoon scats are prevalent in the Historic Entrance Area of Mammoth Cave and in Long Cave. Although scats are less common beyond the Gothic Avenue intersection, they can be found as far into Mammoth Cave as Proctors Arcade.

Evidence for woodrats includes footprints, latrines, scattered fecal pellets, trails, larders, middens, scattered debris, nests, and activity areas. Larders and middens can be informative because woodrats like to collect objects, such as plant materials, bat mummies, raccoon scat, and cultural artifacts.

Now, let's look at significant paleontological discoveries. First we will discuss fossils found near tour routes, then, those located in passages not open to the public and in other caves. For information on the historic use by bats please refer to Chap. 17.

11.5.1 Houchins Narrows

The main entrance passage has been trampled by gazillions of human feet, woodrats, mules, and maybe that legendary bear that ran into Houchins Narrows. It was surprising to find animal remains in this highly trafficked passage. Among the bones found were those of frog, chicken, woodrat, mouse, chipmunk, raccoon, and seven kinds of bats: big-eared bat (*Corynorhinus*), big brown bat (*Eptesicus fuscus*), gray bat, small-footed bat (*Myotis* cf. *leibii*), little brown bat, Indiana bat, and tricolored bat (*Perimyotis subflavus*). Undoubtedly, many of these bones are historic. However, along the wall, paleontologists found intact deposits of sediment and bone that predate the presence of humans. *Myotis* humeri (upper arm bones) were radiocarbon dated to be more than 11,000 years old. Thus, even the most impacted passage has paleontological gems.

11.5.2 Audubon Avenue from the Rotunda to Rafinesque Hall and Lookout Mountain

Audubon Avenue is labeled the "Big Bat Room" on Edmund Lee's 1835 map. The name is a clue to what early visitors and miners saw. But what does the paleontological evidence tell us? The Rotunda and Audubon have been subjected to so much impact that few, if any, fossil resources are left on the surface. However, bat remains in Rafinesque Hall and Lookout Mountain tell a different story. Lookout Mountain has a 10,000-year-old deposit of bat bones. They are too fragmented for species identification, but we think they are remnants of degraded raccoon scat. In another part of Lookout Mountain, accumulations of guano along a wall suggest past summertime use by gray bats. Bat stain on the ceiling in Rafinesque Hall indicates heavy use as a summer roost and/or winter hibernation site. Loose, bone-bearing sediments that predate saltpeter mining contain bones of big brown bat and small-footed bat. These two species prefer hibernation sites that are colder than those used by colonial species of *Myotis*. A former opening at Lookout Mountain would have allowed cold, outside air to descend into the cave.

11.5.3 Little Bat Avenue

Dark stain on the ceiling in Little Bat Avenue reflects repeated, heavy roosting by one or more of the colonial *Myotis* species mentioned above. In fact, historic writings of the nineteenth and early twenty centuries noted the presence of hibernating *Myotis* bats, primarily the Indiana bat, and possibly little brown bat and gray bat. Bats identified from bones found in this passage include big brown bat and Indiana bat, and possibly small-footed bat and gray bat. Other bat bones were identified only as *Myotis* or medium-sized *Myotis*. Raccoon scat and claw marks are also present. However, none of the remains is definitely attributable to prehistoric times, most likely because Little Bat Avenue was cleared for saltpeter mining and tourist activities.

11.5.4 Bunker Hill to Olives Bower

Stain in the dome at Bunker Hill signals that it was used as a summer roost and/or hibernaculum; the area below is scattered with numerous bones and remnants of raccoon scat. A mummified medium-sized *Myotis* found on the surface was radiocarbon dated to be about 2000 years old. The wall profiles in the Borrow Pits have intact sediment deposits that contain bat bones below the charcoal-filled prehistoric and historic cultural layers. Roosting stain in Vespertilio Hall and areas of crushed, bone-filled raccoon scat reflect that this was one of the most intensively used *Myotis* hibernacula in Mammoth Cave. The dome was still actively used in the late 1800s when nearly 700 bats were collected in one night (Hovey 1912). Changes since then have rendered the microclimate in this part of the cave warmer than the temperature range of 3–8 °C (37–46 °F) preferred by hibernating Indiana bats.

11.5.5 Rotunda to Acute Angle

This main thoroughfare was severely impacted by historic activities such as saltpeter mining and trail building. Fortuitously, these activities have exposed bone-bearing sediments in the floor that predate the presence of humans in the cave. An old bat hibernaculum in Cyclops Gateway (a side passage) was identified by heavy, faded roost stain. Sediment mining has removed guano and bone.

The oldest passage in the historic section of Mammoth Cave is high in the B-level. It started at the Double Cellars sink and flowed at ceiling height through Broadway. Today, it is no longer a continuous passage having been severed by breakdown. Remnants of the old passage are the high-level side passages named Blue Springs Branch, Blackall Avenue, Gothic Avenue, Backsliders Alley, and Methodist Church. Despite the heavy human impact of 1812-era saltpeter mining, sediment mining for trail building, and tours to view Gothic Avenue formations, important paleontological discoveries were made in these high-level passages.

Gothic Avenue and Backsliders Alley are located across from each other and intersect Broadway at the Saltpeter Vats. They are joined by a narrow ledge, but are separated by breakdown from other remnants of the high-level passage that lie upstream (Blackall Avenue and Blue Springs Branch) and downstream (Methodist Church). Paleontologists found important fossils and intact sediments, such as in situ sand and gravel layers, tightly packed reddish cave fill, sandy deposits, and thin layers of beige-colored silts and clays. The fine-grained sands and thin laminae of silt and clay were deposited by slow-moving or still waters during episodes of back-flooding. Sediment containing bat bones still adheres to the walls and ceilings. It is evidence from a time when the passage was full of sediment.

The majority of remains in Gothic Avenue date to the Holocene epoch (11,700 years ago to present), but it also has evidence for bat use during late Pleistocene interstadials. Interstadials were warm periods usually of shorter duration than interglacials. Thin strata of old bat guano beneath the gray, sooty cultural surface have been exposed by the tourist trail. In situ guano is 28,000 years old, which corresponds to a short-term warm period called the Farmdalian interstadial. Bats also used Gothic Avenue 12,600 years ago, during the warm, moist Bolling–Allerod interstadial. Extensive roosting stain on the ceiling of Gothic Avenue, particularly beyond the connection to Gratz Avenue, is of an unknown age. It most likely was created by myotis bats. Today, live myotis bats, tricolored bats, red bats, and fresh guano pellets occur in numbers far below that indicated by the heavy stain. Bones of Mexican free-tailed bats found in cave fill were transported in sediment-laden water.

A femur fragment of the extinct bat, Stock's vampire bat (*Desmodus stocki*), and thousands of brittle white bat bones were found with fine-grained fill sediments in Backsliders Alley. Remains of species that still inhabit Kentucky were also identified: big brown bat, up to five species of *Myotis*, tricolored bat, red bat (*Lasiurus borealis*), hoary bat (*L. cinereus*), evening bat (*Nycticeius humeralis*), woodrat, raccoon, mouse, snake, and hellbender salamander (*Cryptobranchus*). It is hypothesized that the Backsliders Alley assemblage is one of the oldest in Mammoth Cave because it is associated with fine, thinly laminated sediment left by back-flooding of the ancient passage. The bones are younger than the gravels deposited in Methodist Church 2.3 Ma, but they failed to yield a radiocarbon date.

This discovery is remarkable not only because Stock's vampire bat is extinct but also because modern vampire bats live outside the USA. Paleontologists use ecological requirements of modern species as an analogue for fossil taxa. Because modern vampire bats are not adapted to cold climates and do not hibernate, they are limited to regions with winters warmer than 10 °C (50 °F) (McNab 1973, 1974) like Mexico, Central America, and South America. Thus, Stock's vampire bat inhabited Kentucky when it was warmer (especially during winter) and conditions ranged from temperate to subtropical. Stock's vampire bat probably lived during one of the many interglacial episodes (long periods of warming between episodes of glaciation) that occurred periodically throughout Mammoth Cave's 10-million-year evolution. The last interglacial took place 130,000 to 71,000 years ago (Lisiecki 2014; Lisiecki and Raymo 2005). Geologists know it as Marine Isotope Stage 5 (MIS 5); it is commonly called the "Sangamon." Temperatures were warmest between 130,000 and 119,000 years ago during the earliest substage, MIS 5e. The peak of warmth was about 123,000 years ago. Sea levels were as much as 6 meters (almost 20 feet) higher than present, and temperatures

Fig. 11.3 Femurs of extinct Stock's vampire bat (*Desmodus stocki*) found in an ancient upper level side passage in Mammoth Cave (*top*) and modern vampire bat (*Desmodus rotundus*) from Illinois State Museum Collection (*bottom*) (*Photo* by Mona Colburn)

were about 2 °C (3.6 °F) warmer than today. *D. stocki* was larger than the modern vampire bat (*D. rotundus*) (Fig. 11.3), which may have allowed it to tolerate cooler conditions than modern forms. We don't know whether a colony of Stock's vampire bats lived in or near Mammoth Cave or whether the specimen was a lone vagrant. Stock's vampire bat fossils have been found in a dozen US localities in Arizona, California, Florida, New Mexico, Texas, and West Virginia. These include both late Pleistocene and Holocene sites that range from 120,000 to 3,000 years old (Grady et al. 2002, Ray et al. 1988).

11.5.6 River Acheron

In March 2010, Park Ecologist Rick Olson found two vertebrae in this lower-level passage that were identified by Dr. Rick Toomey as American black bear (*Ursus americanus*). The bones are jammed into a bedrock crack next to a small waterfall and are presumed to have been transported to their present location by flood water. The bones have not been radiocarbon dated.

11.6 Acute Angle to Cataracts

The Main Cave passage between Acute Angle and S Bend has been impacted by miners, tourists, and patients living in the TB Huts. Bat bones, historic food debris, and numerous soot-covered, mummified bats have been found in less exposed, protected places. A 2600-year-old human paleofeces containing bat bones was an interesting archaeological find. Today, few bats roost in the large area composed of Wrights Rotunda, Black Chambers, and the Dark Room. However, numerous guano accumulations demonstrate that the area was utilized multiple times. Radiocarbon-dated guano indicates that colonial bats have used the area throughout the last 10,700 years of the Holocene. In addition, a large mass of in situ guano covered by a large block of breakdown proved to be the most interesting discovery in Wrights Rotunda. It is a Pleistocene deposit that is 46,000 year old! We don't know what species of bat made this deposit because no identifiable bone was found with the guano. It is likely that large numbers of bats, of multiple species, used various areas of the passage for thousands of years.

Well beyond the bat roost areas—near the terminus of the passage at Sandstone Dome—is an area with loose gray-colored sediment containing burned cane and bones of mink (*Neovison vison*) and big brown bats. A former opening to the surface at Sandstone Dome would have provided a suitable microclimate for the big brown bats, an entrance for bats accessing the roosts in Black Chambers, and perhaps an entrance for prehistoric Native Americans. The evidence suggests that Wrights Rotunda, Black Chambers, and the Dark Room had more bat-friendly microclimates in the past.

Fig. 11.4 *Dark brown* free-tailed bat guano among breakdown exposed in a Chief City excavation unit. *Note* trail dirt that was applied over the breakdown to make a more even walking path (*Photo* by Mona Colburn)

11.6.1 Cataracts to Haines Dome

From Bryans Pass to just past Haines Dome, there are areas with thousands of well-preserved bones and extensive deposits of guano from Mexican free-tailed bats. The largest such accumulation is in Chief City and smaller but sizable deposits occur at Mummy Ledge and Haines Dome. The guano occurs in a variety of forms: fine powder, pellet shapes that disintegrate when touched, dark brown strata exposed in excavation profiles, and solid masses that are ebony-colored and granular. As the Lantern Tour ascends and descends the hill at Chief City, visitors are actually walking on a pile of breakdown and mahogany-colored guano (Fig. 11.4). Originally the guano (discovered by Davies and Chao 1959) was attributed to *Myotis* bats, but later to Mexican free-tailed bats when bones of the latter were found (Jegla and Hall 1962). Some of the bones are larger than modern specimens; they are being studied to determine whether they represent the extinct Constantine's free-tailed bat (*Tadarida constantinei*) found in the American southwest. Bones of big brown bat, hoary bat, red bat, gray myotis, small-footed myotis, tricolored bat, big-eared bat, short-tailed weasel (*Mustela erminea*), deer mouse (*Peromyscus*), and woodrat were also identified from Chief City. The free-tailed bat bones are stained reddish-brown, but most other bones are beige-colored (Fig. 11.5). The color difference may indicate the other animals were not contemporaneous with the free-tailed bats. When free-tailed bats lived in Chief City, high ammonia levels would have rendered the area inhospitable to other mammals; however, the guano would have provided rich nutrients for insect communities.

Beyond Chief City are two high-level side passages that are remnants of the ancient upper passage mentioned earlier. Blackall Avenue and Blue Springs Branch are located across from each other and intersect the Main Cave passage at Ste. Catherine City. Nothing significant was located in Blackall. However, bones and associated guano indicate that a small group of free-tailed bats roosted in Blue Springs Branch. No evidence of them roosting in other areas of the ancient upper level passage has been found.

Based on characteristics of modern free-tailed bat maternity colonies, the Chief City location with its massive guano deposit and high ceiling is thought to have been a maternity roost. The discovery of incompletely fused bones of juvenile free-tailed bats supports this. The configurations of the passages beyond each end of Chief City are not typical roosting sites, but they too contain extensive deposits of free-tailed guano and bones. Juveniles and nursing females would have occupied the high-domed areas where it was warmest, and males and other females would have been on the periphery of the clusters of young and in the adjoining passages.

We don't know the age of the free-tailed bat deposits. Guano has been radiocarbon dated multiple times (>38,000 YBP in the 1950s; >54,000 YBP in the 1990s; and >48,000 YBP in 2010), but each time, the samples were beyond the assessment ability of radiocarbon dating. The absence of water-lain sediments overlying the free-tailed bat deposits

Fig. 11.5 Bones from an excavation unit in Chief City (*Photo* by Mona Colburn)

indicates that bones and guano accumulated after the final flooding of the Main Cave level 2.3 Ma. Thus, the deposits are older than 48,000 years but younger than 2.3 million years.

It is very interesting that Mammoth Cave had a large, long-term population because the species is not considered a resident in present-day Kentucky. Modern populations live in tropical, subtropical, and warmer parts of the temperate zones because they need warmer year-round temperatures and warmer winters. The largest colony today lives in Texas. The subspecies of free-tailed bats found in the southeastern USA do not live in caves and have smaller colonies. Most of the subspecies in the southwestern USA are migratory. In winter, some fly more than 1000 miles (>1600 km) to Mexico, Central America, and South America and return to their summer roosts. Free-taileds in the southeastern USA and some in California do not migrate, they remain year-round. We don't know whether Mammoth Cave's free-taileds were migratory or year-round residents. Free-tailed fossils have been found in Florida and the American Southwest.

11.6.2 Anzers Hall to Violet City

Today, few bats use Anzers Hall, but thin lenses of guano compressed by large slabs of ceiling breakdown indicate that bats roosted here in the past. Numerous bones of small-footed and big brown bats were found in this part of the cave. Species identifications and bone dates suggest that about 14,600 years ago, a former entrance, a different passage configuration, and/or colder surface temperatures made the area colder than today. This time frame coincides with the early years of the Bolling–Allerod interstadial. Between Ultima Thule and the Violet City Entrance is a vast area of watery shafts and pits. Although there are intriguing rooms deep in the breakdown piles, the area is too wet for preservation of organics.

11.6.3 Carmichael Entrance to Frozen Niagara Entrance

Interesting fossils were discovered near the Grand Avenue Tour route. In 1978, CRF cavers and Wilson (1980) collected a "right tibia" of black bear from Sophys Avenue. In 2001, Rick Olson, working with the Illinois State Museum Paleontological Inventory Project, found a left tibia of black bear that was well-camouflaged by limestone rubble and flowstone, under a low overhang. At the time, it was assumed that the two tibiae were from the same individual. In order to resolve the matter, Colburn visited the Carnegie Museum in Pittsburgh, Pennsylvania, to view their collections from Mammoth Cave—both bones are left tibiae! The bears may have denned in the passage, or only some of their bones may have entered by way of Hunts Sink and been transported by water to Sophys Avenue.

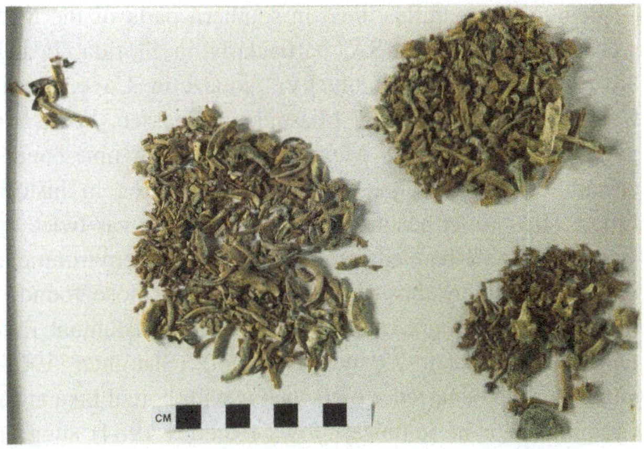

Fig. 11.6 Bones excavated from the Frozen Niagara Section of Mammoth Cave (*Photo* by Gary Andrashko, Illinois State Museum)

In the Frozen Niagara Section, fossils were excavated from beneath a layer of flowstone, which Uranium-thorium analysis indicated to be 125,000 years old (Colburn et al. 2000). The deposit includes bones of frog, salamander, turtle, snake, lizard, bird, bats, small rodents, pocket gopher (*Geomys*), raccoon, deer, Pleistocene horse (*Equus*), beautiful armadillo (*Dasypus bellus*), peccary (*Platygonus*), and Leonard's water rat (*Neofiber leonardi*) (Fig. 11.6). The last four animals are extinct. The following paragraphs describe the most significant taxa.

Pocket gophers do not inhabit Kentucky today. Their bones have been found in other Kentucky caves, including Welsh (Woodford County) and Savage Cave (Logan County). Plains pocket gophers (*G. bursarius*) live in the Great Plains and as far east as western Indiana, and southeastern pocket gophers (*G. pinetis*) live in the southeastern USA in Alabama, Georgia, and Florida. Mammoth Cave's Pleistocene-aged pocket gophers were burrowing herbivores that ate grasses and forbs above ground, and roots and tubers below ground. Depending on the species, they would have lived in an open habitat of grassland and prairie, or in the sandy soils of an open pine forest.

Horses have a long evolutionary history. They appeared in North America about 50 Ma. The early forms were small, about the size of a fox. The earliest horse fossils of the genus *Equus* occur at a 3.5-million-year-old site in Idaho. Paleontologists used to think that there were numerous species of horses, but recent genetic work (Weinstock et al. 2005) suggests there were only two species with great morphological and geographic variation. They were about the size of zebras and ponies and grazed in open grassland or savanna. Native horse species died out in North America by the end of the Pleistocene.

Beautiful armadillos lived in southern parts of the Midwestern and eastern USA, particularly in Florida. Fossils have been found in Kentucky (A-maze-in Cave, Bullitt County), Illinois, Indiana, Missouri, Tennessee, West Virginia, and eastern New Mexico. The modern nine-banded armadillo arrived in the Mammoth Cave area in historic times. The extinct beautiful armadillo (which was twice the size of the nine-banded) may have preferred environments with mild winters that were no colder than those found in modern-day northern-central Texas, and with annual rainfalls of more than 20 inches (0.5 m) (Slaughter 1961). However, the occurrence of *D. bellus* with boreal taxa at the Craigmile site in southwest Iowa (Rhodes 1984) suggests that it may have been more cold tolerant. Modern armadillos burrow, but recent research suggests that the extinct form did not (Jasinski and Wallace 2014).

Extinct peccaries were open habitat herd animals whose remains are common and widespread in the North American fossil record. The flat-headed peccary appears to have used caves intentionally. They were gregarious, and cave localities often contain multiple individuals. For example in Kentucky, at least 31 individuals are represented from Welsh Cave (Woodford County) (Guilday et al. 1971) and 24 from Toolshed Cave (Bullitt County) (Wilson 1985). Other Kentucky localities with bones of flat-headed peccary are Proctor Cave in MCNP; Wells Cave (Boyle County); Chickasaw Bluffs and Hickman (Fulton County); Lone Star Peccary Cave and Granny Puckett Cave (Hart County); Savage Cave (Logan County); Great Saltpeter Cave (Rockcastle County); and an unnamed cave in Wayne County.

The ecological requirements of the extinct Leonard's water rat may have been similar to those of modern *Neofiber alleni* (the Florida water rat or round-tailed muskrat). *N. alleni* is subtropical and lives in shallow, grassy marshes in Florida and extreme southeastern Georgia (Birkenholz 1972). A viable population of Leonard's water rat needed a shallow, fairly permanent body of water, which did not have to be in the immediate vicinity. An owl or carnivore could have brought the bone to an area above ground near the Frozen Niagara location. Fossils have been found in Pleistocene sites in Kansas, Texas, Florida, and West Virginia—but not in Kentucky, until now.

11.6.4 New Discovery

In 1999, paleontologists inventoried Fossil Avenue because it was a location known to have skeletal material of little brown bats (Jegla 1963). The floor of Fossil Avenue is a maze of small rimstone dams that are filled with sediment, limestone grains, cave popcorn, and thousands of bat bones. Bones of little brown bat, Indiana bat, gray bat, and tricolored bat are the most abundant. Bones of spotted skunk, least weasel, raccoon, woodrat, and deer mouse were also identified. Pellets of bat guano are scattered throughout the passage, but no roosting stain was observed. The conclusion is that the bats were not roosting in the passage, but that the bones were deposited by flood waters and trapped by the rimstones. The remains have not been dated.

11.6.5 Proctor Cave

This part of the Mammoth Cave system required hardy cave explorers to discover its Pleistocene fossils. During a 24-hour trip to remote passages in the 1980s, cavers found isolated bones of extinct Pleistocene mammals—giant short-faced bear (*Arctodus simus*), Vero tapir (*Tapirus veroensis*), flat-headed peccary (*Platygonus compressus*), and extinct elephant (Proboscidea). In the 1990s, Illinois State Museum scientists (Rick Toomey and Mona Colburn) and Cave Research Foundation cavers re-explored portions of Proctor. They found saber-toothed cat (*Smilodon* sp.) and additional flat-headed peccary and proboscidean remains. Later, proboscidean expert, Dan Fisher (University of Michigan) identified the tusk fragments as American mastodon. The mastodon fossils were located in the talus rubble of a collapsed sink. The giant short-faced bear, saber-toothed cat, and peccary bones were in lower-level stream channels; they are probably redeposited materials.

Elsewhere in Proctor were bones of smoky shrew (*Sorex fumeus*), American marten (*Martes americana*), and hundreds of gray bats. Smoky shrews live in the park today, but the marten no longer inhabits Kentucky. It prefers habitat with spruce and pine trees and lived as far south as Alabama and Tennessee during the Pleistocene. The gray bat bones indicate that a sizable gray bat roost was located in Proctor. This is interesting because gray bats do not roost in the passage today. Because there is no nearby opening to the surface, the bats may have entered through the sink before it collapsed.

11.6.6 Long Cave

A paleontological survey of Long Cave identified the remains of nine bat species (big brown bat, red bat, possible southeastern bat, gray bat, small-footed bat, little brown bat, Indiana bat, tricolored bat, and big-eared bat). Raccoon (represented by numerous claw marks on the walls and bat bone-filled scat), woodrat, and deer mice (Colburn n.d.) were also identified. All these taxa live in Kentucky today. Woodrats are still active in the cave, but no fresh sign of raccoon was found. A small colony of gray bats roosts here in summer, and *Myotis* species hibernate in winter. Although most radiocarbon-dated remains in Long Cave are less than 3000 years old, they indicate that *Myotis* species hibernated in Grand Avenue, Briggs Avenue, and Lee Avenue in far greater numbers in the past.

11.6.7 Bat Cave

This name seems to be a misnomer for the cave that we see today. In summer, a small colony of gray bats is present. In winter, the hibernating bat population consists of a few hundred individuals composed of tricolored bats and multiple species of *Myotis*. Prehistorically, the case may have been quite different. Bat Cave contains a large deposit of bat bones that measures about 34 feet long by 4 feet wide by 2 feet deep (10.4 m × 1.2 m × 0.6 m) (Jegla 1963). Recent research determined that there are at least 11 natural sediment levels containing bat bones (Colburn et al. 2015b). Identified bones include those of bats (small-footed myotis, little brown bat, Indiana bat, big-eared bat, big brown bat) and other animals (raccoon, deer, mouse-sized rodent, rat-sized rodent, plethodontid salamanders, indeterminate mammal, and indeterminate fish). Medium-sized *Myotis* dominate the assemblage. Radiocarbon dates confirm that multiple depositional events occurred between 2160 and 10,800 years ago.

11.7 What Mammoth Cave's Vertebrate Fossils Tell Us About Past Environments

The Mammoth Cave system has been forming for about 10 million years. During this span of time, different ecosystems, plants, and animals have danced upon the landscape. The oldest fossils suggest ecosystems that were temperate, and perhaps subtropical, and the large Pleistocene mammals indicate glacial conditions. Bats were most abundant during warm periods like interglacials and interstadials.

The fossil assemblage discovered in the Frozen Niagara Section is quite significant not only because it has extinct species, but more importantly because the flowstone is datable. Capped by 125,000-year-old flowstone, the assemblage dates to the earliest substage of the Sangamon interglacial (MIS 5e) and is no older than 130,000 years. Pollen and faunal data from the Pittsburg Basin and Hopwood Farm sites in Illinois indicate that the midcontinent was hotter and drier than present, with mild winters. The Frozen Niagara taxa support these conditions. Water rats indicate warm conditions; extinct peccaries were open habitat herd animals; Pleistocene horses reflect grassland or savanna, and pocket gophers could have inhabited grassland, prairie, or open pine forest. The presence of both Leonard's water rat and pocket gophers in the Frozen Niagara assemblage suggests the area was a mosaic of habitats.

The fossils of extinct Stock's vampire bat and free-tailed bat colonies are very exciting because both species indicate environments that were warmer than present-day Kentucky, with winters above freezing. Bones found in Backsliders Alley (including the Stock's vampire bat femur) appear to be quite old, but we really don't know how old. The free-tailed bat remains display excellent preservation, but we don't know their age, either. The sheer number of free-tailed bones and the thickness and extent of their guano deposits indicate that a very large population used Mammoth Cave for a long time. If the Mammoth Cave free-taileds were migratory, successive generations would have returned to Mammoth Cave each year. Whether they were migratory or year-round residents, Mammoth Cave's large population would have required a lot of food.

Bat Conservation International estimates that the summer colony of 20 million free-tailed bats in Bracken Cave, Texas, may eat 200 tons of insects per night. Thus, a warmer regional climate was necessary not only for warmer cave temperatures but also for the large populations of moths, beetles, and other insects needed to sustain the colony. One or more past interglacials or interstadials could have provided suitably warm climatic conditions for Stock's vampire bats and free-tailed bats. Both may have inhabited central Kentucky during the Sangamon interglacial. However, because modern vampire bats require warmer and more humid conditions than free-tailed bats, Stock's vampire bat and free-taileds were not necessarily here at the same time. Without better dating techniques, we can only say that free-tailed bats inhabited Mammoth Cave for a lengthy, unknown span of time during one or more of the warm periods that occurred between 48,000 YBP and 2.3 Ma. Likewise, the Stock's vampire bat fossil might date to the Sangamon interglacial, but it could be older.

The bones of large extinct Pleistocene mammals (giant short-faced bear, saber-toothed cat, Vero tapir, flat-headed peccary, and mastodon) that were found in Proctor Cave have not been radiocarbon dated. However, they may date to the late Pleistocene time when conditions were cold and dry, like those in modern-day northern Michigan, Wisconsin, Minnesota, or Canada. Trees at that time would have been boreal taxa like spruce, jack pine, firs, birch, elm, and poplar. As environmental conditions warmed, glaciers shrank northward and deciduous trees replaced the cold climate conifers. These mammal species went extinct by the end of the Pleistocene 11,700 years ago.

Other evidence for large numbers of bats in Mammoth Cave during the Pleistocene was found in Wrights Rotunda (dated to 46,000 years ago) and in Gothic Avenue (dated to 28,000 years ago during the Farmdalian interstadial). Big brown bat and small-footed myotis used Anzers Hall about 14,600 years ago, and bats roosted in Gothic Avenue 12,600 years ago; these times coincide with the beginning and the end of the Bolling–Allerod interstadial, respectively. The Bolling–Allerod was followed by a return of cold climate that lasted until the Holocene interglacial circa 11,700 YBP.

Multiple lines of evidence show that several passages in Mammoth Cave were important bat roosts during the Holocene. Colonial bats (especially little brown myotis and Indiana bats) hibernated in Audubon Avenue, Little Bat Avenue, Rafinesque Hall, Lookout Mountain, Bunker Hill, Vespertilio Hall, Gothic Avenue, Cyclops Gateway, Star Chamber, and Proctors Arcade. Most *Myotis* remains in Mammoth Cave are less than 11,320 years old. In addition, thousands of medium-sized *Myotis* bones were deposited at multiple times between 2160 and 10,800 years ago in Bat Cave where they accumulated into a large bone bed. Perhaps regional conditions were not warm enough to support large populations of colonial species until several hundred years into the Holocene. *Myotis* bones associated with the vampire bat sediments and with free-tailed bat deposits are much older and predate the Holocene. Gray bats may have roosted in Lookout Mountain during the Holocene. Large populations of an undetermined colonial bat species used Wrights Rotunda and Black Chambers during the Holocene and the Pleistocene. Big brown bat and mink fossils suggest that Sandstone Dome was open in the past. An entrance would have allowed colonial bats to access Wrights Rotunda and Black Chambers.

11.8 Conclusions

Sandstone and limestone buffer the passages in Mammoth Cave from short-term changes in weather. However, the cave system did not escape major regional and global changes in climate. Changes in paleoclimate, plants and animals, passage configurations, and flood events have created a complex record to interpret. The remains of sixty different kinds of vertebrates discovered in the caves of Mammoth Cave National Park span a broad range of time and indicate the presence of widely varied ecosystems. Remains have been radiocarbon dated to the Pleistocene and Holocene. Bat remains dominate the fossil material and provide evidence that bats have inhabited Mammoth Cave for 130,000 years and perhaps longer. During that time, species composition changed. Millions of free-tailed bats inhabited Mammoth Cave during the Pleistocene. Although evidence shows that free-tailed bats were longtime residents of Mammoth Cave, its fossils have not been found elsewhere in Kentucky. Various *Myotis* species found in present-day Kentucky used to hibernate and roost by the millions in the park caves, but they do so no longer. Based on location and ecological requirements, bones in Backsliders Alley (Stock's vampire bat and other animals) may be the oldest fossils in Mammoth Cave; however, attempts to radiocarbon date bones from this location have been unsuccessful. Mammoth Cave is the only Midwestern location for Stock's vampire bat, and it is the second most northerly occurrence east of the Mississippi River.

The significance of the Frozen Niagara fossil assemblage lies in the ability to date it to the early Sangamon interglacial (MIS 5e). There are fewer than a dozen early Sangamon-aged faunal sites in eastern North America. Chief City is notable for being the largest fossil free-tailed bat site in the eastern USA, and it has been assessed as a maternity colony. Significant taxonomic discoveries include not only the free-tailed bats but also several extinct mammals—mastodon, beautiful armadillo, Pleistocene horse, saber-toothed cat, short-faced bear, Stock's vampire bat, and Leonard's water rat. Mammoth Cave is the only locality in Kentucky where free-tailed bats, Stock's vampire bat, saber-toothed cat, short-faced bear, and Leonard's water rat have been discovered.

References

Birkenholz DE (1972) *Neofiber alleni*. Mammalian Species No. 15:1–4 Amer Soc of Mammalogists

Colburn ML (2005) Paleontological inventory project: vertebrate remains found in select passages and caves at Mammoth Cave National Park, Kentucky. Report submitted to the National Park Service, July 2005, Illinois State Museum Landscape History Program, Technical Report No. 2005-1199-007. On file at Mammoth Cave National Park

Colburn ML (2007) Vertebrate remains from 4 Localities at Mammoth Cave National Park, Kentucky. Master's Thesis in Environmental Studies program, University of Illinois-Springfield

Colburn ML (n.d.) Paleontological inventory project: Vertebrate faunal remains found in Long Cave, Mammoth Cave National Park, Kentucky. Draft manuscript, Illinois State Museum

Colburn ML, Toomey R III, Olson R, Dorale J (2000) An important interglacial fauna from sediments of the Frozen Niagara Entrance, Mammoth Cave National Park—a preliminary report. In: Proceedings of Mammoth Cave National Park's eighth science conference, pp 23–24

Colburn ML, Keenan D, Widga C (2015a) FTIR spectroscopy of Pleistocene-aged Mexican free-tailed bat guano from Chief City, Mammoth Cave, KY. Poster presented at the 2015 National Speleological Society Convention, Biology Session, July 15, Waynesville, MO. Available on ResearchGate.net

Colburn ML, Toomey R III, Widga C, Olson R (2015b) Holocene paleontology of Bat Cave, Edmonson County, Kentucky, USA. J of Cave and Karst Stud 77(2):91–98

Davies WE, Chao ECT (1959) Report on sediments in Mammoth Cave: U.S. Department of Interior, U.S. Geological Survey Administrative Report

FAUNMAP Working Group (1994) FAUNMAP: a database documenting late Quaternary distributions of mammal species in the United States. Illinois State Museum Scientific Papers 25(1–2): 1–690

Gradstein FM, Ogg JG, Smith AG (eds) (2004) A geologic time scale. Cambridge University Press, Cambridge

Grady F, Arroyo-Cabrales J, Garton ER (2002) The northernmost occurrence of the Pleistocene Vampire Bat *Desmodus stocki* Jones

(Chiroptera: Phyllostomatidae: Desmodontinae) in Eastern North America. In: Robert J. Emry (ed.), Cenozoic Mammals of Land and Sea: Tributes to the Career of Clayton E. Ray Smithsonian Institution Press, Washington, DC, pp 73–75

Granger DE, Fabel D, Palmer AN (2001) Pliocene-Pleistocene incision of the Green River, Kentucky, determined from radioactive decay of cosmogenic ^{26}Al and ^{10}Be in Mammoth Cave Sediments. Geo Soc of America Bul 113(7):825–836

Guilday JE, Hamilton HW, McCrady AD (1971) The Welsh Cave peccaries (*Platygonus*) and associated fauna, Kentucky Pleistocene. Annals Carnegie Museum 43(9):249–320

Hovey HC (1912) Mammoth Cave of Kentucky with an account of Colossal Cavern. John P. Morton & Company, Louisville

International Commission on Stratigraphy (ICS) (2015) International chronostratigraphic chart version 2015/01, http://www.stratigraphy.org Retrieved 7 Dec 2015

Jasinski SE, Wallace SC (2014) Investigation into the paleobiology of *Dasypus bellus* using geometric morphometrics and variation of the calcaneus. J of Mammalian Evolution 21(3):285–298

Jegla TC (1963) A recent deposit of *Myotis lucifugus* in Mammoth Cave. J Mammology 44(1):121–122

Jegla TC, Hall JS (1962) A Pleistocene deposit of free-tailed bat in Mammoth Cave, Kentucky. J Mammalogy 43:444–481

Johnson WK (2002) A study of the Chesterian Girkin Formation Gasperian. Geo Soc of Amer, Abstracts with Programs 34(2):40

Lisiecki LE (2014) Ages of MIS boundaries. Retrieved 6 July 2014 from http://lorraine-lisiecki.com

Lisiecki LE, Raymo ME (2005) A Pliocene-Pleistocene stack of 57 globally distributed benthic delta18O records, Paleoceanography 20, PA1003, doi:10.1029/2004PA001071

McNab BK (1973) Energetics and the distribution of vampires. J of Mammalogy 54:131–144

McNab BK (1974) The behavior of temperate cave bats in a subtropical environment. Ecology 55:943–958

Ogg JG, Ogg G, Gradstein FM (2016) A concise geologic time scale. Elsevier

Pohl ER (1970) Upper Mississippian deposits in South Central Kentucky. Kentucky Academy of Sciences Transactions 31

Qi Y, Wang Z (2005) Serpukhovian conodont sequence and the Visean-Serpukhovian boundary in South China. Riv Ital Paleontol Stratigr 111(1):3–10

Ray CE, Linares OJ, Morgan GS (1988) Paleontology. In: Greenhall AM, Schmidt U (eds) Natural History of Vampire Bats. CRC Press, Boca Raton, pp 19–30

Rexroad C, Liebe R (1962) Conodonts from the Paoli and equivalent formations in the Illinois Basin. Micropaleontology 8(4):509–514

Rhodes RS II (1984) Paleoecology and regional paleoclimatic implications of the Farmdalian Craigmile and Woodfordian Waubonsie mammalian local faunas. Illinois State Museum Reports of Investigations 40

Schmidt VA (1982) Magnetostratigraphy of sediments in Mammoth Cave, Kentucky. Science 217(4562):827–829

Schultz CB, Tanner LG, Whitmore FC Jr, Ray LL, Crawford EC (1963) Paleontological investigations at Big Bone Lick State Park, Kentucky: A preliminary report. Science, New Series 142(3596): 1167–1169

Slaughter BH (1961) The significance of *Dasypus bellus* (Simpson) in Pleistocene local faunas. Texas J of Science 13(9):311–315

Stuiver M, Reimer PJ, Reimer R (2014) CALIB Radiocarbon calibration. http://calib.org/calib

Thompson B, Thompson J (2003) Mammoth Cave and the Kentucky Cave Region. Arcadia Publishing, Charleston

Toomey RS III, Colburn ML, Schubert BW (1998) Late Holocene utilization by bats of the Mammoth Cave Historic Entrance area, Mammoth Cave National Park, Kentucky. Report submitted to Mammoth Cave National Park in partial fulfillment of Cooperative Agreement No. 1443 CA5530 97 001

Toomey RS III, Ward RH, Olson RA, Colburn ML (1995) Cooperative cave paleontological resource management, Mammoth Cave National Park, Kentucky. 1995 National cave management symposium proceedings. Indianapolis, In: Karst Conservancy, pp 291–296

Weinstock J, Willerslev E, Sher A, Tong W, Ho SYW, Rubenstein D, Storer J, Burns J, Martin L, Bravi C, Prieto A, Froese D, Scott E, Xulong L, Cooper A (2005) Evolution, systematics, and phylogeography of Pleistocene horses in the New World: a molecular perspective. PLOS Biol 3:1373–1379

Widga C, Colburn M (2015) Paleontology and paleoecology of guano deposits in Mammoth Cave, Kentucky, USA. Quaternary Research 83(3):427–436

Willman HB, Atherton E, Buschbach TC, Collinson C, Frye JC, Hopkins ME, Lineback JA, Simon JA (1975) Handbook of Illinois Stratigraphy. Illinois Sate Geological Survey Bul 95:140–158

Wilson RC (1980) Letter dated 13 June 1980 to John E. Guilday and catalogue [a list] of bones collected in Mammoth Cave National Park between 1978 and 1980 by the Carnegie Museum of Natural History

Wilson RC (1981) Extinct vertebrates from Mammoth Cave. Proceedings of the Eighth International Congress of Speleology, p 339

Wilson RC (1985) Vertebrate remains in Kentucky caves. Caves and Karst of Kentucky. Kentucky Geological Survey, Special Publication 12:168–175

12 Landscape Ecology of Mammoth Cave: How Surface and Cave Ecosystems Influence Each Other

Rickard A. Olson

Abstract

At first, the surface of Mammoth Cave National Park and the caves below may seem like two different landscapes or even different worlds, but these surface and underground spaces interact in ways that are sometimes surprising. The roughly 645 documented kilometers (400 miles) of Mammoth Cave are part of a regional karst landscape, which is defined by subterranean drainage. From the southeast to the northwest corners of the park, there is a gradient of decreasing extent of karst development, which corresponds to the regional dip of the bedrock. The major cave-bearing carbonates (limestone and dolomite) are barely exposed in the northwestern part of the park, and so cave development there is in the earliest stages. The hydrogeology and landforms of this highly varied karst landscape have profound effects upon vegetation ecosystems. Surface rivers, cave aquatic, and cave terrestrial ecosystems are all largely dependent upon food energy from photosynthesis in the vegetation ecosystems. Therefore, this chapter will examine physical aspects of the landscape that determine what type of vegetation is found where, the status of park vegetation, and address how all the ecosystems interact.

12.1 Karst Development South of Green River

Karst-linked ecosystems in and near the park south of Green River can most reasonably be divided into two sections near Turnhole Spring on Green River (Fig. 12.1). The major karst development east of Turnhole Spring includes the Mammoth Cave System plus other major caves. Due to underground drainage, perennial surface streams are almost nonexistent. West of Turnhole Spring and south of Green River, the terrain is in a much earlier stage of karst development. A "layer cake" of inter-bedded limestone and sandstone locally called the "cap rock sequence" causes surface streams to be present on a segmented basis as water sinks into limestone, and reappears on sandstone at springs. Only minor cave development in the underlying massive carbonate beds has taken place.

On top of Mammoth Ridge, the perched karst system in the cap rock is quite different from the sinking springs found elsewhere. Most of the cap rock sequence has eroded away, leaving the Big Clifty Sandstone and patches of the Haney Limestone. Often associated with these remnants are upland swamps, which may have been sinkholes in the Haney Limestone before most of this stratum weathered away (Fig. 12.2). These upland swamps are an interesting aspect of the karst landscape because they provide habitat for plant species not found elsewhere in the park such as pin oak, and they are vital also for amphibians such as Jefferson's salamanders (*Ambystoma jeffersonianum*). Upland swamps are not common north of Green River, and there are fewer upland springs in the northwest part of the park because the cap rock sequence was partially eroded away by a river back in the Pennsylvanian period.

R.A. Olson (✉)
Cave Research Foundation, Cave City, KY, USA
e-mail: rickardolson16@gmail.com

H.H. Hobbs III et al. (eds.), *Mammoth Cave*, Cave and Karst Systems of the World,
DOI 10.1007/978-3-319-53718-4_12

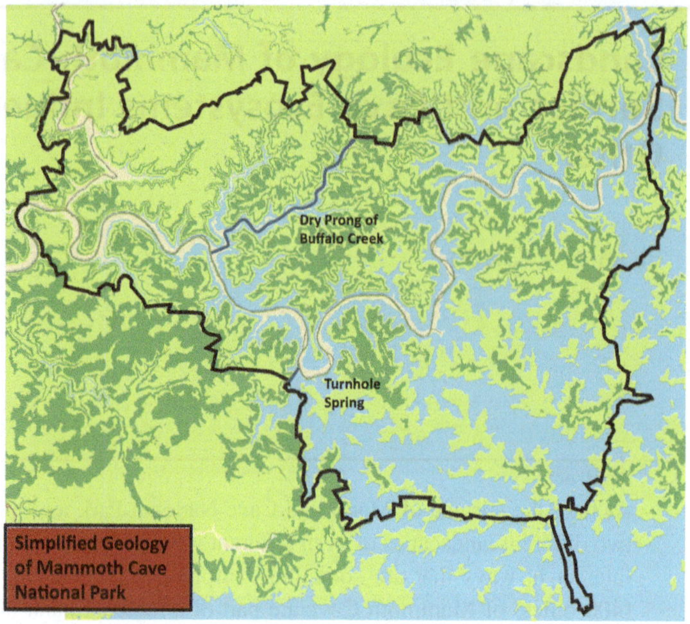

Fig. 12.1 A simplified geology map of the park showing boundaries between major and minor karst development north of the Chester Escarpment and on either side of Green River. North of the river, minor karst is found west of the Dry Prong of Buffalo Creek, with major caves found under the Dry Prong and to the east. South of the river, the dividing line is near Turnhole Spring, with major karst to the east. *Yellow* cap rock sandstone, *green* cap rock limestone, and *blue* major cave-bearing limestone

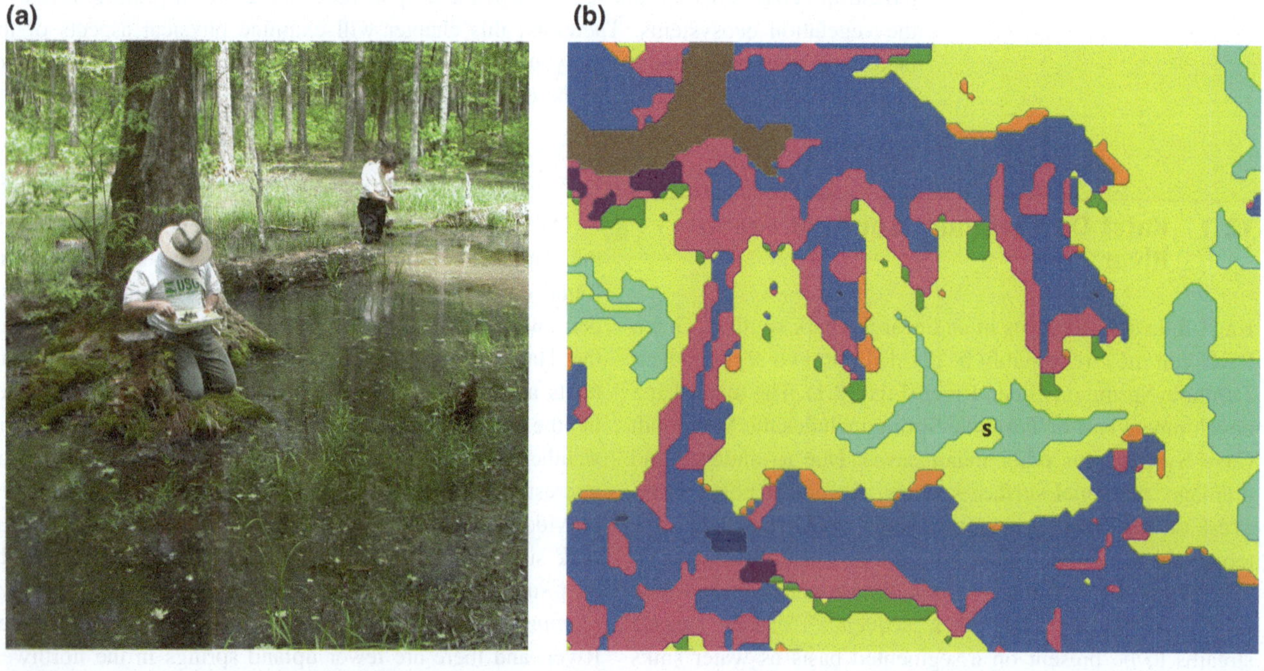

Fig. 12.2 **a** Photograph of an upland swamp on Jim Lee Ridge. At *left* is a visiting researcher with Kurt Helf at right. **b** A portion of the park habitat map showing how this upland swamp on Jim Lee Ridge is nearly surrounded by limestone. Long ago, this depression may have been a sinkhole in the Haney limestone. Today there is no visible inflow, yet there is substantial outflow, so it can be viewed as a special kind of Haney spring. *Yellow* Big Clifty sandstone and *pale blue* Haney limestone

12.2 Karst Development North of Green River

Karst-linked ecosystems in the park north of Green River are most easily described in two sections divided along the Dry Prong of Buffalo Creek (Fig. 12.1). From the Dry Prong east, there is significant karst and cave development within the major carbonates beneath the cap rock sequence. Buffalo Creek, Bat, Ganter, Running Branch, and Wilson caves are all extensive and provide important habitat for cave life. West of the Dry Prong, karst development is limited in much the same way as described for the area west of Turnhole Spring, south of Green River. In addition, perennial surface streams that are tributaries to the Nolin River, such as Bylew and Second Creeks, dissect the landscape more deeply than in any other area of the park.

12.3 Vegetation Habitat Types in the Park

Taking regional geography and hydrogeology into account, a vegetation habitat classification was developed for Mammoth Cave National Park (Olson and Franz 1998). This habitat classification combines bedrock geology, slope, and aspect in the park's geographic information system (GIS) with a spatial resolution of 30 m (Fig. 12.3, Table 12.1). The underlying rationale is that for a given climate, bedrock geology largely determines soil type and whether surface or subsurface (karst) drainage prevails. The effects can be striking as shown in Fig. 12.4 where trees growing on limestone along the Dripping Springs Escarpment had their leaves die during a severe drought in 1999, but trees growing on the sandstone above were still green. The name of the escarpment hints at the explanation for this because the sandstone at the top of the escarpment is a perched aquifer. Soils on calcareous bedrock are pH buffered as the underlying rock dissolves; soils on non-calcareous rocks tend to be more acidic due to the lack of buffering. Because of this, the Kentucky State Nature Preserves Commission classifies habitats as "calcareous" or "acid" based upon bedrock type, and I follow this convention here. Due to the tendency for subsurface drainage to develop in calcareous bedrock, virtually any surface site with a given set of landform characteristics will be more xeric (dry) than an equivalent situation underlain by insoluble rocks such as sandstone or shale. The magnitude of this general difference appears to be less on the steepest exposures due to rapid surface drainage.

One significant attribute of the habitat map is that natural physical influences on vegetation types are made clear in a way that they are not by direct study of geologic quadrangle maps. This is especially important in conducting ecological restoration, given the complex history of cultural disturbance over the past two centuries since settlement, and the profound impact on vegetation patterns seen today. For example, the vast majority of coniferous forest stands in the park today are successional following pre-park agriculture; Virginia pine dominates in old fields underlain by sandstone, and Eastern red cedar is most abundant in old fields underlain by carbonate rock. The potential natural vegetation (climax community) for these old fields can be estimated by looking at what vegetation is present on less disturbed sites with the same habitat type as the old field in question.

12.4 Vegetation Community Types

You may wonder why so much space is given to vegetation in a book about Mammoth Cave. For the river and connected aquatic cave ecosystem, vegetation determines the amount and quality of water, sediment, and organic matter that enter. For the terrestrial cave ecosystem, the types and quantities of insects, fungi and plants available to bats, woodrats, and cave crickets are largely determined by major vegetation types.

The most recent vegetation map for Mammoth Cave National Park (MCNP) was derived from a 2008 LANDFIRE map (Fig. 12.5). In this exercise, 24 vegetation categories were consolidated into four vegetation types useful for fire management (Olson et al. 2013). These vegetation types are Oak Forest–Woodland, Coniferous–Deciduous Successional Forest, Mesic Hollow–Floodplain Forest, and Disturbed Lands, which are developed areas and roadsides. Other vegetation types added to the GIS coverage included Barrens, Prairie Plantations, and Forest Canopy Gaps linked to both storms and prescribed fires. Species typical for each vegetation type and areal coverage for the park are summarized in Table 12.2.

When comparing geographic patterns of potential natural vegetation in Table 12.1 to mapped coverage in Table 12.2, there are some discrepancies. For example, if we add up all the habitat types in Table 12.1 that support Oak Forest–Woodland in the park, then we arrive at just under 14,160 ha (35,000 acres), assuming half of Calcareous Subxeric habitats are not in Barrens. However, in Table 12.2, Oak Forest–Woodland was mapped at only a bit over 8900 ha (22,000 acres). This shortfall is due to two types of ecological distortion—disturbance and the spread of mesic species due to lack of fire. The disturbance is from pre-park farm fields, and more recent damage from a major ice storm in 2009, and prescribed fires in 2010. So the 8900 ha of Oak Forest–Woodland plus half the Coniferous–Deciduous Successional Forest (the other half may have been Barrens in the karst valleys) at 1645 ha (4064 acres), plus the Canopy Gap acreages at 1990 ha (4920 acres), to the total is a little over 12,545 ha (31,000 acres).

The spread of mesic species such as beech and maple into oak habitats shows similar discrepancies. In Table 12.1, there is listed 4860 ha (12,000 acres) of habitat types that

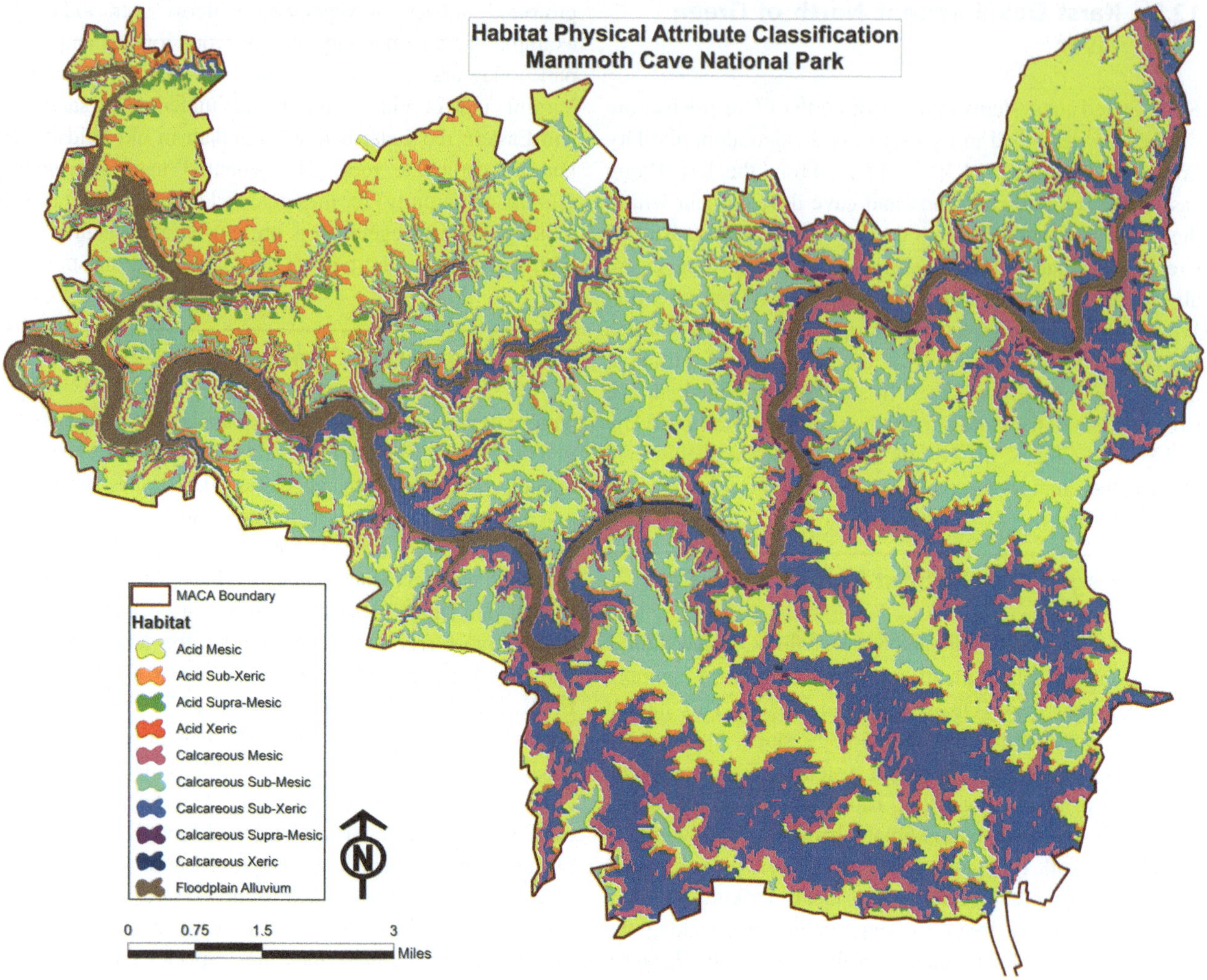

Fig. 12.3 Vegetation habitat map of the park based upon bedrock geology, slope, and aspect. Soils in the park are almost wholly determined by parent material and are therefore of secondary importance as controlling factors for plant habitats

Table 12.1 Areal Extent of Habitat Classes in the Park Habitat types in red and bold are capable of carrying fire during the spring and fall fire seasons, and account for approximately three-fourths of the park

Habitat type	Potential natural vegetation	Acres-hectares
Calcareous Xeric	*Cedar-Oak Slope Glade*	*150–61*
Calcareous Subxeric	**Forest-Woodland**	**9240–3739**
Calcareous Sub-Mesic	**Forest-Woodland**	**9530–3850**
Calcareous Mesic	Mesic Hollow Forest	9050–3662
Calcareous Supra-Mesic	Mesic Hollow Forest	130–53
Acid Xeric	*Pine Cliff Edge Forest*	*60–24*
Acid Sub-Xeric	**Forest-Woodland**	**2500–1012**
Acid Mesic	**Forest-Woodland**	**18000–728**
Acid Supra-Mesic	Hemlock-Yellow Birch Forest	1000–405
Alluvium	Floodplain Forest	2700–1092

Blue and underlined indicates non-fire-adapted vegetation, and green in italics indicates where caution must be exercised in determining whether fire could possibly be beneficial

Fig. 12.4 During a severe drought in the summer of 1999, leaves of trees growing on limestone along the Dripping Springs Escarpment died. Trees growing on the sandstone above were still *green* because of water retained and seeping from the sandstone

would support Mesic Hollow or Floodplain Forest in three categories. However, in Table 12.2 more than 6880 ha (17,000 acres) is indicated for Mesic Hollow–Floodplain Forest. This discrepancy of about 2025 ha (5000 acres) is mostly accounted for by the spread of mesic species such as beech and maple into oak habitat due to lack of fire. If these 2025 ha are added to the previous subtotal of 12,545 ha (31,000 acres) of Oak Forest–Woodland (mapped and currently disturbed), then this brings us to 14,570 ha (36,000 acres), which is in pretty good agreement with the 14,165 ha (35,000 acres) predicted by the habitat model. It must be understood that the vegetation map and the habitat model are both best estimates with the limited resources available at the time, and this is why numbers for areas are rounded.

Next consider the Barrens, which is the local name for grasslands. Those areas that had few trees and were dominated by grasses were referred to by settlers as Barrens. Once again, if acreages of potential Barrens in Table 12.1 are compared to mapped coverage in Table 12.2, there is a significant difference. Even if we assume only half of the Calcareous Subxeric habitats in the park would support Barrens, there are a little under 5,000 acres that have the same characteristics as the historically documented Barrens on the Sinkhole Plain south of the park. However, in Table 12.2, Barrens were mapped at only 120 acres. The Barrens were lost to agriculture and fire suppression both on the Sinkhole Plain and probably in the drier parts of big karst valleys in the park.

12.5 The Role of Fire in Some Vegetation Habitat Types

In pre-settlement times, fire played a role of variable importance in habitat types that are moderately well drained and sunny. Obviously, shaded moist sites are the least fire prone. For example, on the shady sides of sandstone cliffs we find Acid Supra-Mesic habitats, which in places support stands of hemlock and yellow birch. However, nature is full of surprises, and the two driest or xeric habitat types do not support fire-adapted communities. On the sun-baked, southwest-facing limestone hillsides (Calcareous Xeric), dry conditions limit the species that can live there, and the Eastern red cedars that dominate are not fire-adapted. The combined effect of internal drainage down through limestone and afternoon sun in summer months results in prickly pear cactus being able to grow in a climate that averages 127 cm (50 in) of rain per year! Virginia pine (*Pinus virginicus*) holds forth on the sunny edges of sandstone cliffs (Acid Xeric) where again, dry conditions keep out competitors, not fire. Virginia pine is easily killed by fire.

Acid Mesic habitat covers huge swaths of oak–hickory-forested ridge tops north and south of Green River where it was not cleared for agriculture. This community type requires fire in order to prevent invasion by more moisture-tolerant species such as beech (*Fagus*) and maple (*Acer*). We can infer from Indian torch remains in Salts Cave that oak–hickory stands either had significant open fields

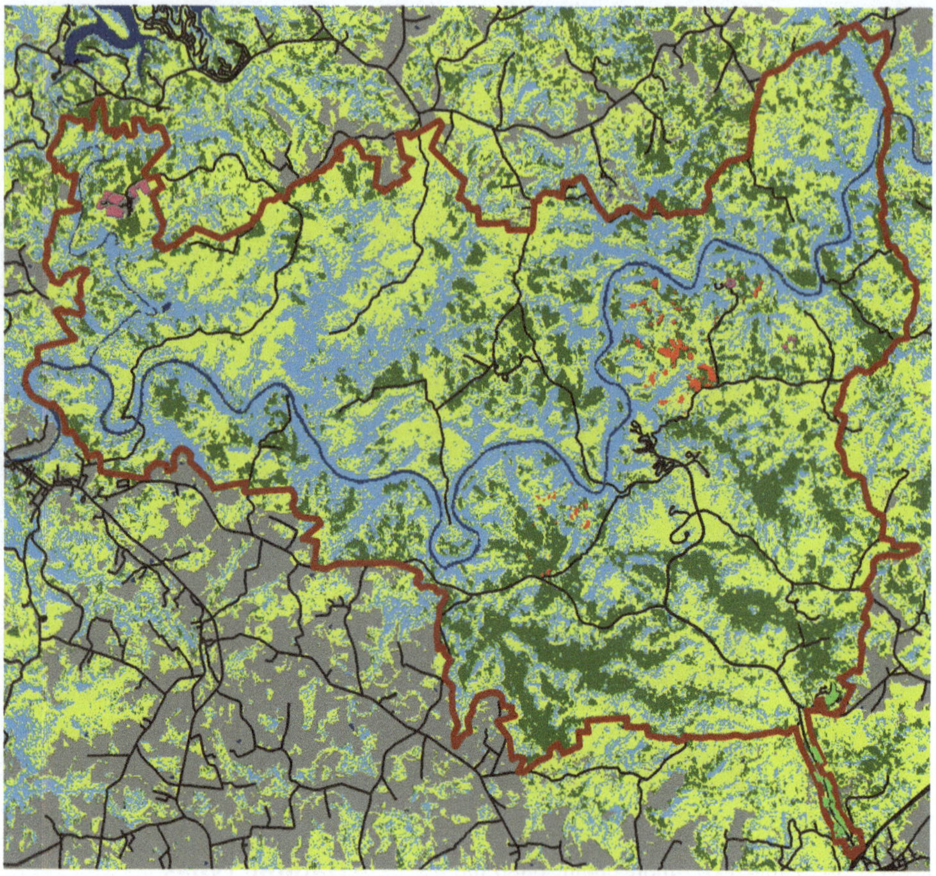

Fig. 12.5 Vegetation map of the park. *Yellow* Oak Forest–Woodland, *blue* Mesic Hollow Floodplain-Forest, *green* Coniferous-Deciduous Successional Forest, *red* Fire-Killed Coniferous Forest, *fuchsia* Planted Prairie, and *lime green* Barrens

Table 12.2 Community vegetation types, representative species, and coverage in the park. Due to the low precision inherent in mapping large areas with complex vegetation, acreages/percentages have been rounded

Vegetation type	Typical species	Acres–hectares
Oak Forest—Woodland	**Post, blackjack, chestnut, black, white, red, and chinquapin oak**	22,300–9024
Barrens	**Prairie grasses, forbs, plus shingle , post, and blackjack oak**	120–49
Prairie Plantation	**Prairie grasses and forbs**	110–45
Mesic Hollow—Floodplain Forest	Sugar maple, beech, tulip poplar, box elder, sycamore	17,100–6920
Coniferous—Deciduous Successional Forest	*Eastern red cedar , Virginia pine, red maple, tulip poplar, dogwood, sweetgum*	8130–3290
Disturbed Lands	*Developed areas in fescue, road sides*	150–61
Forest Canopy Gap—Storm Linked	*Downed pines, early successional and invasive plants*	800–324
Forest Canopy Gap—Fire Linked	*Downed and standing dead pines, successional and invasive plants*	4120–1667

Red and bold indicates fire-adapted vegetation, blue and underlined indicates non-fire-adapted vegetation, and green in italics indicates where fire is not welcome such as developed areas, or where caution must be exercised in determining whether fire can help move successional vegetation toward desired future conditions

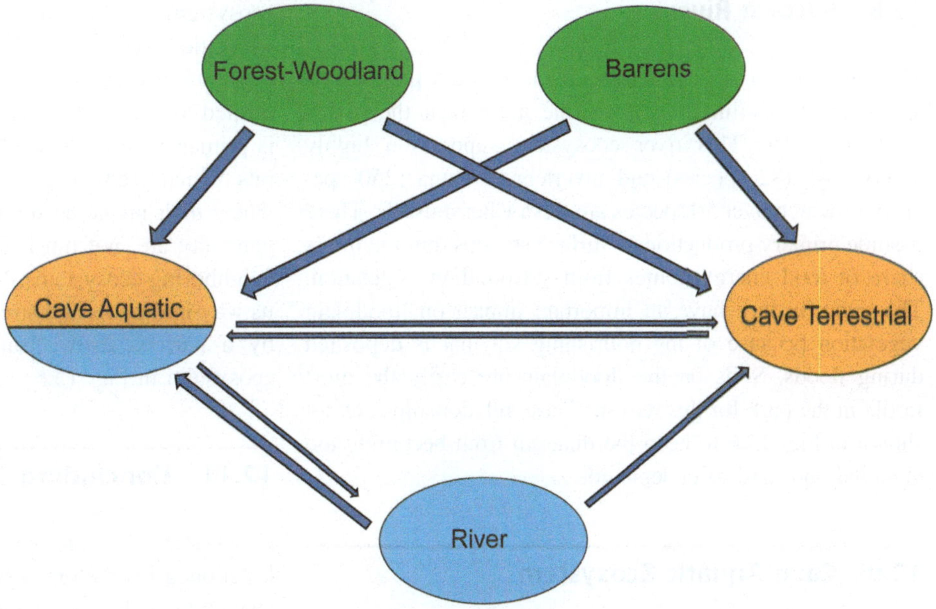

Fig. 12.6 Ecosystems within the regional karst landscape, and how they relate. *Arrows* indicate direction of food energy transfer with *wider arrows* meaning greater transfers. Relationships between vegetation and rivers are not shown in order to keep the figure understandable. Figure courtesy of Shannon Trimboli

where the Indians grew crops, or that there existed open oak woodland. This is possible because false foxglove (*Aureolaria virginica*), the most abundant torch material in Salts Cave (Watson 1969), must attach its roots to those of oak trees, and yet must have a lot of light in order to grow tall enough for torch use (Olson 1998). This is important to know for setting restoration targets (desired future conditions) for vegetation, and properly restoring vegetation is important for the long-term well-being of both the aquatic and terrestrial cave ecosystems.

12.6 How Ecosystems in the Landscape Interact

Within the regional karst landscape, there is one historical and four functioning ecosystems (Fig. 12.6). Historically, Barrens bordered by woodland covered large portions of the Sinkhole Plain located south of the park. Karst valleys within the dissected upland of the Mammoth Cave Plateau offer a similar habitat type (Calcareous Subxeric) to the Sinkhole Plain, and there is a shared indicator species, called shingle oak (*Quercus imbricaria*), which grows only on this habitat type in the region. Unfortunately, no historical descriptions of pre-settlement vegetation in the karst valleys have been found. Soon after settlement in the late 1700s, the grasslands described were gradually converted to agriculture. Except for the loss of woodland portions of the forest ecosystem due to fire suppression, other ecological components of the karst landscape in and near the park are reasonably intact.

12.7 Forest–Woodland and Barrens

I will discuss the vegetation ecosystems in the Mammoth Cave area together because the plant species list cannot be separated easily due to extensive overlap, and the photosynthetic productivity of all are important to the cave ecosystems. Plant diversity in the region is exceptional with over 1100 species, including 82 kinds of trees. An unknown percentage of currently forested land in the park would have been maintained as woodland and some Barrens via natural and American Indian set fires. Barrens in the park are currently limited to small areas, each no greater than 28 ha (70 acres). Only those growing on Calcareous Subxeric habitat, and that developed from the natural seed bank, can be considered linked to actual grassland remnants from pre-settlement times. Even so, these areas are rich in prairie grasses and forbs such as big bluestem (*Andropogon gerardii*) and tall coreopsis (*Coreopsis tripteris*). These patches of grassland serve as refuges for species marginalized by conversion of former prairie on the sinkhole plain to agriculture and by fire suppression within and beyond park boundaries (Seymour 1996). Diversity of plant species in the park has almost certainly suffered due to forest succession in the absence of fire. Sorted among vegetation communities according to habitat preferences, 203 species of birds, 43 species of mammals, 29 species of amphibians, and 38 species of reptiles have been reported in the park. The species data for Mammoth Cave National Park are maintained in NPSpecies (https://irma.nps.gov/NPSpecies), a database developed by the National Park Service's Inventory and Monitoring Program.

12.8 Surface Rivers

The Green River runs east to west through the park and is joined by the Nolin River from the north near the park's western border. This river ecosystem supports a highly diverse fish (82 species) and invertebrate fauna (250 species), of which over 50 species are freshwater mussels. There is some primary production in surface streams, but the lion's share of food energy comes from surrounding vegetation. The rivers in turn have an important impact on floodplain vegetation because of the nourishing silt that is deposited during floods. Soils on the floodplain are easily the most fertile in the park for this reason. These relationships are not shown in Fig. 12.6 to keep the diagram from becoming too cluttered and hard to understand.

12.9 Cave Aquatic Ecosystem

Functionally, since sinking streams and cave streams are tributaries of base-level rivers by way of springs, they are all part of the river continuum, with the important distinction that the middle section is underground. The cave aquatic ecosystem is almost wholly energetically supported by dissolved organic carbon from the Forest–Woodland and Barrens ecosystems. Food transport is usually down gradient, but natural back flooding from Green River through springs into the lower cave streams is also important. Ruhl (2005) found that a great diversity and abundance of larval fish are transported into base-level cave streams during back-flooding events. The magnitude of food transport into the cave was a surprise, and this natural process is an extremely important food source for aquatic cave life in these base-level streams. The aquatic ecosystem in Mammoth Cave is quite diverse with 16 obligate species spread across four basic community types (see Chaps. 14 and 15). The cave aquatic ecosystem cannot contribute much energetically to other ecosystems due to lack of significant primary productivity. Still, when cave streams flood during heavy rains, a film of organic matter is deposited on cave walls, and this is fed upon by terrestrial organisms such as springtails, which are tiny invertebrates closely related to insects. This is shown in Fig. 12.6 by a narrow arrow from the aquatic to terrestrial cave ecosystem.

12.10 Terrestrial Cave Ecosystem

As surface rivers carve deeper into the rock and lower their channels, cave streams follow and leave dry upper levels. These passages become habitat for the terrestrial cave ecosystem, which is also mostly dependent upon the surface vegetation ecosystems for its food base. The component communities are highly diverse with 32 terrestrial species limited to caves (see Chaps. 13 and 15). The import of food is primarily accomplished by cave crickets, bats, and packrats which feed on the surface and use caves for refuge where their guano accumulates. Being energy poor as caves generally are, not much food energy can be contributed to neighboring ecosystems. Notwithstanding, guano does make its way into pools and streams in caves, and this is indicated by a narrow arrow from the terrestrial to aquatic cave ecosystem in Fig. 12.6.

12.11 Concluding Thoughts

Ecosystems are very good at self-regulation and certainly functioned fine before people came on the scene. The major challenge with karst landscapes is to prevent inadvertent distortions (linked with human activities) from causing irreversible loss in any of the component ecosystems. For information on ecosystem management issues, see Chap. 18. One little studied energy input to both aquatic and terrestrial ecosystems in Mammoth Cave is hydrogen sulfide and hydrocarbons (see Chap. 14 for more on this). The magnitude of these inputs is not known but deserves greater attention (Olson 2013).

References

Olson R, Franz M (1998) A vegetation habitat classification for Mammoth Cave National Park. In: Proceedings of Mammoth Cave National Park's seventh science Conference. Mammoth Cave National Park, KY, pp 19–25

Olson R (1998) Torch fuels used by prehistoric Indian cavers: their utility and botanical significance. Proceedings of Mammoth Cave National Park's seventh science conference. Mammoth Cave National Park, KY, pp 5–8

Olson R, Scoggins L, Toomey RS, Burton J (2013) 2011 Vegetation map for Mammoth Cave National Park. In: Mammoth Cave National Park's 10th Research Symposium, 14–15 Feb 2013, pp 4–8

Olson R (2013) Potential effects of hydrogen sulfide and hydrocarbon seeps on Mammoth Cave ecosystems. In: Mammoth Cave National Park's 10th Research Symposium, 14–15 Feb 2013, pp 25–30

Ruhl M (2005) Flow reversal events increase the abundance of nontroglobitic fish in the subterranean rivers of Mammoth Cave National Park, Kentucky. Masters Theses & Specialist Project, http://digitalcommons.wku.edu/theses/475

Seymour R (1996) A contribution to the recorded flora of Mammoth Cave National Park. In: Proceedings of the fifth annual Mammoth Cave National Park science conference, 2

Watson PJ (1969) The prehistory of salts cave, Kentucky. Reports of investigations No. 16, Illinois State Museum, pp 12, 33, 34, 36, 56

Terrestrial Cave Ecology of the Mammoth Cave Region

13

Thomas L. Poulson

Abstract

Terrestrial communities of cave-adapted invertebrates are found where surface organic material is deposited: cave entrances, upper-level terminal breakdowns, upper-level vertical shafts, cave stream banks, and cave base-level rivers. The high diversity of cave-adapted species in the Mammoth Cave region is partly due to the variety of food inputs from the surface. Surface organic material is imported into caves passively via gravity, water, or wind, or actively through transport by organisms that use caves for shelter or live inside them most of their lives. Though many organisms defecate inside cave entrances, both feces and eggs deposited by cave crickets (*Hadenoecus subterraneus*) support the most diverse communities of cave invertebrates in the region. *H. subterraneus*, relative to the two species of surface crickets with which it shares cave entrances, is long lived and feeds on the surface throughout the year. Scientific studies have partially revealed the life histories of many invertebrates supported by *H. subterraneus,* but there are many aspects that remain to be discovered.

13.1 Introduction

The Mammoth Cave region, with its large number of cave-adapted species, is a wonderful place to study terrestrial cave ecology. For geological reasons, Mammoth Cave itself has a huge lateral yet quite modest vertical extent with many passage levels and characteristics (see Chap. 6 this vol.). This contrasts with caves in the Sinkhole Plain, which are shorter and less interconnected.

The large number of species in the combined caves of the Sinkhole Plain and Mammoth Cave Plateau is high for biogeographic reasons as well as time. There has been perhaps ten million years for species to evolve locally, and other species have dispersed to the area since it is geographically at the junction of several regional cave faunas (Barr 1967, see Chap. 15 this vol.).

Finally, the region has a variety of food input types. This is important because caves are typically food limited due to the absence of green plant producers. Food input results from physical factors such as percolating and flowing water and biological factors such as deposition of feces by species that feed outside and return to the cave periodically. Biological inputs dominate in Mammoth Cave Plateau caves, whereas physical inputs dominate in Sinkhole Plain caves.

In this chapter, I will concentrate on cave-adapted species called **troglobionts**. They are generally blind with very reduced eyes and pigment. I will compare troglobionts that have been isolated in caves for different durations of evolutionary time. I will also compare them to facultative species called **troglophiles** that can complete their life cycles outside caves. Aside from cave crickets, I will not consider the adaptations of species that use caves as shelter or for reproduction, but cannot complete their life cycles in caves; these are called cave guests or **trogloxenes**. They include wood rats and bats (See Thomas and Toomey Chap. 17 this vol.). Their importance for this chapter is as a source of feces that are the basis for some terrestrial cave communities.

13.2 Aquatic Versus Terrestrial Cave Habitats

Compared to terrestrial species, aquatic cave species face conditions that are more extreme (rigorous), are more variable, and have less predictable chemical, physical, and food

T.L. Poulson (✉)
Cave Research Foundation, Laurel, USA
e-mail: tomandliz05@gmail.com

H.H. Hobbs III et al. (eds.), *Mammoth Cave*, Cave and Karst Systems of the World,
DOI 10.1007/978-3-319-53718-4_13

conditions (Poulson 2012). Food rigor in aquatic habitats is higher due to lower quantity and quality than in terrestrial habitats. Aquatic temperature, water flow, and food are more variable especially with seasonal floods.

The main reason for low aquatic food quality is that organic matter progressively decreases in size and amount and is diluted and leached of the most soluble and digestible nutrients as it passes from the surface through the epikarst and into the cave via drip water or enters the cave directly in sinking streams.

Terrestrial cave food types are higher in quality because they usually remain where deposited and are rarely leached by water. Examples are the feces of trogloxene raccoons, wood rats, and bats.

With increased evolutionary time of isolation in caves, all terrestrial and aquatic troglobionts have progressive reduction in eyes and pigment and, where studied, reduction in circadian activity and circadian metabolism. For both terrestrial and aquatic troglobionts, the degree of reduction in traits has been used as an index of evolutionary time isolated in caves (Culver 1982).

13.3 Mammoth Cave Terrestrial Environment

Figure 13.1 shows an idealized cross section of Mammoth Cave adjacent to the Green River. It shows the variety of entrances and passage types that manifest great differences in winter to summer range of temperature, wind, and moisture. In addition, the passages have a wide range of substrate types, moisture, and food supply.

In some large entrances connected to the extensive cave system (*e.g.,* Historic Entrance), it is too rigorous in winter for terrestrial troglobionts. Dry, cold, winter winds penetrate over 1000 m, and subfreezing temperatures can even cause frost-wedging and ceiling rock falls. Only bats that seek cold microclimates to hibernate are found at such entrances in winter (see Chaps. 10, 17).

The highest diversity and abundance of terrestrial troglobionts occur in microclimatically buffered small caves not connected to the over 640 km (400 miles) of Mammoth Cave. Examples on the Mammoth Cave Plateau are White Cave and Little Beauty Cave. There are many more examples on the Sinkhole Plain. The terrestrial troglobionts in these caves have thin exoskeletons with little waxy waterproofing and cannot tolerate humidity much below 97–100%.

Another habitat with a high number of species and stable environmental condition is the terminal breakdown of upper-level passages at the edge of valleys. An easily accessible example is Rafinesque Hall where Barr and Kuehne (1971) documented greatly increased inflow of dissolved organic matter and bacteria during unusually heavy rains.

Beyond the entrances to the Mammoth Cave System, the only habitats with enough food for the most energy efficient troglobionts are upper-level passages around vertical shafts. Adjacent areas under the sandstone cap rock are very dry, may have gypsum speleothems, and typically harbor few visible organisms (Poulson 1992).

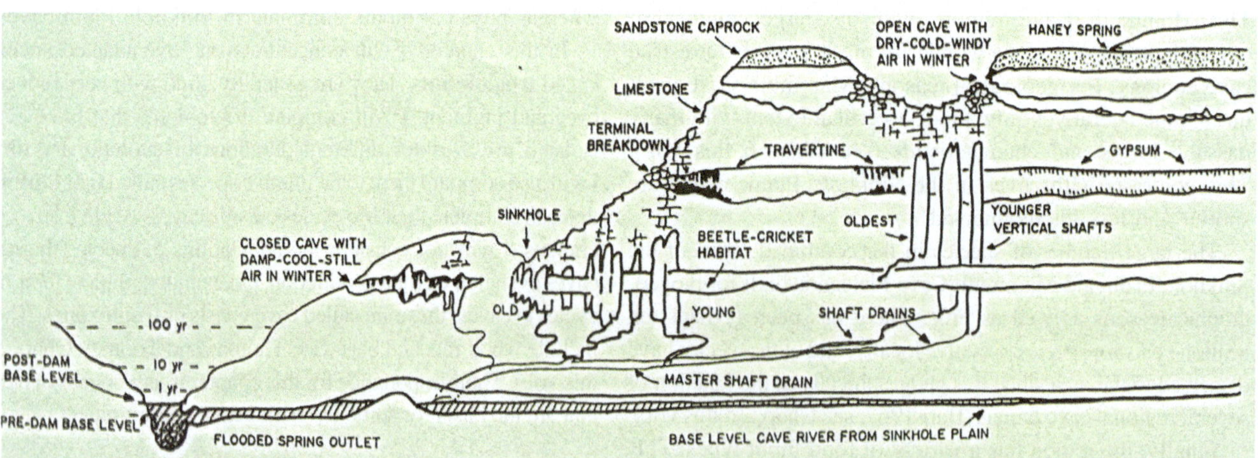

Fig. 13.1 Cross section of typical passages of Mammoth Cave adjacent to Green River Cave types range from "closed caves with damp–cool–still air in winter" that are microclimatically buffered (e.g., White Cave) to "open caves with dry–cold–windy air in winter" that are not microclimatically buffered (e.g., Historic Entrance). Intermediate situations are "terminal breakdowns" (e.g., Rafinesque Hall) and sinkholes above vertical shaft complexes that have periodic organic in-wash. Some of the in-wash of particulate organic matter from vertical shaft complexes reaches the silt–sand banks of base-level cave rivers (e.g., Styx–Echo river) via master shaft drains. Other shaft drains rarely flood and have very little particulate organic matter (e.g., Eyeless Fish Trail). Finally, there are the dry and lifeless upper-level passages under the sandstone cap rock with gypsum speleothems. Out from under the cap rock, passages have both carbonate speleothems and silt–sand substrates that are beetle–cricket habitat

Even the nearly ubiquitous cave cricket, *Hadenoecus subterraneus*, does not roost and rest where cold, dry winter air enters entrances to Mammoth Cave. Where cave crickets "sleep," they are roosting, digesting food, and defecating. Cricket entrance roosts are important since feces of cave crickets provide a reliable food source for many troglobionts.

Along a gradient of temperature and moisture in winter, different species of bats hibernate in species-specific microclimates (see Chap. 17 this vol., Mohr and Poulson 1966). In summer, substrates in such entrances often remain too dry for troglobionts. Most Sinkhole Plain caves are small, and many have sinking streams or active stream passages near the entrances. Often entrances have regular input of coarse particulate organic matter (CPOM) during rainfall events. As a result, stream banks have all sizes of organic matter and many troglophilic earthworms. With such predictable organic input, Sinkhole Plain caves have many more troglophiles than troglobionts, have high densities of organisms, and have high species richness (e.g. McKinney 1975).

Mammoth Cave stream banks have much lower species richness and abundance than those on the Sinkhole Plain. The reason is that very few streams in Mammoth Cave have direct input from the surface. Their sources are typically drains of vertical shaft complexes that rarely get fresh input of particulate organic matter input. A corollary is that their stream banks never have high rigor of low temperatures and low humidity and do not flood very often.

Three stream-bank communities that I followed for more than a decade differ in the diversity, frequency, and amount of organic matter input and the diversity and kinds of cave animals (Poulson 1992).

Unlike streams that drain vertical shafts, the large base-level rivers of Mammoth Cave flood every year. An example is the Natural Bridge *aka* Orpheus Pass near Styx and Echo Rivers with 10–20 m floods. These floods leave dissolved organic matter (DOM) and fine particulate organic matter (FPOM) at the high water mark. Every decade or so, there may also be in-wash of coarse particulate organic matter (CPOM) of sticks and leaves when a nearby surface channel unplugs and allows flushing into the huge Mammoth Dome sinkhole. This CPOM input results in a transient mix of many species. As the organic matter is depleted, a succession of species occurs. The soil millipede, *Chaetapsis fragilis,* survives only a year with no reproduction and the troglophilic snail, *Carychium,* a few years longer with limited reproduction. Troglobiont predators such as Carabid beetles (*Pseudanophthalmus striatus)* and opilionids (*Phalangodes*) remain in low numbers for up to five years. Troglobiont detritivores such as bristletails (*Litocampa*) and millipedes (*Scoterpes*) may remain for almost a decade.

13.4 Comparative Biology of Species that Are not Closely Related

Even though not in the same genus, comparisons of the ecology and degree of cave adaptation of pairs of species in the same family may still be useful. Perhaps as a result of the past or present competition, the species I discuss are mainly separated by habitat and food supply, often around entrances rather than in deep cave habitats.

We have most intensely studied those species that have high importance value that are easily and regularly found. Importance value of a species is the sum of its frequency of occurrence in time and space, density where it occurs, and impact per individual in terms of body size and metabolic rate. On all of these criteria of importance value, the cave cricket, *Hadenoecus subterraneus,* outranks all other species found in the cave today. It inhabits virtually all cave entrances, it has dense roosts just inside cave entrances, and it is large for an insect and has a relatively high metabolic rate.

13.5 Two "Crickets" in the Family Rhaphidophoridae

Crickets that live in caves have some traits that preadapt them to life in caves. All are nocturnal with relatively long antennae, labial palps, and cerci. These organs at least allow them to touch, and smell, taste, and detect air movements so that the crickets can navigate and forage in the dark. All are omnivorous and will eat a wide variety of foods.

Here I compare a trogloxene, the camel cricket *Ceuthophilus stygius,* to a possible troglophile, the cave cricket *Hadenoecus subterraneus.* The camel cricket is certainly a trogloxene because it can never complete its life cycle in caves. The status of the cave cricket is much less clear. It could be called a troglophile because it could complete its life cycle inside caves if there is a lot of food. But I could also argue that it is a habitual trogloxene because it does virtually all of its feeding outside caves and shelters inside cave entrances. I could even argue that it is an incipient troglobiont. First, we have never seen a hint that it reproduces outside of caves. Second, it is especially compromised by low humidity and warm temperature. Third, it has highly elaborated sense organs that would allow it to find scarce food inside a cave.

The two crickets also differ daily and seasonally in their use of habitats. During the day in summer, the camel cricket can be found tightly clustered just inside some cave entrances and in cave-like cellars and spring houses. In summer, the cave cricket always roosts inside caves in loose clusters. It leaves the cave to forage outside whenever its crop is empty of stored food, but it never remains outside.

In winter, some immature camel crickets hibernate inside cave entrances in cracks and crannies and never leave the cave in winter. A few cave crickets leave the cave to forage in winter if it is warm and wet.

The two crickets differ in their ability to survive in caves. Cave crickets are partially cave adapted, whereas camel crickets have little hint of the adaptations needed for efficient location of scarce food that would be necessary to evolve to live in caves (Lavoie et al. 2007). The outline drawings in Fig. 13.2 compare the relative sizes and shapes of bodies, legs, crops, crop distension, and sensory appendages. Figure 13.3 shows a drawing from life that shows the differences in color.

Levy called the cave cricket a "scavenger on stilts." She hypothesized that adults are especially efficient at finding smelly organic matter. A cricket effectively detects smelly food from far away by being far off the ground with long legs and effectively even further off the ground with exceedingly long antennae.

Adult crickets came only to the smelliest baits in the cave. Their long legs and long antennae probably allowed a more

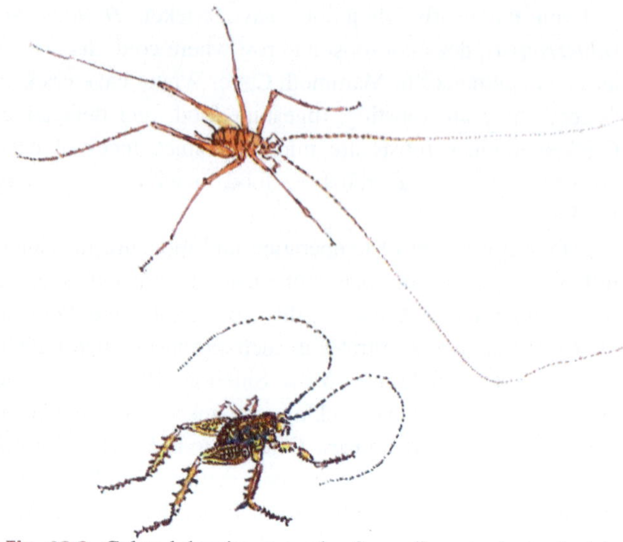

Fig. 13.3 Colored drawings to scale of a medium-sized camel cricket *Ceuthophilus stygius* and an adult cave cricket *Hadenoecus subterraneus*

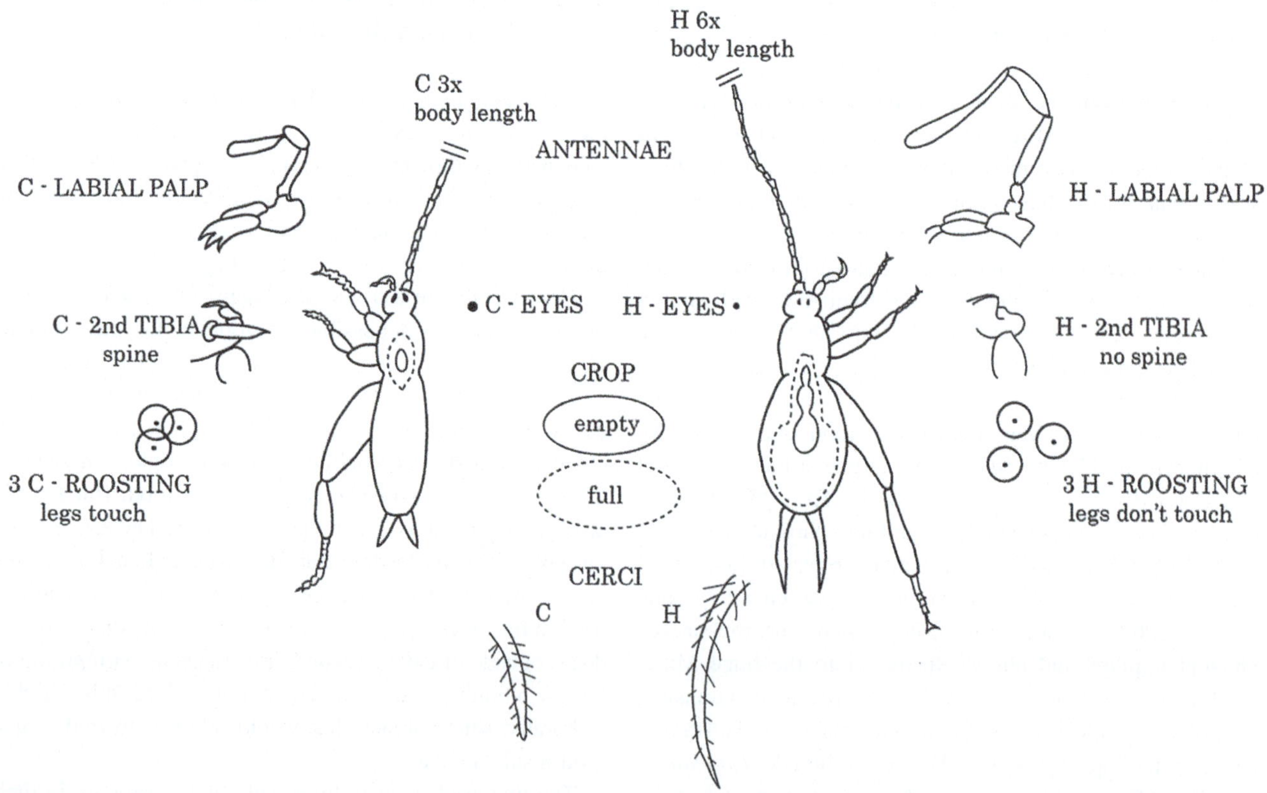

Fig. 13.2 Outline drawings of the camel cricket *Ceuthophilus stygius* (C) and cave cricket *Hadenoecus subterraneus* (H) C and H are shown at the same body length to best show differences in body build and relatively longer lengths of all sense organs and legs for H. Only part of the antennae are shown but H's are six times its body length while C's are only three times its body length. Antennae have sense cells that at least detect odor and touch. The labial palps have both taste receptors on the tips and "teeth" used to chew food. C's cerci are relatively short with few, short hairs that detect air movement. H's longer cerci with longer hairs are much more sensitive to detect air movement. Inside the bodies of the crickets in the center, note the differences in crop size when empty (*solid line*) and full (*dotted line*). H's larger and more distensible crop allows it to eat much larger meals when foraging outside. H can stay in the cave ten days before it needs to risk going outside to forage again. C must forage every two days

effective triangulation. Small crickets with short legs and antennae came to all baits. We infer that they either could not discriminate at a distance or just ate the first bait encountered.

13.5.1 Trade-offs of Roosting in Cave-like Habitats

Both cave and camel crickets have advantages from roosting in caves or cave-like habitats by day. It is wet and cool, and there are either few or no predators. As they leave shelter to feed at night, they are vulnerable to mice and salamanders that congregate around the cave entrance. But with its much bigger and more distensible crop, cave crickets need to go out to feed and face predation risk only every 8–11 days, whereas camel crickets must go out nearly every night.

Most camel crickets do not roost inside cave entrances in summer and so are exposed to warmer and drier conditions in their outside foraging and hiding areas than cave crickets. The camel cricket's lower risk of desiccation is associated with a thicker exoskeleton that includes a shiny, waxy cuticle. Camel crickets have 5–6 times lower rates of evaporative water loss compared with cave crickets and also a higher tolerance for cold and warm temperature (Lavoie et al. 2007).

13.5.2 Lifespans and Growth Rates of Cave and Camel Crickets

Camel crickets are a short-lived annual species, whereas cave crickets may live as long as 6–10 years. A correlate of shorter life is faster growth with more frequent molting. On any date in summer cave roosts, 5–10% of camel crickets are molting. Consistent with a lower growth rate, a maximum of 0.1–1.0% of cave crickets in cave entrance roosts are molting at any one time. Another indicator of longer life for cave crickets is that all sizes are in entrance roosts. In contrast, camel crickets live only a year. The evidence is that there is only one size cohort that grows to adult size during the summer in entrance roosts.

13.5.3 Reproduction of Cave and Camel Crickets

Cave crickets and camel crickets have major differences in seasonal timing and location of reproduction.

In fall, camel crickets copulate and then lay eggs. Nymphs hatch in fall, and at least some intermediate size nymphs enter cave entrances where they appear to hibernate. In spring, nymphs forage outside and grow to adult size by the end of summer.

In winter, some adult cave crickets leave entrance roosts and travel to sandy reproductive areas deeper in the cave. There many can be seen copulating, and many females can be seen laying eggs in the sand substrate.

13.5.4 Cave Crickets Are a Keystone Species in the Mammoth Cave Region

The cave cricket is a keystone species that supports a high diversity and abundance of other species. Some keystone species, like the beaver and the alligator, foster species diversity as ecosystem engineers whose activities create new habitats. Others are keystone predators that increase diversity by eating a species that otherwise monopolizes space, for example, the *Pisaster* sea stars that eat mussels and sea otters that eat sea urchins. Still other keystone species increase local species diversity as what have been called key industry species that many others depend on for food. Examples are krill in the Antarctic and cave crickets in Mammoth Cave.

I suggest that, as a keystone species, the cave cricket is the main reason for the high terrestrial species richness in Mammoth Cave. Cave crickets provide food in an otherwise food-poor environment directly by their feces and their eggs. In entrance areas, feces are deposited by dense roosts of crickets as thin layers of guano. Elsewhere in the cave, sparse feces are deposited by transient crickets especially on the way to and from reproductive areas. In reproductive areas, cricket eggs are eaten by the Carabid beetle *Neaphaenops tellkampfii.* After eating an egg, a beetle hides under a rock and leaves feces as it digests the egg (Griffith and Poulson 1993).

Beetle feces and cricket feces are the bases for cave communities along with feces of raccoons and cave wood rats, plus in-washed leaf and twig litter (Poulson 2012).

13.5.5 A Spider, a Salamander, and Cave Crickets

A large orb-weaving spider, *Meta ovalis,* and the cave salamander, *Eurycea lucifuga,* both depend mainly on cave crickets as prey.

Mammoth Cave entrances that have spiders have on average 13–30 adults with half again that number of subadults. Entrances with salamanders have 12–19 adults and virtually no subadults. Spiders are found year around, but in winter salamanders go to wet areas away from entrances to mate and lay eggs.

13.5.6 *Meta ovalis* Spiders

Good caves for *M. ovalis* spiders have relatively narrow cave passages through which crickets have to travel. This makes crickets vulnerable to capture in spider webs.

It is a bit of a mystery how any newly hatched spiders and small instars get enough food of small enough size to reach adult size. My main hypothesis is cannibalism among spiders. Other small enough prey is simply absent this far from the cave entrance. There are no flying midges. And at 5 mg, rare first instar cave crickets are too big to catch for even the biggest early instar spider of about 1–2 mg.

Poor caves for *M. ovalis* spiders have tall ceilings and large passages. They lack narrow places where spiders could spin webs to reliably intercept crickets as crickets leave the cave to feed at night or move deeper into the cave to reproduce. I have observed *M. ovalis* only once in the deep cave areas where cave crickets mate and lay eggs, and this is an astounding story. For six years, I observed one adult female *M. ovalis* in the same ceiling crevice at the back of Great Onyx Cave. She must have caught an occasional cricket to survive, but produced no egg cases. So she had never mated and stored sperm. Then, a male must have arrived because the first egg case appeared. Within a year, the population of spiders bloomed to over 100 with mostly subadults and a few adults with egg cases. Two years later, the spider population crashed, and for at least 10 years, the population was steady at 12–25 adult and subadult spiders.

13.5.7 *Eurycea lucifuga* Salamanders

The best cave entrances for salamanders are those where cave crickets have to exit to feed at night through a small opening. During the day in the Great Onyx entrance blockhouse, 17–22 salamanders are on the walls and under rocks. As dusk approaches, they move and line up along a narrow crevice that crickets must traverse on their way outside. Thus, the crickets have to run a predation gauntlet.

13.6 Two Troglobiont Linyphiid Spiders

Phanetta subterranea is less cave adapted than *Anthrobia monmouthia*. The latter has more reduction in eyes, pigment, exoskeleton thickness, layers of silk in egg cases, thinner and longer legs, and a smaller and more slender body. It is more energy efficient as measured by slower rate of weight loss and longer life. Additionally, it is more reproductively efficient as measured by fewer yet larger eggs (Poulson 1981).

P. subterranea is much more flexible in growth, web density, web spacing, reproductive rate, and local migration than *A. monmouthia*. When the outside weather is mild, *P. subterranea* moves to food-rich entrance slopes with abundant CPOM and reproduces rapidly. In winter, many entrance slopes are too cold and dry for live spiders. They either die or enter a hibernation-like state. Their eggs survive within an egg case with three silk layers and hatch in spring when it is cool and humid again.

13.7 A Tale of Three Millipedes: An Accidental, a Troglophile, and a Troglobiont

The accidental ***Oxidus gracilis*** is an 18–23 mm long, robust millipede that is native to southeast Asia and is often found in greenhouses in the USA. It is found in cave entrances only when polluted wastewater or commercial lighting results in considerable algal growth. In recent years, it has been found occasionally around several Mammoth Cave entrances. There it has reached incredible densities of as many as 100 per m^2.

The troglophile *Chaetapsis fragilis* is mainly a surface-dwelling soil millipede adapted to living in small spaces. It is only 8–9 mm long, thin in width, and has short legs. It is found only in shallow caves overlain by rich forest soils. From the forest soil in the Mammoth Dome Sink area, it often invades via roots into the back of White Cave. If organic matter is dense, then they can attain densities of 2 per m^2 and reproduce. Once in White Cave on dense cricket guano slopes, I observed 80% of an adult population in copulo. Weeks later, the adults were gone, and there were scores of tiny baby millipedes.

The troglobiont millipede *Scoterpes copei* is found only in areas with low food supplies. In exceptional cases around old organic matter that has low nutrient content, it can reach densities as high as 5 per 10 m^2. In one such situation, a few females 10–13 mm long actually laid eggs because a year later I found six newly hatched young 2.5–3.0 mm long. These same six millipedes grew to only 6–8 mm in the next three years! It lives at least 8–10 years.

As detailed by Poulson (1992), the three millipedes differ in life history and local population sizes as expected for accidental versus troglophile versus troglobiont.

13.8 A Tale of Two Ubiquitous Troglobiont Detritivores, a Beetle and a Dipluran

The beetle *Ptomaphagus hirtus* and the dipluran *Litocampa cookei* are found in almost all habitats in the cave and are the only two species always attracted to smelly baits placed by biologists. Both are widely distributed in karst areas of the eastern USA. Neither is highly cave adapted compared to their closest surface relatives.

We know much more about the beetle than the dipluran. This is because *P. hirtus* reproduces whenever adults encounter high-nutrient organic matter. We never see local areas where *L. cookei* reproduces since we never see many small individuals in an area. A corollary of this contrast is that several workers have raised *P. hirtus* easily in the laboratory, but nobody has been able to raise *L. cookei*. This is consistent with the observation that the beetle is an opportunistic species with high reproductive potential. In contrast, *L. cookei* is never aggregated around natural patches of organic matter in the cave and has a low reproductive potential.

It is frustrating that we know so little about *L. cookei,* which is such a common species. Its size–frequency distribution, mainly large individuals 6–10 mm length, suggests infrequent reproduction. However, we catch many of these large sizes in baited traps, so either adult must be very common or they sense baits from incredibly long distances. They move very fast compared to most of our other troglobionts. When we observe them, they are walking rapidly in the open and are only sometimes under rocks.

13.9 Component Communities of Six Food Resources

In the late 1970s, I first hypothesized that six different food types have different communities of cave animals (Poulson and Kane 1981). Of 44 species, 30 had >95% of their occurrence on one food, seven on two, four on three, and two on four of the food types. Thus, there appeared to be specialization by at least 30 species. Species composition was most distinct on raccoon and rat feces and least distinct on leached wood and organic mud–silt of stream banks.

I have discussed all six communities in detail elsewhere (Poulson 1992, 2012; Poulson and Kane 1981). Below, I briefly summarize the communities associated with feces from four different animals. The different feces are used by different diversities and species of cave animals because of differences in digestibility, successional decomposition, frequency and predictability of occurrence, and susceptibility to drying. They also differ in size of fecal unit and dispersion (Poulson 2012). With a large unit in a small pile we have a tangle of 13,000 mg turds for raccoon, a piling of 60 mg pellets for cave rat, a thin spreading of 5 mg splotches for crickets, and a sprinkling of 0.1 mg specks for sand beetles.

13.9.1 Raccoon Feces

Procyon lotor feces have the greatest useable calories and most possible succession, but few species use it because of its rarity and problems of drying. At times in the past, paleo-feces show that raccoons entered the seasonally cold and dry Historic Entrance presumably to eat hibernating bats. The preservation of feces is explained because today raccoon feces in entrances are too dry for animals, but if a little damp show a succession of fungi starting with Phycomycetes, proceeding to Ascomycetes, and ending with Basidiomycetes (mushrooms) and Fungi Imperfecti.

If raccoon feces are wet enough, they can be used by a few fast-reproducing species. The larvae of the troglobiont fly *Spelobia cavernarum* can completely use a turd. Those turds not dominated by fly larvae are found be the beetle *Ptomaphagus hirtus* whose larvae are in local aggregations. Larvae of Psychodid and Phorid flies are occasionally found but always in low numbers.

13.9.2 Cave Rat Feces

Neotoma magister feces potentially have the highest diversity of cave species because of the varied diet of rats and the microclimate buffering of piles of pellets that occur near nests in latrines where the rats also urinate. But rats usually nest in areas too dry for animal decomposers, and often we cannot find latrines even when we know a rat is resident.

Animals and fungi decrease in density and species diversity from freshest pellet at the surface to oldest pellets deep in the pile. Animals and fungi may compete for the fecal resources. Larvae of a small and a large species of Sciarid fly can graze abundantly on a felt-work of Phycomycete fungi on fresh pellets unless a Staphylinid beetle finds and eats most of them. At the bottom of the pile, the species of mite and springtail depend on whether the remaining fecal fragments are on sand, mud, or rock.

13.9.3 Cave Cricket Feces

Hadenoecus subterraneus feces support the greatest variety of cave species because they occur in two dispersion patterns. Feces occur as guano veneers under dense entrance roosts. And they occur as individual splotches where individual crickets leave them while commuting from entrance roosts to deep cave reproductive areas in winter.

The sparse feces are important for our most troglobiont species. Among detritivores, they include the *Litocampa* bristletail and the *Scoterpes* millipede. Among predators, they include the *Anthrobia* spider and the *Phalangodes* harvestman.

The veneers of guano under roosts at entrances support different species at different moisture levels at one time and at different frequencies of renewal at different times. From the early 1970s to the early 1990s in White Cave, the density and species diversity of cave animals declined with droughts and cold winters, both of which restricted the nighttime foraging of crickets outside and so decreased the deposition of feces (Poulson et al. 1995). The snail *Carychium* and the millipede *Chaetapsis* were found only under the best of conditions, whereas the more troglobiont millipede *Scoterpes* and Dipluran *Litocampa* were always found. Even under the best of conditions, the greatest species diversity is found where moisture conditions are just right, not too wet and not too dry.

13.9.4 Sand Beetle Feces

Neaphaenops tellkampfii feces have the least diverse but most consistent community of troglobionts (Poulson 1992). The supported species are least diverse and most energy efficient because beetle feces, even though concentrated under rocks, have very low energy availability and are very small. The communities of detritivores (different springtails and mites) and their predators (mainly the spider *Anthrobia monmouthia* and pseudocorpions *Kleptochtonius* spp.) are consistently different with sand and mud substrates that maintain the same micro–zonation over many years.

13.10 Competition Among Closely Related Species of Carabid Beetles

Darwin predicted that the most closely related species will be the strongest competitors and so might diverge to avoid competition by evolving differences in habitat and/or feeding niche. This prediction is supported for the five Carabid beetle species in the Mammoth Cave region. Barr (1967) reviewed their differences in size, general habitat, and rarity. McKinney (1975) showed that two species of the same size on the Sinkhole Plain live in different microhabitats with different prey as food. In addition, Kane and Poulson (1976) showed that two species of different sizes differ in microhabitat and degree of food specialization with season.

13.11 Mechanisms and Consequences of Competition in the Sand Beetle

Griffith and Poulson (1993) have shown that the sand beetle *Neaphaenops tellkampfii* competes for its main cricket egg prey by both direct completion *aka* interference and indirect competition *aka* exploitation.

Exploitation competition was shown by showing, in the field and laboratory, that the number of and depth or holes dug for buried eggs decreased with increased density of beetles.

Interference competition included chasing, nipping, and fighting especially at holes that were dug to the depth of buried egg. The context is that beetles spend only 7% of their time budget searching for eggs but 74% of their energy budget digging up eggs. The advantage of defending one cricket egg is high since the egg is the same mass as a beetle.

13.12 Mysteries Suggest Further Studies

13.12.1 Where Are Larval *Neaphaenops tellkampfii*?

We do not find the number of larvae and pupae needed to account for the yearly presence of very many new-hatched teneral adults. The larvae do not occur in tiny cracks and crannies or in beetle holes. It is a mystery as to what they eat.

13.12.2 Why Is the Recapture Rate of Marked Adult Cave Crickets so Low?

Researchers have marked hundreds of cave crickets at several entrance roosts and get a maximum of 10% recapture the first day. They are not retreating to cracks and crannies where we cannot census them, and we see very low predation rates when they forage outside at night.

13.12.3 Why Are Cave Cricket Populations so Often Sink Populations?

Sink populations have too few early instars to infer that reproduction is occurring locally. Only two of nine study entrances are source populations with size distributions that indicate that reproductive areas are close to the entrance roosts. In addition, size distributions at the nine entrances have been stable from 1990 to 2015.

13.12.4 Why Are Some Organic Resources in Caves not Used?

Lamp flora such as algae and moss around electric lights are not used by troglobionts. Likewise, organic lint from people along tour trails is not used. Finally, colonies of Actinomycete bacteria are not consumed. Actinomycetes are chemically protected and may be indigestible, but this is not the case for algae.

13.12.5 Why Are Some Species Found Abundantly but Apparently not Eaten?

Heliomyzid flies are dense on some flat cave ceilings near entrances in winter, but seem not to be part of any cave food chains. They do not come to bait attractants, and populations are stable for months. Similarly, Enchaetryid worms are dense in flood zone mud of base-level rivers, but seem not to be eaten.

Research Resulting in Unpublished Theses and Dissertations

Cohen J (1985) Competition of cave beetles (*Ptomaphagus hirtus*) and fungi for dung. Masters Thesis, Biological Sciences, University of Illinois at Chicago

Griffith DM (1990) Ecology and evolution of predatory behavior in the Carabid cave beetle *Neaphaenops tellkampfii*. Ph.D Dissertation, Biological Sciences, University of Illinois at Chicago

Helf KL (2003) Foraging ecology of the cave cricket *Hadenoecus subterraneus*: Effects of climate, ontogeny, and predation. Ph.D. Dissertation, Biological Sciences, University of Illinois at Chicago

Kane TC (1974) Studies of simple cave communities: Predation strategies of two co-occurring Carabid beetles. Ph.D. Dissertation, University of Notre Dame, South Bend

Lavoie KH (1982) The influence of biotic and abiotic factors on the successional decomposition of cave rat dung. Ph.D Dissertation, Biological Sciences, University of Illinois at Chicago

Levy ES (1976) Aspects of the biology of *Hadenoecus subterraneus* with special reference to foraging behavior. Masters Thesis, Biological Sciences, University of Illinois at Chicago

Richards PM (1992) Predatory effects. life cycle, and husbandry of the Staphylinid beetle (*Quedius erythrogaster*) in the cave rat latrine. Masters Thesis, Biological Sciences, University of Illinois at Chicago

VanZant TR (1976) Coexistence of a multi-species complex of Carabid beetles: Micro-distribution and life history characteristics. Masters Thesis, Biological Sciences, University of Illinois at Chicago

References

Barr TC Jr (1967) Ecological studies in the Mammoth Cave System, Kentucky. I: The biota. Inter J of Speleology 3:147–203

Barr TC Jr, Kuehne R (1971) Ecological studies in the Mammoth Cave System, Kentucky II: The ecosystem. Annales de Speleologie 26:47–96

Culver DC (1982) Cave life: evolution and ecology. Harvard University Press, Cambridge, Massachusetts

Griffith DM, Poulson TL (1993) Mechanisms and consequences of intraspecific competition in a Carabid cave beetle. Ecology 74 (5):1373–1383

Kane TC, Poulson TL (1976) Foraging by cave beetles: spatial and temporal heterogeneity of prey. Ecology 57(4):793–800

Lavoie KH, Helf KL, Poulson TL (2007) The biology and ecology of North American cave crickets. J of Cave and Karst Studies 69 (1):114–134

McKinney T (1975) Studies on the niche separation in two Carabid cave beetles. Inter J of Speleology 7:65–78

Mohr CE, Poulson TL (1966) The life of the cave. McGraw Hill, New York

Poulson TL (1981) Variations in life history of Linyphiid cave spiders. In: Proceedings 8th international congress of speleology 1:60–62

Poulson TL (1992) The Mammoth Cave ecosystem. In: Camacho A (ed) The natural history of biospeleology. Museo Nacional de Ciencias Naturales, Madrid, Spain, pp 569–611

Poulson TL (2012) Food sources. In: White WB, Culver DC (eds) Encyclopedia of Caves, 2nd edn. Elsevier, Amsterdam pp 323–334

Poulson TL, Kane TC (1981) How food type determines community organization in caves. In: Proceedings 8th international congress of Speleology 1:56–59

Poulson TL, Lavoie KH, Helf KL (1995) Long-term effects of weather on the cricket (*Hadenoecus subterraneus*, Orthoptera, Rhaphidophoridae) guano community in Mammoth Cave National Park. Amer Midland Natur 134:226–236

Subsurface Aquatic Ecology of Mammoth Cave 14

Kurt L. Helf and Rickard A. Olson

Abstract

The hidden nature of Mammoth Cave National Park's (MCNP) cave aquatic ecosystem may suggest it is disconnected from events occurring on the surface. However, it is part of a continuum of water that begins with rain falling within the Green River watershed upstream of Mammoth Cave. Locally, water drains through Mammoth Cave's 184 km^2 (71 square mile) watershed and ends with the master stream for south-central Kentucky, the Green River. Rain falling in an area underlain by insoluble rock, such as sandstone on the Chester Upland, flows overland as runoff until it reaches a crevice, where it feeds a sinking stream or vertical shaft below. Rain falling on an area underlain by epikarst, the layer of highly weathered limestone and soil just beneath the surface common in the Mammoth Cave region, either percolates relatively slowly through interconnected vertical and horizontal channels (Fig. 14.1) or flows rapidly through a sinkhole. Whether water moves slowly or rapidly through this unsaturated zone of partially water-filled channels it ultimately reaches the water table and flows out of springs along the Green River (Fig. 14.2). When the Green River rises during floods, springs and base-level cave streams temporarily reverse their flow. A combination of back-flooding and local water influx can cause water levels in the cave to rise as much as 18 m (60 feet). During water's journey through the cave aquatic ecosystem, it transports organic matter, and often contaminants, from the surface. This is a greatly simplified description of the Mammoth Cave region's hydrogeology and vulnerability to pollution (see Chaps. 6, 8 and 18 for more information), but it is important for understanding the interplay between the nonliving components of the cave aquatic ecosystem, the organisms living there, the structure of their communities, and the impact of human activities.

It takes generosity to discover the whole through others. If you realize you are only a violin, you can open yourself up to the world by playing your role in the concert.
—Jacques Yves Cousteau

14.1 Habitat Types and Organisms

Absent the effects of human activities on surface water quality, the reliability of water is one of the major limiting factors for cave aquatic communities in upper level habitats in the Mammoth Cave area (Poulson 1992) and generally increases from the unsaturated zone to the water table (Fig. 14.3).

K.L. Helf (✉)
Department of Biology, Western Kentucky University, Bowling Green, KY, USA
e-mail: kurthelf@gmail.com

R.A. Olson
Cave Research Foundation, Cave City, KY, USA
e-mail: rickardolson16@gmail.com

H.H. Hobbs III et al. (eds.), *Mammoth Cave*, Cave and Karst Systems of the World,
DOI 10.1007/978-3-319-53718-4_14

Fig. 14.1 Icicles flowing from underground channels at a roadcut in MCNP demonstrate the vertical and horizontal flow of water through the epikarst. *Photograph* by Kurt Lewis Helf

14.1.1 Epikarst

Located at the top of the unsaturated zone, the epikarst is an ecotone, or transitional habitat, between surface and cave ecosystems. As such, ecotones exhibit a blend of environmental conditions and organisms from these very different ecosystems. Forest and grasslands on the surface supply the epikarst with tiny particles and molecules of organic matter, but the perpetual darkness in epikarst is a feature of subterranean habitats. Epikarst is a permanent habitat for many small surface and cave aquatic organisms (Pipan et al. 2006; Pipan and Culver 2013).

Epikarst drainage into the cave also creates a wide variety of isolated, diverse temporary habitats such as thin films of water on cave speleothems, drip pools (Fig. 14.4), and seeps at collapsed areas (Barr and Kuehne 1971). Aquatic organisms found in these habitats may include those from the surface such as cladocera, copepods, earthworms, fungi, nematodes, ostracods, and protozoa (Culver and Pipan 2009; Pipan et al. 2006). These temporary habitats also harbor a number of stygobionts, organisms only found in cave aquatic habitats, including the eyeless, unpigmented isopod *Caecidotea stygia*, the amphipod *Crangonyx barri*, which can be eyeless and unpigmented or not, the eyeless, unpigmented amphipod *Stygobromus vitreus*, the aquatic earthworms *Aeolosoma* and *Chaetogaster*, and the unpigmented flatworms *Sphalloplana percoeca* or *Sphalloplana buchanani* (Hubricht 1943; Gittleson and Hoover 1970; Barr and Kuehne 1971; Kenk 1977; Lewis 1988; Zhang and Holsinger 2003; Culver et al. 2010). Some permanent residents of the epikarst are unlikely to survive long-term if they are deposited in temporary cave pools. Hypothesized sources of mortality for epikarst organisms in drip pools include predation, lack of suitable habitat, reduced/absent reproduction, and competition for relatively scarce dissolved organic carbon (Pipan et al. 2010). The organic carbon in drip pools supports a thin film of bacterial colonies coating the substrate which largely feeds grazers such as flatworms (Simon et al. 2003). Water levels in these temporary habitats will decrease considerably and often dry out completely during droughts. However, rains that follow eventually replenish pools with water, organic matter, and organisms from the epikarst.

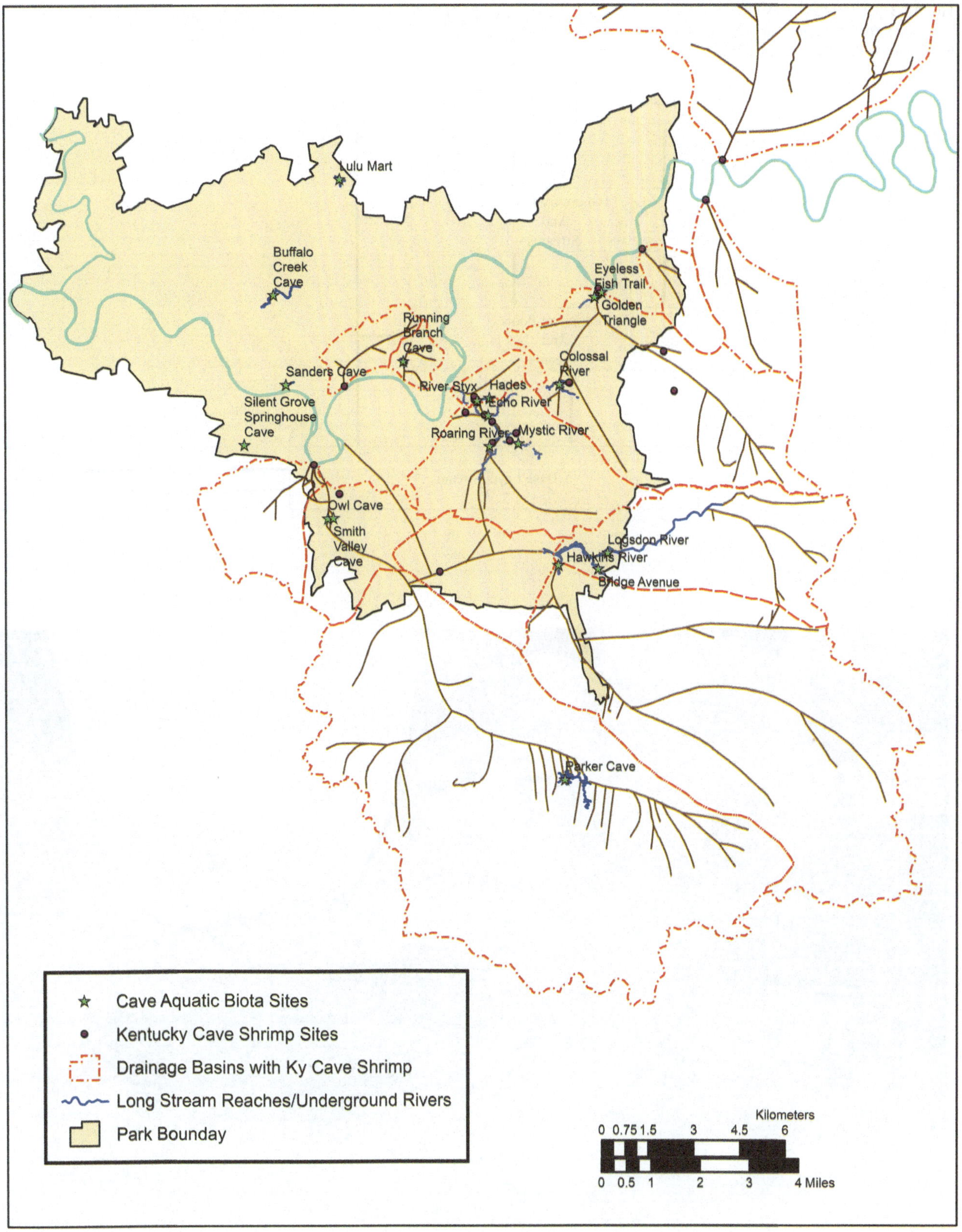

Fig. 14.2 Major karst groundwatersheds/basins of Mammoth Cave National Park, springs, and locations of surveyed cave aquatic communities in human accessible cave streams/rivers. Note the difference among basins in the proportion of land outside MACA boundaries. Map courtesy of Rickard Toomey

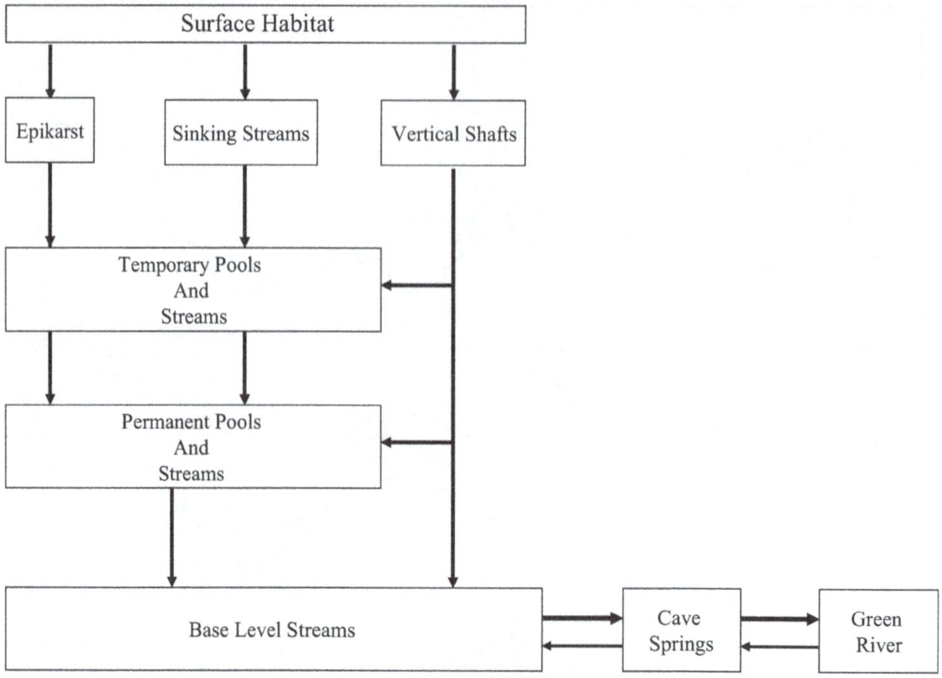

Fig. 14.3 Conceptual model of water's journey from the surface to the Green River in the MCNP region

Fig. 14.4 Epikarst drip is the source of water for this pool in White Cave. *Photograph* by Kurt Lewis Helf

14.1.2 Shafts

Water flowing off the sandstone caprock onto the limestone sinks and may form shafts where water can plummet directly to mid or lower levels (Fig. 14.5). Being storm driven, these sources of water flowing into the cave are more variable than epikarst flow. The community of aquatic organisms found in streams and pools fed by shaft drains is transitional in that stygobionts become more prevalent than epikarst organisms. The small eyeless, unpigmented amphipod *Stygobromus exilis* (Fig. 14.6), is a generalist widely distributed among cave aquatic habitats and replaces the epikarst specialist *S. vitreus* (Lewis 1988; Culver et al. 2010). Another amphipod associated with cave streams, *C. barri*, may also be found. The isopod *C. stygia* typically inhabits upper level shaft drain streams, whereas in lower levels both *Caecidotea bicrenata* and *C. stygia* may be occasionally be found together in disturbed areas (J.J. Lewis personal communication). The blind, unpigmented cave crayfish *Orconectes pellucidus* are occasionally found in isolated pools. These crayfish are known to be scavengers but are also predators (Hobbs III and Daniel 1977).

Fig. 14.5 Water flow into the Maelstrom; an example of a vertical shaft. *Photograph* by Rickard A. Olson

Fig. 14.6 Epikarst specialist amphipod (*Stygobromus vitreus*). *Photograph* by Rickard A. Olson

14.1.3 High-Gradient Streams Above Base Level

Water from shaft drains and overflowing drip pools collect to form permanent flowing streams. Examples in Mammoth Cave include much of Mystic River, all of Logsdon River (Figs. 14.2 and 14.7), Eyeless Fish Trail, and River Acheron. Stygobionts found in these high-gradient streams are more adapted to food poor conditions compared with their closest surface relatives. Stygobionts display enhanced sensory organs, which enable them to find scarce or low-quality sources of food more easily. For example, the Southern cave fish *Typhlichthys subterraneus*, a blind, unpigmented species found in these high-gradient streams, possesses higher numbers of sensory organs and corresponding expansions in its brain, than its closest surface relative the cave springfish *Forbesichthys agassizii* (Niemiller and Poulson 2010). The Southern cave fish is highly sensitive to distant vibrations in the water and can more easily locate the patchily distributed copepods, isopods, amphipods, small crayfish, and salamander larvae or even smaller cave fish on which it feeds. These streams may also support large numbers of cave crayfish *O. pellucidus*. The highest density was 376 individuals per 5000 m^2 in a section of Logsdon River (Pearson and Jones 1998). These crayfish are wide ranging foragers and can even move overland between isolated pools as long their gills do not dry out (Fig. 14.8).

In nutrient-enriched cave streams, the surface crayfish *Cambarus tenebrosus* can be highly abundant and may even be able to reproduce and, in this context, likely out-competes *O. pellucidus* since it is larger and stouter-bodied. However, in the most recent surveys of Park cave streams, densities of these crayfish never rose above 29 per 5000 m^2 (Pearson and Jones 1998). Thus, in cave streams with low food availability, the cave crayfish's significantly longer antennae enable them to locate food more efficiently than the surface crayfish (Ziemba et al. 2003; Taylor et al. 2010). Other cave invertebrates commonly found in these high-gradient streams include the isopod *C.*

Fig. 14.7 Logsdon River, an example of a high-gradient cave stream above base-level, flows into P. Strange Falls. *Photograph* by Gary Berdeaux

bicrenata (occasionally, with *C. stygia*), the amphipod *S. exilis*, and *Sphalloplana* sp. flatworms.

14.1.4 Base Level

Ultimately, all water moving through these habitats reaches the water table, or base level, which is equivalent to the elevation of the Green River near that part of the cave (Fig. 14.9). One of the more highly adapted cave fishes is the blind, unpigmented Northern cave fish *Amblyopsis spelaea*. It is a top predator that reaches its highest abundance in Roaring River, an excellent example of a base-level stream in the park (Figs. 14.2 and 14.10). Another highly adapted organism observed at base level is the unpigmented, eyeless Kentucky cave shrimp *Palaemonias ganteri*, a federally listed endangered species (Fig. 14.11). Its highest estimated density to date (i.e., 1308/5000 m^2) was documented during a survey of Mystic River, a tributary of Roaring River on the south side of the Green River, only slightly above base level (Pearson and Jones 1998). Since *P. ganteri* is found in cave streams on both sides of the Green River (Table 14.1), a potential geographic barrier to genetic exchange between populations, it is possible there is more than one species. These shrimp are thought to feed on the microorganisms living in the sediment, which it has been observed filtering

Fig. 14.8 Cave Crayfish (*Orconectes pellucidus*) can walk overland between cave pools and streams. *Photograph* by Kurt Lewis Helf

through its mouthparts (Cooper and Cooper 2010). Roaring River is also the habitat of the eyeless, unpigmented shaggy cave snail *Antroselates spiralis*. Other cave aquatic biota found in base-level streams include the isopod *C. bicrenata*, the amphipod *S. exilis*, *Sphalloplana* sp. flatworms, the cave crayfish *O. pellucidus*, the Southern cave fish *T. subterraneus*, occasionally, the cave spring fish *F. agassizi,* and the sculpin *Cottus carolinae*.

The diversity of aquatic organisms in base-level streams, particularly those associated with spring outlets, is partially attributable to aquatic invertebrates from the surface. Whitman (1989) samplers in base-level cave sediments and Barr and Kuehne's (1971) plankton collections from base-level streams found myriad surface aquatic organisms such as diatoms, filamentous green and blue algae, flatworms, and roundworms. Barr (1967) collected surface rotifers in park cave streams, whereas Whitman (1989) speculated some rotifer species he found might have been cave adapted and new to science. Oligochaetes, or segmented worms, such as *Aelosoma*, *Chaetogaster*, tubificids, and enchytraeids were reported by both Barr and Kuehne (1971) and Whitman (1989) in cave stream sediments. Whitman also found the larvae of at least five different genera of midges, occasionally in high densities, living in the sands of cave streams such as Echo River. While it is unknown whether adult midges can survive or reproduce in the cave, it is clear that at least their larval stages play some role in the park's cave aquatic ecosystem. Barr and Kuehne (1971) and Whitman (1989) regularly found cladocera and copepods in their zooplankton samples; and Barr and Kuehne (1971) observed both water fleas and copepods bearing young and egg sacs. They attributed winter increases in zooplankton to the influx of water and detritus from percolating ground water, sinking streams, and backflow from the Green River. They also speculated the increased zooplankton density they found in summer and fall were due to secondary microbial production in cave pools and streams, based on detritus carried in by floods.

Fig. 14.9 Roaring River; an example of a base-level cave stream. *Photograph* by Rickard A. Olson

Relatively recent biological monitoring in Roaring and Echo/Styx rivers (Pearson and Jones 1998) found aquatic vertebrates from the surface such as salamanders, frogs (*Rana palustris*, *R. clamitans*), toads (*Bufo woodhouse fowleri*), and surface fishes. There is no good evidence surface aquatic organisms make a significant contribution to the cave aquatic community as predators. However, two fish species, the Spring Cavefish (*F. agassizii*) and the Banded Sculpin (*C. carolinae*), are regularly observed, though in low densities, in base-level streams associated with springs in the park (Pearson and Jones 1998; Niemiller and Fitzpatrick 2012). *F. agassizii* may be found in both surface and cave streams which classifies them as stygoxenes (Culver and Pipan 2009; Niemiller and Fitzpatrick 2012). Gut contents indicate *F. agassizii* in cave aquatic habitats feeds on amphipods, midge larvae, and worms (Niemiller and Poulson 2010). Gut contents from *C. carolinae* in surface streams indicate they feed on aquatic insects, crayfish, isopods, amphipods, snails, and other fish (Poly and Boucher 1996; Tumlinson and Cline 2002).

14.2 Energy Input

Inputs of photosynthetically derived organic matter from surface ecosystems, such as dissolved organic matter leached from the vegetation litter–soil interface (think of water percolating through coffee grounds to produce coffee), appear to be the dominant source of energy input to the cave aquatic ecosystem in the Mammoth Cave region (Fig. 14.12). Indeed, temperature, precipitation, and forest biomass, indicators of primary productivity on the surface and the availability of organic carbon, were all found to be important factors in predicting the presence of cave organisms (Christman et al. 2016). Flow reversals and back-flooding from the Green River into cave springs also transport

Fig. 14.10 Northern Cave Fish (*Amblyopsis spelaea*). *Photograph* by Matt Niemiller

Fig. 14.11 Kentucky cave shrimp (*Palaemonias ganteri*). *Photograph* by Michael Durham

Table 14.1 Drainage basins, caves, and cave streams in which Kentucky cave shrimp (*Palaemonias ganteri*) have been found. Localities based on United States Fish and Wildlife Service (1988) and data in Mammoth Cave National Park files

Drainage basin	Cave/Spring	Site
McCoy Bluehole		
	McCoy Bluehole Spring	
Suds Spring		
	Suds Spring	
Mile 205.7		
	Mile 205.7 Spring	
Pike Spring		
	Northtown Cave	Lower level stream
	Roppel Cave	Grand Central Sump
	Colossal Cave	Colossal River
	Unknown Cave	Eyeless Fish Trail
	Unknown Cave	Golden Triangle
	Pike Spring	
River Styx		
	Mammoth Cave	Hades
	Mammoth Cave	River Styx
Echo River		
	Mammoth Cave	Echo River submerged passage
	Mammoth Cave	Echo River
	Mammoth Cave	Roaring River
	Mammoth Cave	Shrimp Pools
	Mammoth Cave	Mystic River
	Mammoth Cave	Lucy's Dome Drain
Running Branch		
	Running Branch	Reccius River
Ganter Bluehole		
	Ganter Cave	
Turnhole		
	Lee Cave	Snake River
	Whigpistle Cave	Red River
Turnhole—Double Sinks		
	Sandhouse Cave	Sandhouse Cave Spring

organisms and photosynthetically derived organic matter into the cave aquatic ecosystem.

14.2.1 Green River Flow Reversals and Back-Flooding into Cave Streams

Flow reversal events are a normal and vital part of cave aquatic ecosystem function in Mammoth Cave. When the Green River rises above nearby cave streams the river flows into, rather than out of, cave springs. This backflow will continue until surface and subsurface water levels equalize and normal flow out of cave springs resumes (see Chap. 8 for more information on this phenomenon). These flow reversals carry particulate organic matter and myriad surface aquatic organisms (e.g., surface fishes) that greatly contribute to food energy input. A recent study of surface fish captured in Mammoth Cave's base-level streams yielded 22 species from nine families over two field seasons (Ruhl 2005).

An unusual back-flooding relationship exists at times when Green River water enters River Styx Spring and exits Echo River Spring, a straight line mile (1.6 km) to the south. In a recent three-year study, water temperature was used as a proxy to determine the direction of flow (Trimboli et al. 2016). The authors reasoned that during flow reversal events, water temperatures in Styx and downstream Echo Rivers, typically above or below that of the Green River,

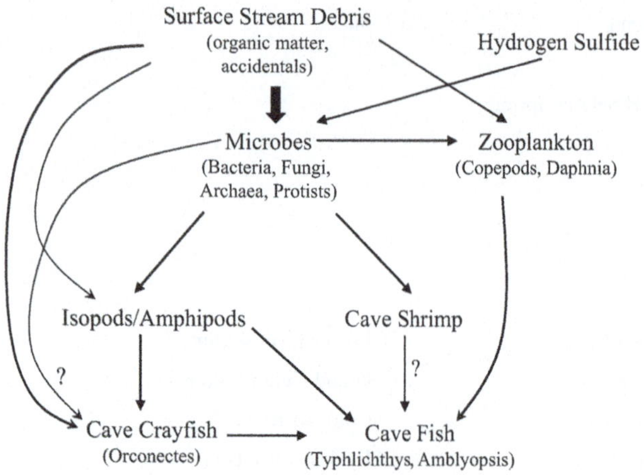

Fig. 14.12 Hypothesized food web diagram for cave aquatic habitats in MCNP region. *Arrows* indicate the direction of energy flow. Modified from Barr and Kuehne (1971)

would be similar to Green River water temperatures. Their data showed only River Styx underwent dramatic deviations from its mean temperature of 13.5 °C ± 2.5 reaching a maximum of 23.8 °C and a minimum of 3.6 °C (Trimboli et al. 2016). Water temperature in Echo River, upstream of the area affected by flow coming from River Styx, remained relatively stable at 13.4 °C ± 0.6. Over the three-year study, their temperature data showed that these flow reversal events occurred an average of five times each year.

14.2.2 Shifting Paradigm

The long-standing paradigm among cave ecologists was that the cave aquatic ecosystem was almost exclusively supported by particulate organic matter (POM: leaves, twigs, etc.) washed in from the surface. Logically, increased POM input should support more stygobionts. Yet data collected at Mammoth Cave over many years show no discernible relationship between POM and stygobiont density. For example, cave streams with low POM (Logsdon River/Bridge Avenue) had high stygobiont densities and cave streams with high POM (Mystic River) overall, with the exception of cave shrimp, had lower densities of stygobionts (Table 14.2). Cave streams have been generally thought to have low POM supply compared with surface streams. However, data from MCNP indicate there are a few exceptions to the general assumption of the old paradigm. For example, Pearson and Jones' (1998) POM data from nine cave stream reaches in Mammoth Cave ranged from 12.7% at Mystic River, remarkably nearly double Whitman's (1989) data from surface streams, to 1.4% at Echo/Styx (Table 14.2). However, POM from the surface is mostly processed by consumers (e.g., amphipods and isopods) near its point of entry into cave streams, and dissolved organic carbon is a more important source of carbon in deep cave habitat (Simon and Benfield 2001; Simon et al. 2007).

14.3 Food Web

Compson (2004) analyzed ratios of carbon and nitrogen isotopes in the tissue of biota from Mammoth Cave's surface streams, springs, and cave streams to evaluate their food sources and feeding relationships. Because carbon is relatively stable between trophic levels, differences in the ratios of carbon isotopes in animal tissue are used to determine an organism's food sources. Nitrogen, however, is enriched as it moves through successive trophic levels, and so increased ratios of nitrogen isotopes in an organism's tissues can differentiate between producers and consumers among the ecosystem's constituent organisms. He concluded periodic back-flooding events through cave springs likely contribute substantial pulses of nutrients to the cave stream community. Surface fish trapped in cave streams are a clear example of food input from the surface because after their inevitable death, they become food for stygobiont scavengers. Similarly, the surface crayfish *C. tenebrosus*, while not abundant in MCNP cave streams, is frequently encountered in cave streams and might subsist on detritus washed in from the surface. Compson's nitrogen isotope data place *C. tenebrosus* in a low trophic level and its carbon isotope data are close to that of detritus and fungal mycelia.

Compson's (2004) data on carbon and nitrogen isotopes offer some insight into the cave stream community's ultimate food source and its feeding relationships. Carbon isotope data clearly show the ultimate food source of MCNP's stygobionts is derived from bacteria (Fig. 14.13), likely as bacterial biofilms, and so provides support for the new paradigm. Carbon isotopes in stygobiont tissues were enriched relative to detritus, which suggests it is not the food source of their prey. Since the cave crayfish *O. pellucidus* is known to be at least partially predatory, it is not surprising its enriched nitrogen isotope levels place it in one of the upper trophic levels. Interestingly, the nitrogen isotope data for the cave isopod, presumably *Caecidotea* sp., place it near *O. pellucidus* trophic level suggesting that it, too, is partially predatory. Finally, enriched nitrogen isotope levels in the Southern cave fish (*T. subterraneus*) indicate that it, of the stygobionts tested, occupied the highest trophic level and so is one of two top predators in MCNP's cave stream communities. Presumably the Northern cave fish *A. spelaea*, though its tissue was not tested, occupies a similar position. These trophic relationships, however, are generalizations of what are likely much more complicated interrelationships within cave stream communities (Fig. 14.12). As yet, we have limited data regarding the origins of the dissolved organic matter that fuel the bacterial communities driving the

Table 14.2 Density of stygobiotic and Stygophilic fauna in Mammoth Cave area subsurface streams as a function of their length, coarse particulate organic matter, and microbial biomass

Reach	Basin	Reach length (m)	Coarse particulate organic matter (% weight LOI)	Microbial biomass (m mol/g)[d]	Mean density of stygobiotic fauna/5000 m^2	Mean density of Stygophilic fauna/5000 m^2
Echo and Styx Rivers	Echo Spring and River Styx	835	1.3–2.1[a]/1.4[c]	–	12.3	9.4
Mystic River	Echo Spring	1548	0.6–1.3[a]/12.7[c]	956	75.8	1.8
Roaring River/Shrimp Pools	Echo Spring	1371	2.2[c]	441	125.4	16.5
Colossal River	Pike Spring	1116	3.5[b]/3.4[c]	–	63.8	8
Eyeless Fish Trail	Pike Spring	726	1.6[b]/3.3[c]	726	67.8	3.7
Logsdon River/Bridge Ave.	Turnhole	913	2.2[c]	–	232.3	0.3
Logsdon River/Hawkins River	Turnhole	570	6.1[c]	1370	90	3.2
Owl Cave	Turnhole	11	–	2849	0	1
Brown River	Turnhole (Parker Cave)	100	–	–	900	–
Parker River	Turnhole (Parker Cave)	200	–	–	125	–
North Creek	Turnhole (Parker Cave)	150	–	–	1133	–
Sulphur River	Turnhole (Parker Cave)	225	–	–	405	20

[a]Whitman (1989)
[b]Poulson (1992)
[c]Pearson and Jones (1998)
[d]Fowler, unpublished

regional cave aquatic ecosystem. However, their origin and energetic content can be inferred based on the bacterial community found in area cave streams.

DNA analysis of bacterial biofilms grown on artificial substrates in MCNP's cave aquatic habitats (i.e., Owl cave, Hawkins/Logsdon confluence, Charon's Cascade, Mystic River, and Eyeless Fish Trail) suggests a diverse phylogenetic groups of bacterial communities are able to exploit a wide range of environmental conditions due to a wide range of metabolic processes (Fowler et al. 2009*). Proteobacteria, a group that includes both consumers and producers, were the dominant phyla and made up greater than 50% of all bacterial DNA found among all cave stream sites. Alphaproteobacteria, which are known to grow at very low nutrient levels, were the dominant group at most cave stream sites. DNA from gamma- and deltaproteobacteria, groups that include common gut fauna in animals, predators on other bacteria, and contributors to the sulfur cycle as producers of hydrogen sulfide under anaerobic conditions, was also found at most cave stream sites. Intriguingly, DNA from betaproteobacteria, a group that includes chemoautotrophs, was also found at all cave stream sites. The presence of alpha- and betaproteobacteria at most MCNP cave streams sites suggests they do not utilize POM subsidies from the surface as their primary energy source but instead rely on likely energy sources such as dissolved organic matter or chemoautotrophy. Cave stream sites such as Hawkins/Logsdon and Owl Cave, however, appear to be organically enriched due to agricultural input from watersheds

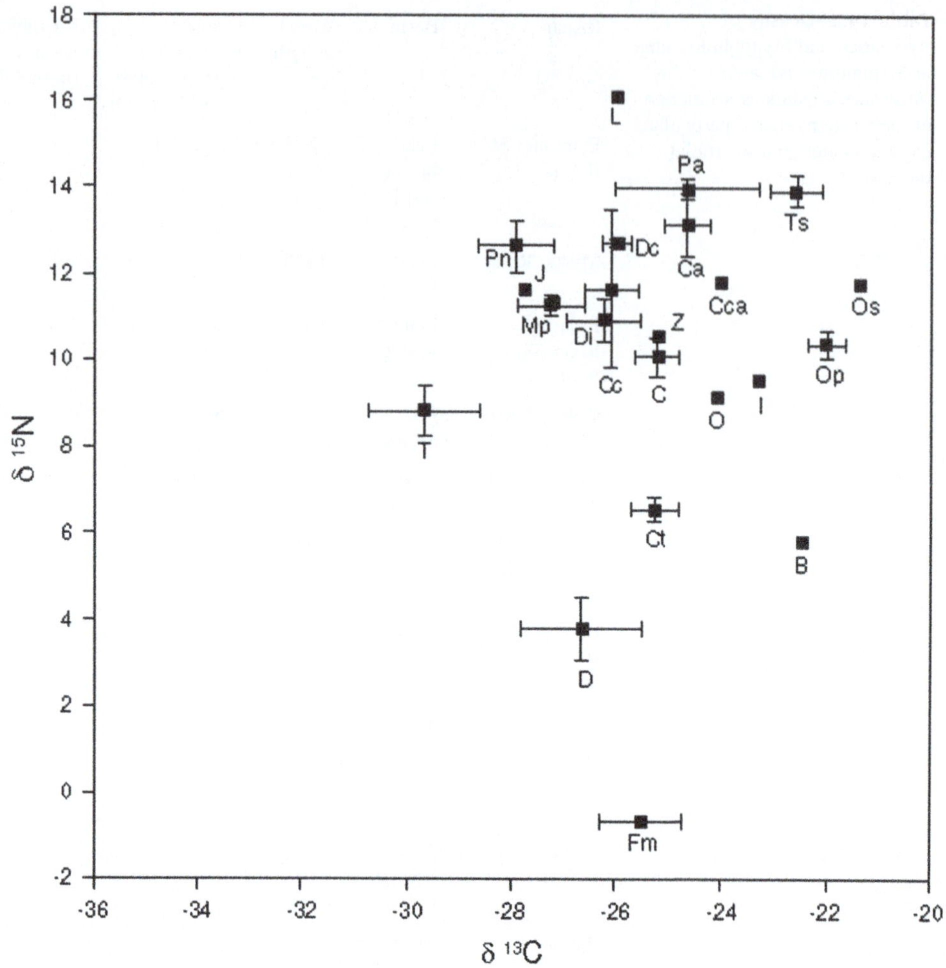

Fig. 14.13 Cave composite graph of temporally and spatially pooled $\delta^{13}C$ and $\delta^{15}N$ data (Compson 2004). Bacteria (B); *Cambarus tenebrous* (Ct); Chironimidae (C); *Forbesichthys agassizi* (Ca); *Cottus carolinae* (Cc); *Cyprinus carpio* (Cca); detritus (D); Diptera (Di); *Dorosoma cepedianum* (Dc); fungal mycelia (Fm); Isopoda (I); larval fish (L); *Micropterus punctulatus* (Mp); Oligochaeta (O); *Orconectes pellucidis* (Op); Ostracoda (Os); *Pimephales notatus* (Pn); *Pomoxis annularis* (Pa); tadpole (T); *Typhlichthys subterraneus* (Ts); and Zooplankton (Zo). Used with permission

outside the park. Indeed, their data showed total microbial biomass on artificial substrates in MCNP's deep cave aquatic habitats decreased from organically enriched cave streams, with input from watersheds outside MCNP's boundaries, to "pristine" cave streams whose watersheds are entirely within park boundaries: Owl cave ≫ Hawkins/Logsdon > Roaring River ≫ Eyeless Fish Trail > Mystic River. There may also be chemoautotrophic energy inputs provided by sulfur oxidizing bacteria and possibly some energy provided by hydrocarbon oxidizing bacteria. The relative magnitudes of the latter two energy inputs are unknown at this time.[1]

[1]An unpublished poster titled "Concentration and Diversity of Bacteria in Clastic Sediments and Limestone Biofilms of Mammoth Cave, Kentucky" by Rick Fowler, Rick Olson, Hazel Barton, and Shivendra Sahi, a progress report to the National Cave and Karst Research Institute in Carlsbad, NM. The poster is stored in the Mammoth Cave National Park Curatorial Facility, accession number 818.

14.4 Chemoautotrophy and Potential Support of Troglobionts via Hydrocarbon Energy Inputs

Cutting across the Central Kentucky Karst is a warp in the bedrock along an east–west band where sulfurous brine rises under artesian conditions. Parker Cave is developed within this structural feature and has three streams that receive sulfurous brine (Fig. 14.2). Sulfur oxidizing bacteria use hydrogen sulfide to make organic carbon through a process called chemosynthesis. Although chemosynthesis is analogous to photosynthesis, the difference is that it occurs regardless of light level or season. Of the three streams with sulfides in Parker Cave, Sulphur River is the most studied (Angert et al. 1998; Olson and Thompson 1988; Roy 1988; Thompson and Olson 1988). These organic-rich cave streams drain into Mammoth Cave and so provide biomass that would not otherwise exist. Other sites along the warp in the bedrock may also provide energy to Mammoth Cave, but they are not yet documented. Indirect methods of investigation further downstream may help gauge the relative contributions to cave streams from photosynthetic versus chemosynthetic sources. For more discussion on this subject see Chap. 16.

Hydrogen sulfide-laden fresh water is common regionally due to sulfate minerals within the St. Louis Limestone. Unlike the brines rising in Parker Cave, these sulfides are shallow and several streams in Mammoth Cave are vertically less than 100 feet and maybe as little as 50 feet (30.5 − 15 m) above this sulfide rich zone. To migrate up, all they need is a fracture or a fault, and there are plenty of these. Artesian conditions in this sulfate zone are known to occur, and it makes perfect sense due to the regional northwest dip of the bedrock housing Mammoth Cave. These hydrogeological conditions and their potential biological significance to Mammoth Cave ecosystems are just now being considered (Olson, in press).

A sulfur spring was reported within Mammoth Cave in the mid-1800s (Bullitt 1845), and Hebes Spring, reported by Hovey (1912), is likely the same feature. These seeps, located in Marianne's Pass, contain low concentrations of hydrogen sulfide, which support bacterial mats typical of sulfur oxidizing bacteria. To date no similar seeps have been found in the cave, but Cave Research Foundation explorers have not been trained to recognize them, and there are thousands of tiny passages where similar seeps might exist. Such training for cavers has been identified as a top priority by microbiologists (Barton 2006). Both hydrogen sulfide and hydrocarbons are very abundant in the Mammoth Cave region which is one hypothesis helping explain the high diversity of troglobionts in the region (Olson 2013). This could represent another paradigm shift regarding Mammoth Cave in the views of biospeleologists. For details on biodiversity in Mammoth Cave see Chaps. 1 and 15. Hydrocarbon odors are associated with the sulfurous seeps in Mariannes Pass, and such sources of organic carbon could also be an auxiliary source of energy to Mammoth Cave ecosystems (Olson, in press) as it is in the Edwards Aquifer. For more discussion of hydrocarbons in Mammoth Cave see Chap. 10 (Meteorology).

14.4.1 A Case Study of Recovery from Severe Cave Stream Pollution

The strongest data available regarding the ecological effects of nonpoint and chronic point source pollution on a cave aquatic community in the Mammoth Cave region, including its post-mitigation succession and recovery, are available from long-term monitoring data in Hidden River Cave (Jones and Pearson 1997; Lewis 1995). Located beneath Horse Cave, KY Hidden River Cave was a tourist attraction and water source for the town in the early twentieth century until groundwater pollution ended the latter practice. In 1944, a local creamery began discharging its waste into the cave and both Horse Cave and Cave City began discharging their sewage effluent into the cave; the former town's effluent containing a mixture of domestic and industrial sewage (Lewis 1995). In the early 1980s, Lewis (1982) began monitoring the cave's aquatic community finding large numbers of aquatic organisms indicative of high nutrient loading: sewage fungus, sewage bacteria, and tubificid worms. Subsequent surveys every few months following the dedication of a new sewage regional treatment facility in the winter of 1989 showed little change in the community though, interestingly, Lewis (1995) observed a single individual of the surface crayfish *C. tenebrosus* and the stygobiotic isopod *C. bicrenata* (Table 14.3). In late 1991, *C. tenebrosus* were observed in high abundance, a condition which continued for the next several years, indicating a cave stream habitat still enriched enough to support large numbers of surface crayfish. In March 1993, the first stygobiotic cavefish (i.e., *T. subterraneus*) and several cave salamanders (*Eurycea lucifuga*), indicating further recovery of the cave aquatic community, were observed. In October 1993, Lewis finally observed *O. pellucidus* and *T. subterraneus* in abundances greater than surface stream organisms. Nearly a decade later, Lewis et al. (2015) observed diversity in Hidden River's Cave's stream community remained relatively low (Table 14.3). Jones and Pearson (1997) speculated the high abundance of *T. subterraneus* they found was due to a reproductive event in 1993, presumably due to high numbers of observed juveniles, where slight decreases in later years reflected local population "adjustments". Their final observations in October 1995, which included the amphipod *C. barri*, suggest the cave aquatic community was near full recovery; though a few typical accidentals (i.e., a green frog *R. clamitans*, a surface fish, and a salamander larva) were also present (Table 14.3). The utility of long-term monitoring data, collected by trained professionals via systematic biological surveys, to resource management at MCNP cannot be overstated.

14.5 Future Directions in Long-Term Monitoring

The overall purpose of natural resources monitoring in parks is to provide scientifically sound information on the current status and long-term trends in the composition, structure, and function of park ecosystems to park resource managers. Use of monitoring information increases confidence in managers' decisions, improves their ability to manage park resources, and enables them to confront and mitigate threats to the park and operate more effectively in legal and political arenas. Critical to resource managers is information on whether the true absence of a species of concern from specific habitats indicates the habitat is simply unoccupied, marginal, or

Table 14.3 Successional recovery of Hidden River Cave stream community after sewage diversion/treatment

Species	1982 (November)	1989	1990 (March)	1990 (Oct)	1990 (Dec)	1991 (Oct)	1993 (March)	1993 (July)	1993 (Oct)	1994 (Aug)	1994 (Oct)	1995 (Oct)	2013 (June)	2013 (Sept)	2013 (Nov)	2014 (Jan)	2014 (June)	2014 (July)	2014 (Aug)	2014 (Sept)
Tubificid Worms	X	X	X		X															
Sewage Fungus	X	X																		
Sewage Bacteria	X	X	X	X	X															
Isopod				1			1	21	18	62	3	1		14			51	8	17	
Amphipod								2	3	3	0	1							1	
C. tenebrosus			1	1		45	29		33	39	14	3	1	2	1		1		1	
O. pellucidus								12	70	109	7	122	1	21	17	8	22	1	15	52
E. lucifuga							3													
T. subterraneus							1	5	29	22	1	19		2						2
Surface Fish												1								
Salamander								1	17	2		1								
Frog									1			1								

degraded (Peterman et al. 2013; Peterson et al. 2013). For example, one of the arguments made by Lisowski and Poulson (1979) for removing the recently failed Lock and Dam #6, just downstream of the Green River segment flowing through MACA, was that with the dam in place-specific cave aquatic habitats saw increased siltation, decreased habitat heterogeneity, and reduced abundance of the federally endangered Kentucky cave shrimp *P. ganteri*. However, the failure to detect a species is not necessarily an indication it is absent from the community. Indeed, a monitoring method that fails to distinguish between whether a species is present and undetected or absent severely limits its utility to resource managers. Because these two states are not distinguishable, the likelihood of a species being associated with particular habitats (e.g., ecotonal cave spring habitat), even when it is not detected, must be estimated.

Future monitoring will utilize rigorous methods to determine cave aquatic organisms' habitat associations and their area of occurrence. State-of-the-art statistical modeling will be used to analyze counts of organisms and data on their presence/absence and so provide resource managers with valuable information regarding whether the absence of a target species from specific habitats indicates the habitat is simply unoccupied, marginal, or impacted (Peterman et al. 2013). Implementing a rigorous monitoring protocol for cave aquatic biota and their habitat covariates provides an excellent opportunity to gather baseline data on current habitat associations among cave aquatic biota before these changes occur as well as test prior predictions based on past research.

References

Angert ER, Northup DE, Reysenbach AL, Peek AS, Boebel BM, Pace NR (1998) Molecular phylogenetic analysis of a bacterial community in Sulphur River, Parker Cave, Kentucky. Am Miner 83:1583–1592

Barr TCJ (1967) Ecological studies in the Mammoth Cave System of Kentucky. I. The biota. Int J Speleol 3:147–203

Barr TCJ, Kuehne RJ (1971) Ecological studies in the Mammoth Cave system of Kentucky. II. The ecosystem. An de Spéléologie 26: 48–96

Barton HA (2006) Introduction to cave microbiology: a review for the non-specialist. J Cave Karst Stud 68:43–54

Bullitt A (1845) Rambles in the Mammoth Cave during the year 1844 by a visitor. Morton and Griswold, Louisville, Kentucky, p 88

Christman MC, Doctor DH, Niemiller ML, Weary DJ, Young JA, Zigler KS, Culver DC (2016) Predicting the occurrence of cave-inhabiting fauna based on features of the earth surface environment. PLoS ONE 11:e0160408

Compson ZG (2004) An isotopic examination of cave, spring and epigean trophic structures in Mammoth Cave National Park. M.S. Western Kentucky University

Cooper JE, Cooper MR (2010) Observations on the biology of the endangered stygobiotic shrimp *Palaemonias alabamae*, with notes on *P. ganteri* (Decapoda: Atyidae). Subterranean Biol 8:9–20

Culver DC, Holsinger JR, Christman MC, Pipan T (2010) Morphological differences among eyeless amphipods in the genus Stygobromus dwelling in different subterranean habitats. J Crustac Biol 30:68–74

Culver DC, Pipan T (2009) The biology of caves and other subterranean habitats. Oxford University Press, New York

Gittleson S, Hoover R (1970) Protozoa of underground waters in caves. Ann Speleol 25:91–106

Hobbs HH III, Daniel MA (1977) A review of the troglobitic decapod crustaceans of the Americas. Smithson Contrib Zool 244:1–183

Hovey H (1912) Mammoth Cave of Kentucky. John P. Morton & Company, Louisville, Kentuck, p 87

Hubricht L (1943) Studies in the Nearctic Freshwater Amphipoda, III. Am Midl Nat 29:683–712

Jones TG, Pearson WD (1997) The recovery of the aquatic biological community in Hidden River Cave, Horse Cave, Kentucky. In: Sixth annual Mammoth Cave National Park science conference. National Park Service, Mammoth Cave, Kentucky, p 231

Kenk R (1977) Freshwater Triclads (Turbellaria) of North America, IX: the genus *Sphalloplana*. Smithson Contrib Zool 246:38

Lewis JJ (1988) The systematics, zoogeography and life history of the troglobitic isopods of the Interior Plateaus of the eastern United States. University of Louisville, Kentucky

Lewis JJ (1995) The devastation and recovery of caves and karst affected by industrialization. In: 1995 National Cave management symposium. National Cave and Karst management symposia. Spring Mill State Park, Indiana, pp 214–227

Lewis JJ, Lewis TM, Eckstein J (1982) A biological reconnaissance of a polluted cave stream: the hidden river groundwater basin. Cave Research Foundation 1982 Annual Report: 9–10

Lewis JJ, Lewis S, Nims P (2015) Observations on the ecosystem of hidden river cave bioinventory: community census and water quality analysis. Kentucky Heritage Land Conservation Fund

Lisowski EA, Poulson TL (1979) Impacts of Lock and Dam Six on base level ecosystems in Mammoth Cave. Cave Research Foundation 1979 Annual Report: 48–54

Niemiller M, Fitzpatrick B (2012). Status and life history of the Amblyopsid Cavefishes in Kentucky. Kentucky Department of Fish and Wildlife Resources

Niemiller M, Poulson TL (2010) Subterranean fishes of North America: Amblyopsidae. In: Trajano E, Bichuette M, Kapoor B (eds) Biology of subterranean fishes. Science Publishers, Enfield, pp 169–282

Olson R (2013) Potential effects of hydrogen sulfide and hydrocarbon seeps on Mammoth Cave ecosystems. In: Mammoth Cave National Park's 10th research symposium. Mammoth Cave, KY, pp 25–30

Olson R (in press) Sulfides in the Mammoth Cave area. In: Klimchouk A, Palmer AN, Audra P, DeWaele J, Auler A (eds) Hypogene Karst areas and caves of the world. Springer International Publishing, Cham, Switzerland

Olson R, Thompson B (1988) Scanning electron microscopy and energy dispersive X-ray analysis of artificial and natural substrates from the Phantom flowstone of Sulphur River in Parker Cave, Kentucky. Bull Natl Speleol Soc 50:47

Pearson WD, Jones TG (1998) A final report based on faunal inventory of subterranean streams and development of a cave aquatic biological monitoring program using a modified index of biological integrity. National Park Service, Mammoth Cave National Park, p. 139

Peterman WE, Crawford JA, Kuhns AR (2013) Using species distribution and occupancy modeling to guide survey efforts and assess species status. J Nat Conserv 21:114–121

Peterson EE, Ver Hoef JM, Isaak DJ, Falke JA, Fortin M-J, Jordan CE, McNyset K, Monestiez P, Ruesch AS, Sengupta A, Som N, Steel EA, Theobald DN, Torgersen CE, Wenger SJ (2013)

Modelling dendritic ecological networks in space: an integrated network perspective. Ecol Lett 16:707–719

Pipan T, Christman MC, Culver DC (2006) Dynamics of epikarst communities: microgeographic pattern and environmental determinants of epikarst copepods in Organ Cave, West Virginia. Am Midl Nat 156:75–87

Pipan T, Culver DC (2013) Forty years of epikarst: what biology have we learned? Int J Speleol 42:215–223

Pipan T, Holt N, Culver DC (2010) How to protect a diverse, poorly known, inaccessible fauna: identification and protection of source and sink habitats in the epikarst. Aquat Conser Mar Freshw Ecosyst 20:748–755

Poly WJ, Boucher CE (1996) Nontroglobitic fishes in caves: their abnormalities, ecological classification and importance. Am Midl Nat 136:187–198

Poulson TL (1992) The Mammoth cave ecosystem. In: Camacho A (ed) The natural history of biospeleology. Museo Nacional de Ciencias Naturales, Madrid, pp 569–611

Roy W (1988) Preliminary assessment of the solution equilibria of Sulphur River in Parker Cave, Kentucky. Bull Nat Speleol Soc 50:37–41

Ruhl M (2005) Flow reversal events increase the abundance of nontroglobitic fish in the subterranean rivers of Mammoth Cave National Park. Western Kentucky University, Kentucky

Simon KS, Benfield EF (2001) Leaf and wood breakdown in cave streams. North Am Benthol Soc 20:550–563

Simon KS, Benfield EF, Macko SA (2003) Food web structure and the role of epilithic biofilms in cave streams. Ecology 84:2395–2406

Simon KS, Pipan T, Culver DC (2007) A conceptual model of the flow and distribution of organic carbon in caves. J Cave Karst Stud 69:279–284

Taylor MS, Blechle BE, Pobst BS (2010) Morphological divergence between cave and surface populations of the digger crayfish, *Fallicambarus fodiens* (Cottle, 1863) (Decapoda, Cambaridae). Crustaceana 83:1303–1313

Thompson B, Olson R (1988) A preliminary survey of the protozoa and bacteria from Sulphur River in Parker Cave, Kentucky. Bull Nat Speleol Soc 50:42

Trimboli S, Toomey III RS, Weber K, Ryan S (2016) An overview of the reverse flow patterns of River Styx in Mammoth Cave, Kentucky: 2009–2012. In: Mammoth Cave National Park's 11th research symposium: celebrating diversity of research in the Mammoth Cave Region. Western Kentucky University, Mammoth Cave National Park, pp 128–129

Tumlinson R, Cline GR (2002) Food habits of the Banded Sculpin (*Cottus carolinae*) in Oklahoma with reference to predation on the Oklahoma salamander (*Eurycea tynerensis*). Proc Okla Acad Sci 82:111–113

United States Fish and Wildlife Service (1988) Kentucky cave shrimp recovery plan. United State Fish Wildlife Service, Atlanta

Whitman RL (1989) Meiofaunal sampling at Mammoth Cave National Park, draft report. National Park Service, Indiana Dunes National Lakeshore

Zhang J, Holsinger JR (2003) Systematics of the freshwater amphipod genus Crangonyx (Crangonyctidae) in North America. Virginia Mus Nat Hist Mem 6:274

Ziemba RE, Simpson A, Hopper R, Cooper RL (2003) A comparison of antennule structure in a surface- and a cave-dwelling crayfish, genus Orconectes (Decapoda, Astacidae). Crustaceana 76:859–869

Biodiversity of Mammoth Cave

15

David C. Culver and Horton H. Hobbs III

Abstract

The first cave-limited species found in Mammoth Cave was described in 1842 by DeKay. By 1888, 19 stygobionts and troglobionts were known. At present, there are 32 described troglobionts (terrestrial obligate cave dwellers) and 16 stygobionts (aquatic obligate cave dwellers) from Mammoth Cave. It is the third richest cave worldwide in terms of terrestrial species. Globally, the aquatic species richness is unremarkable, but it is one of the most diverse aquatic caves in the USA. Possible reasons for the high terrestrial species richness include high levels of available organic matter, a unique geographic position at the intersection of several major cave fauna regions, and its immense size. In comparison with Vjetrenica, a global hot spot of terrestrial species richness in Bosnia and Herzegovina, taxonomic diversity in Mammoth Cave is similar, although Vjetrenica has more families of beetles.

15.1 Introduction

Since Darwin's time, and even before, the fauna of Mammoth Cave has figured prominently in any discussion of cave life, especially concerning adaptation to the cave environment. The most famous quotation from Darwin's *On the Origin of Species by Means of Natural Selection* (1859) relating to cave fauna is

> It is well known that several animals which inhabit the caves of Carniola [Slovenia] and Kentucky are blind.... As it is difficult to imagine that eyes, though useless, could be in any way injurious to animals living in darkness, their loss may be attributed to disuse. (p. 137)

The quotation is noteworthy because it shows that Darwin was influenced by the now discredited ideas of Lamarckism, in particular the inheritance of acquired characters. What is of special interest in the context of Mammoth Cave is that in 1859, the two areas known to Darwin to have blind, depigmented cave animals were Mammoth Cave and Slovenia (in the nineteenth century, part of Slovenia was called Carniola, part of the Austro-Hungarian empire). The most famous Slovenian cave animal was the salamander *Proteus anguinus*, originally described by Laurenti in 1768, and was well known among biologists by Darwin's time (Shaw 1999). The first species with reduced eyes and pigment described from Mammoth Cave was the fish *Amblyopsis spelaean*, described in 1842 by James DeKay (Packard 1888). By 1850, six additional blind, depigmented species had been described that were known from Mammoth Cave, including an enigmatic, little known mite with eyes but known only from Mammoth Cave, described in 1804 (Table 15.1). Among the biologists visiting Mammoth Cave from the 1820s to the 1850s were Rafinesque (probably the first), DeKay, Tellkampf, Wyman, and Motschulsky (Barr 1967a). Darwin even knew about a nonexistent species of cave rat (*Neotoma*), supposedly with large non-functional eyes, which defied Darwin's explanation of eyelessness (Romero 2009). What can be charitably described as a faulty account of the Mammoth Cave (Allegheny) woodrat was given by Silliman in 1851.

With the publication of *On the Origin of Species by Means of Natural Selection*, interest in the connections

D.C. Culver (✉)
American University, Washington, D.C., USA
e-mail: dculver@american.edu

H.H. Hobbs III
Wittenberg University, Springfield, Ohio, USA

H.H. Hobbs III et al. (eds.), *Mammoth Cave*, Cave and Karst Systems of the World,
DOI 10.1007/978-3-319-53718-4_15

Table 15.1 List of all described cave-limited species (terrestrial troglobionts and aquatic stygobionts) from Mammoth Cave

Phylum/Class	Order	Family	Species	Authority	Endemic to Mammoth Cave?	Synonyms
TERRESTRIAL						
Arachnida	Acari	Belbidae	*Belba bulbipedata*	(Packard 1888)	**Endemic**	*Damaaeus bulbipedata*
Arachnida	Acari	Cocceupodidae	*Linopodes mammouthia*	Banks 1897		
Arachnida	Acari	Galumnidae	*Galumna alata*	(Hermann 1804)		*Oribata alata*
Arachnida	Acari	Laelaptidae	*Laelaps cavernicola*	Packard 1888		*Gamasus cavernicola*
Arachnida	Acari	Macrochelidae	*Macrocheles troglodytes*	(Packard 1887)		
Arachnida	Acari	Rhagidiidae	*Traegaardhia holsingeri*	(Zacharda 1980)		*Rhagidia cavernarum, Acarus cavernarum*
Arachnida	Araneae	Linyphiidae	*Anthrobia mammouthia*	Tellkampf 1844		
Arachnida	Araneae	Linyphiidae	*Bathyphantes weyeri*	(Emerton 1875)		
Arachnida	Araneae	Linyphiidae	*Phanetta subterranea*	(Emerton 1875)		
Arachnida	Araneae	Linyphiidae	*Porrhomma cavernicola*	(Keyserling 1886)		
Arachnida	Opiliones	Phalangodidae	*Phalangodes armata*	Tellkmapf 1844		
Arachnida	Pseudoscorpiones	Chernetidae	*Hesperochernes mirabilis*	(Banks 1895)		*Pseudozaona mirabilis*
Arachnida	Pseudoscorpiones	Chthoniidae	*Kleptochthonius cerberus*	Malcolm & Chamberlin 1961	**Endemic**	
Arachnida	Pseudoscorpiones	Chthoniidae	*Kleptochthonius hageni*	Muchmore 1963		*Kleoptochthonius packardi*
Arachnida	Pseudoscorpiones	Chthoniidae	*Tyrannochthonius hypogeus*	Muchmore 1996	**Endemic**	
Diplopoda	Chordeumatida	Trichopetalidae	*Scoterpes copei*	(Packard 1871)	**Endemic**	
Diplopoda	Polydesmida	Macrosternodesmidae	*Chaetaspis fragilis*	(Loomis 1943)		
Hexapoda	Collembola	Arrhopalitidae	*Pygmarrhopalites altus*	(Christiansen 1966)	**Endemic**	*Smynthurus* sp.
Hexapoda	Collembola	Entomobryidae	*Pseudosinella espanita*	Christiansen & Bellinger 1996	**Endemic**	
Hexapoda	Diplura	Campodeidae	*Litocampa cookei*	(Packard 1871)		
Insecta	Coleoptera	Carabidae	*Neaphaenops tellkampfi tellkampfi*	(Erichson 1844)		*Anophthalmus tellkampfi*
Insecta	Coleoptera	Carabidae	*Pseudoanophthalmus audax*	(Horn 1883)		
Insecta	Coleoptera	Carabidae	*Pseudoanophthalmus inexpectatus*	Barr 1959	**Endemic**	
Insecta	Coleoptera	Carabidae	*Pseudoanophthalmus menetriesii*	Motschulsky 1862		*Anophthalmus menetreisii*
Insecta	Coleoptera	Carabidae	*Pseudoanophthalmus pubescens*	Horn 1868		
Insecta	Coleoptera	Carabidae	*Pseudoanophthalmus striatus*	Motschuleky 1862		*Pseudanophthalmus interstitialis*
Insecta	Coleoptera	Leptodiridae	*Ptomaphagus hirtus*	(Tellkampf 1844)		*Adelops hirtus*

(continued)

Table 15.1 (continued)

Phylum/Class	Order	Family	Species	Authority	Endemic to Mammoth Cave?	Synonyms
Insecta	Coleoptera	Staphylinidae	*Batrisodes henroti*	Park 1956		
Insecta	Diptera	Sphaeroceridae	*Spelobia tenebrarum*	(Aldrich 1897)		
Mollusca	Stylommatophora	Carychiidae	*Carychium stygium*	Call 1897	**Endemic**	
Mollusca	Stylommatophora	Endodontidae	*Helicodiscus punctallus*	Morrison 1942	**Endemic**	
Mollusca	Stylommatophora	Zonitidae	*Glyphyalinia specus*	Hubricht 1965		
AQUATIC						
Crustacea	Amphipoda	Crangonyctidae	*Crangonyx barri*	Zhang & Holsinger 2003		
Crustacea	Amphipoda	Crangonyctidae	*Stygobromus exilis*	Hubricht 1943		
Crustacea	Amphipoda	Crangonyctidae	*Stygobromus vitreus*	Cope 1972		*Crangonyx vitreus*
Crustacea	Cyclopoidea	Cyclopoidae	*Megacyclops donnaldsoni donnaldsoni*	Chappuis 1929		
Crustacea	Decapoda	Atyidae	*Palaemonias ganteri*	Hay 1903		
Crustacea	Decapoda	Cambaridae	*Orconectes pellucidus*	(Tellkampf 1844)		*Cambarus pellucidus*
Crustacea	Harpacticoidea	Canthocamptidae	*Attheyella pilosa*	Chappuis 1929		
Crustacea	Harpacticoidea	Canthocamptidae	*Bryocamptus morrisoni elegans*	Chappuis 1929		
Crustacea	Isopoda	Asellidae	*Caecidota stygia*	Packard 1871		
Crustacea	Podocopida	Entocytheridae	*Sagittocythere barri*	(Hart & Hobbs 1961)		
Crustacea	Podocopida	Entocytheridae	*Sagittocythere stygia*	Hart & Hobbs 1966	**Endemic**	
Mollusca	Mesogastropoda	Hydrobiidae	*Antroselates spiralis*	Hubricht 1963		
Pisces	Percopsiformes	Amblyopsidae	*Amblyopsis spelaea*	DeKay 1842		
Pisces	Percopsiformes	Amblyopsidae	*Typhlichthys subterraneus*	Girard 1860		
Turbellaria	Tricladida	Kenkiidae	*Sphalloplana buchanani*	(Hyman 1937)		*Speophila buchanani*
Turbellaria	Tricladida	Kenkiidae	*Sphalloplana percoeca*	(Packard 1879)		*Dendrocoelum percoecum*

between species and their evolutionary relationships intensified. During the late nineteenth century, attention to cave organisms increased, largely as the result of the growth of neo-Lamarckianism. The loss of structures such as eyes and pigment in cave animals was most easily explained by the inheritance of acquired characters, and hence neo-Lamarckians such as A.S. Packard studied the cave fauna extensively. Packard co-founded the journal *The American Naturalist* (Romero 2009), nowadays one of the premier journals of evolutionary ecology. In the late nineteenth century, Packard and others published numerous articles both describing species from Mammoth Cave as well as speculating about evolutionary patterns.

In 1888, Packard summarized information on the North American cave fauna in the monograph, *The Cave Fauna of North America with Remarks on the Anatomy of the Brain and Origin of the Blind Species*. In fact, in 1888, most of what was known about the North American cave fauna was known from Mammoth Cave, which Packard repeatedly visited beginning in 1874. He begins the monograph with a description of Mammoth Cave (including Hovey's 1882 map, see Chap. 5) and produces the first species list for the cave, which he divided into permanent and temporary inhabitants. Among the permanent inhabitants are 31 species (Table 15.2), some, but not all of which we now consider obligate cave dwellers (aquatic stygobionts and terrestrial troglobionts). He reported six stygobionts and 13 troglobionts. At the time, it was by far the most species-rich cave in the USA (it still is), but Packard reported only on a few other caves known outside the Mammoth Cave region, especially Indiana, Virginia, and Utah. Also noteworthy is that Packard collected in relatively few habitats in Mammoth Cave (Table 15.2), but these habitats were species rich, relative to most cave habitats in the USA.

With the collapse of neo-Lamarckism came the near collapse of biological interest in the fauna of Mammoth Cave until the mid-1950s when T.C. Barr Jr., a beetle taxonomist, evolutionary biologist, and speleologist began a decades-long study of the fauna of Mammoth Cave, especially its beetles (Barr 1967b). He produced the first fauna overview since Packard (Barr 1967a), and it remains the definitive source of information on the fauna of the cave. Another prominent speleobiologist, T.L. Poulson, also commenced a decade-long study of the cave fauna of Mammoth Cave in the 1960s, but his emphasis was ecological and evolutionary in nature, rather than faunistic and systematic (see Poulson 1992). Surprisingly, even Barr did not create a tabular list of stygobiotic and troglobiotic species in the cave, and the one included here is the first one since Packard (Table 15.1).

15.2 The Fauna

The list of all the species found in a cave is nearly endless, including as it does accidentals, facultative dwellers, and obligate cave dwellers. We focus here on the obligate cave-dwelling species (stygobionts and troglobionts), which are enumerated in (Table 15.1). A few ecologically important facultative cave dwellers are also mentioned. Barr (1967a) provides a more thorough treatment of these species.

A total of 32 troglobionts and 16 stygobionts are known. Of these, nine troglobionts and two stygobionts are endemic to Mammoth Cave.

15.2.1 Terrestrial Species

The mite fauna of Mammoth Cave is rich. There are six troglobionts known from the cave and at least that number of troglophiles (Barr 1967a). Unfortunately, they have been little studied since Packard's time (see Zacharda et al. 2010), and even appropriate generic assignments are in doubt. The most conspicuous species is *Traegaarhdia holsingeri* (=*Rhagidia cavernarum*). It is a predator of microarthropods, especially Collembola and other mites.

The most conspicuous spider in Mammoth Cave is a troglophile, the pigmented and eyed cave orb weaver *Meta Americana*, and it is common near cave entrances. The four troglobiotic spider species are all relatively widespread in caves in the eastern USA, and Barr considers them to be Pleistocene relicts. One troglobiotic opilionid—*Phalangodes armata*—is widespread, but not especially common in the cave. Of the four troglobiotic pseudoscorpions, two—*Kleptochthonius cereberus* and *Tyrannochthonius hypogeus*—are endemic to Mammoth Cave. Cave pseudoscorpions typically have very restricted ranges. One species—*Hesperchernes mirabilis*—is most common near entrances, whereas the others are deep-cave species. Typical prey for pseudoscorpions is Collembola.

Among myriapods, two troglobiotic millipedes are known from the cave. Of the two, *Scoterpes copei* is also endemic to Mammoth Cave and is commonly found on rotting wood, wet flowstone covered with cricket guano, and debris left in the cave.

Barr (1967a) reports a total of ten species of Collembola from Mammoth Cave, including two troglobionts—*Pygmarrhopalites altus* and *Pseudosinella espinata*—which are both endemic to Mammoth Cave. Numerous troglophilic species are also present, including *Tomocerus bidentatus, Pseudosinella argentea*, and *Pygmarrhopalites pygmaeus*, all of which are rather common. Likewise, some troglobiotic species found in nearby caves are absent, including *Pseudosinella hirsuta* and *Sinella cavernarum*. One troglobiotic dipluran species, *Litocampa cookei*, is known from wet flowstone, cricket guano, silty areas in upper level galleries, and damp silt along rivers (Barr 1967a).

Even though there are no troglobiotic Orthoptera in Mammoth Cave (and very few in any cave), camel crickets (Rhaphidophoridae), which includes what are commonly called cave crickets, are common, periodically leaving the cave to feed and returning during the day and to reproduce. Their guano and eggs are a major carbon source for the terrestrial community (Lavoie et al. 2007). The most common species is *Hadenoecus subterraneus*.

Among terrestrial species, the most diverse, in terms of the number of species per genus is the carabid beetle genus *Pseudanophthalmus*, with five species known from the cave (Barr 1967b). They occur in a variety of habitats and are generalized predators. The closely related *Neaphaenops tellkampfi* is a major predator of cricket eggs. It is especially common in sandy-floored passages in upper levels. Two other troglobionts are known—one Staphylinidae and one Leptodiridae.

The final troglobiotic insect known from the cave is the dipteran *Spelobia tenebrarum*. This troglobiont has one of the largest known ranges of any troglobiont, occurring throughout the eastern USA. Many other troglophilic Diptera, especially in the families Heleomyzidae, Mycetophilidae, and Phoridae, are common along cave passage ceilings and walls.

Table 15.2 Packard's list of permanent inhabitants of Mammoth Cave, excluding the Protista (Infusoria in Packard's list). Obligate cave-dwelling species are indicated by an asterisk. From Packard (1888)

Species	Site in cave
VERMES:	
*Dendrocoelum percoecum** Packard	Shaler's Brook, Richardson Spring, and other pools
a nematoid parasite of *Adelops*	
CRUSTACEA:	
Lumbricus[1]	Dead Sea
*Canthocamptus cavernarum*Packard	River Hall, Richardson's Spring
*Caecidotea stygia** Packard	Shaler's Brook' River Styx
*Crangonyx vitreus** (Cope)	Wandering Willie's Spring and Labyrinth
*Cambrus pellucidus** (Tellkampf)	River Styx
ARACHNIDA:	
*Acarus cavernarum** Packard	River Hall, Richardson's Spring
*Gamasus cavernicola** Packard	Labyrinth
*Chthonius packardii** Hagen	River Hall
*Oribates bulipedata** Packard	Richardson's Spring
*Phalangodes armata** Tellkampf	River Hall
*Anthrobia mammouthia** Tellkampf	Labyrinth
Coelotes juvenalis Keys	
Liocranoides unicolor Keys	Keyserling collection
MYRIDAPODA:	
Scoterpes copei (Packard)	Richardson's Spring and Labyrinth
INSECTA:	
Lipeura sp.	Richardson's Spring
Isotoma sp.	River Hall, under stones
Degeeria sp.	Devil's Cooling Tub
*Smythurus** (white species)	Labyrinth
*Campodea cookei** Packard	River Hall, Labyrinth, Richardson's Spring
Machilis cavernicola Tellkempf	Wandering Willie's Spring
Hadenoecus subterraneus Scudder	River Hall
Atropos divinatoria Mueller	Rotunda (Hubbard collection)
Typerates tessulatus Hagen	
*Adelops hirtus** Tellkampf	Richardson's Spring, River Hall, Labyrinth
*Anopththalmus tellkampfi** Erichson	Richardson's Spring, River Hall, Labyrinth
*Anophthalmus menetreisii** Motschulsky	Richardson's Spring
*Anophthalmus interstitialis** Hubbard	Washington's Hall (Hubbard collection)
Blepheroptera defessa Osten Sacken	Near the entrance
VERTEBRATA:	
*Typhlichthys subterraneus** Girard	River Styx
*Amblyopsis spelaea** De Kay	River Styx

[1]Almost certainly a misplacement of the earthworm genus *Lumbricus*

Three terrestrial troglobiotic gastropods are known from Mammoth Cave (Table 15.1), the most common being *Carychium stygium*. It can be commonly found on wet cricket guano. Numerous troglophiles have also been reported, but none are common (Barr 1967a).

15.2.2 Aquatic Species

Mammoth Cave is the type locality (but not the only locality) for two stygobiotic flatworms (Table 15.1)—*Sphalloplana buchanani* and *S. percoeca. S. percoea* occurs in large drip pools, essentially being an epikarst habitat (see Pipan 2005). They probably can survive drying by encystment (Barr 1967a) *S. buchanani* occupies stream gravels.

A single aquatic stygobiotic gastropod is known from Mammoth Cave—*Antroselates spiralis*. It occurs primarily under large stones in shallow riffles and is uncommon. It is observed also in Echo River Spring.

Barr (1967a) states that sixteen species of copepods have been reported from streams and pools in the cave, but some of these are likely accidental species that got washed into the cave. Three of the species are stygobionts, although none are limited to Mammoth Cave. Completely unstudied is the epikarst habitat, which at least in Europe harbors a rich stygobiotic fauna with more than ten stygobiotic species found in a single cave (Pipan 2005). A study by Pipan and Culver (2005) in Organ Cave, West Virginia found about half that many. The epikarst of Mammoth Cave is a very promising habitat in which to find new species of microcrustaceans. Two other microcrustaceans—*Sagittocythere barri* and *S. stygia*—are ectocommensals of the stygobiotic crayfish, *Orconectes pellucidus*.

Major stygobiotic constituents of the streams in Mammoth Cave are the isopod *Caecidotea stygia* and the amphipod *Crangonyx barri*. Two other amphipods in the genus *Stygobromus* are common in drip pools that drain the epikarst. An endemic shrimp—*Palaeomonias ganteri*—occurs in a number of subterranean basins in MCNP (Leithauser and Holsinger 1985). The largest stygobiotic crustacean in Mammoth Cave is the crayfish, *Orconectes pellucidus*.

Perhaps the most interesting component of the aquatic fauna is the fish fauna. Two stygobiotic species are known from the cave—*Amblyopsis spelaea* and *Typhlichthys subterraneus*. *A. spelaea* has a very restricted range in the cave and is typical of deeper pools; *T. subterreanea* is more common and often found in shallower pools and streams.

15.3 Species Richness in Mammoth Cave

In 2000, Culver and Sket published a list of caves and karst wells that were reported to have more than 20 species of stygobionts and troglobionts combined. Worldwide, they found twenty such sites, including three in North America—San Marcos Spring in Texas, Shelta Cave in Alabama, and Mammoth Cave. Based on the numbers available, Mammoth Cave ranked sixth in terms of overall stygobiotic plus troglobiotic species richness. Culver and Sket's approach is compelling because it only requires species lists of a relatively few high diversity caves. However, it has severe limitations because (a) it does not take into account regional diversity, (b) it is not possible to estimate list completeness, and (c) no confidence intervals can be placed on the numbers. More detailed analyses circumvent these problems, but these are much more data intensive and require detailed analysis (Culver et al. 2012, Zagmajster et al. 2008, 2010). The "quick and dirty" approach of Culver and Sket does allow comparison among many regions, including ones for which less data are available.

Culver and Pipan's (2013) latest tabulation of the most species-rich caves is shown in (Table 15.3) separated into stygobionts and troglobionts. With better and more complete data, the cutoff for global hot spots of subterranean biodiversity is now at 25 stygobionts or 25 troglobionts. According to this criterion, the terrestrial fauna of Mammoth Cave is a global subterranean biodiversity hot spot, tied for third among the world's caves. The stygobiotic fauna is not among the richest caves and is generally unremarkable in its richness.

A series of hypotheses can be put forward to explain the high species richness in Mammoth Cave (see Barr 1967a; Olson 2013; Poulson 1992). The first is the frequently expressed concern that cave species richness patterns largely reflect sampling intensity, rather than real differences. In an analysis of a very large dataset from the Dinaric karst of southeastern Europe, Zagmajster et al. (2008) show that cave beetle richness in the Dinaric karst is strongly influenced by collection intensity. However, in a more thorough analysis, Zagmajster et al. (2010) showed that, when sampling bias was removed, the same pattern of species richness persisted.

More interestingly, there is a biological explanation that has been put forward for the location of terrestrial cave biodiversity hot spots globally (Culver and Sket 2000; Culver et al. 2006). This hypothesis can be evaluated with respect to Mammoth Cave.

Based on an extensive analysis of hundreds of caves in relatively small ($\sim 10{,}000$ km^2) regions, Culver et al. (2006) concluded that there was a ridge of high subterranean terrestrial biodiversity in southern Europe that corresponded to a ridge of high actual primary productivity, and hence to allochthonous input in caves. Similarly, Culver and Sket (2000) concluded that terrestrial cave biodiversity hot spots were ones with high productivity, either chemoautotrophic or allochthonous. Mammoth Cave may have both. Olson (2013) points out that considerable chemoautotrophic productivity occurs in Mammoth Cave and vicinity, the best known of which are Sulfur River in Parker Cave, and the sulfur–hydrocarbon seeps in Marianne's Pass. Rickard Toomey (pers. comm.) points out that Mammoth Cave has had large bat populations in the past, including free-tailed bats (*Tadarida*) and gray bats (*Myotis grisescens*), with colonies perhaps in the millions. Today there are still significant populations of bats present in Mammoth Cave and nearby caves (see Chap. 17). Both the current colonies and the guano piles from the old, historical colonies provide considerable organic material that is allochthonous in origin.

Table 15. 3 Cave withmore than 25 stygobionts (A) or 25 troglobionts (B). From Culver and Pipan (2013), with additional data from Oromi (pers. comm.)

Site name	Country	Number of Species	Remarks
A. Stygobionts			
Postojna Planina Cave System	Slovenia	48	Dinaric Karst
Vjetrenica	Bosnia &Hercogovina	40	Dinaric Karst
Walsingham Cave	Bermuda	37	Anchialine cave
Triadou Aqujifer well	France	34	Phreatic
Robe River well	Australia	32	Phreatic
Križna jama	Slovenia	29	Dinaric Karst
Logarček	Slovenia	28	Dinaric Karst
Šica-Krka System	Slovenia	27	Dinaric Karst
Edwards Aqjuifer well	Texas, USA	27	Phreatic
B. Troglobionts			
Postojna Planina Cave System	Slovenia	36	Dinaric karst
Cueva de Felipe Reventón	Canary Islands, Spain	34	Lava tube
Mammoth Cave	**Kentucky,USA**	**32**	**Longest cave**
Cueva del Viento	Canary Islands, Spain	32	Lava tube
Vjetrenica	Bosnia & Hercecovina	32	Dinaric karst
Peştera Movile	Romania	29	Chemoautotrophic

The large cave-cricket population that regularly leaves and reenters the cave certainly increases the carbon and nutrient flux, but many other caves in the region also have such a system (Lavoie et al. 2007).

Three other explanations for high terrestrial biodiversity unique to Mammoth Cave have been proposed. One is that Mammoth Cave lies near the intersection of two biogeographic provinces (Pennyroyal Plateau and Bluegrass) resulting in high dispersal rates into Mammoth Cave (Barr 1967a). The Mammoth Cave fauna thus has elements of the fauna of the Pennyroyal and Cumberland Plateaus, as well as an endemic fauna. A second explanation is the result of the high habitat diversity as a result of the preservation of fossil passages under the protective sandstone caprock overlying the soluble limestones (see Chaps. 6 and 7). Poulson (1992) discusses the habitats present and emphasizes the different food types (especially various feces) present. Finally, the immense size of Mammoth Cave may allow for high levels of endemism (Barr 1967a).

A detailed comparison with another terrestrial subterranean biodiversity hot spot—the cave Vjetrenica in Bosnia and Herzegovina—is possible because there is also a published species list for Vjetrenica (Sket 2003, supplemented by Ozimec and Lučić 2009), the only other published hot spot list as far as we can determine. Both caves have 32 described troglobionts. There are more species of Arachnida in Mammoth Cave than any other group, and there are more species of Coleoptera in Vjetrenica than any other group (Fig. 15.1). In both caves, these two groups together make up more than 50% of the species. Somewhat surprisingly, Hexapoda (e.g., Collembola) are minor components of both fauna. The level of endemism is roughly the same in the two caves—Mammoth Cave has eight species endemic to the cave, and Vjetrenica has seven species endemic to the cave. The most obvious difference between the two caves is that in Vjetrenica, five genera are represented by two species; all others are represented by one. In Mammoth Cave, only two genera have more than one species, but one of these (the beetle *Pseudanophthalmus*) has five. If adaptive radiation, that is, the diversification of species through natural selection, were important (see Fišer et al. 2012 for a subterranean example), more examples of multiple species in the same genus in the same cave would be expected.

Nearly all of the global high diversity aquatic sites in (Table 15.3) are either in the Dinaric Mountains or are relatively deep phreatic aquifers. Mammoth Cave is neither. Sket et al. (2004) suggest that part of the explanation for high subterranean species diversity in the Balkans, and the Dinaric Mountains in particular, is its long and complex geological and evolutionary history, especially the proximity to the Adriatic Sea, which was a source of subterranean colonists during the Messinian salinity crisis, when the Adriatic and Mediterranean Seas dried-up. In the USA, only the stygofauna of the Edwards Aquifer of Texas has a component of marine origin that likely increased species richness (Culver et al. 2009).

Of course, the species counts for Mammoth Cave are likely to increase in the future. New species are being

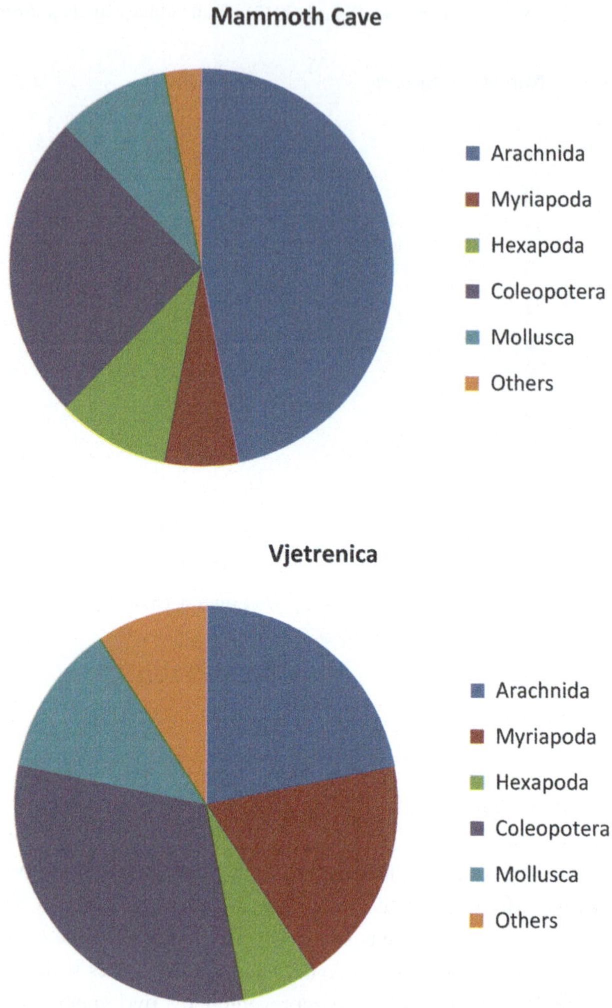

Fig. 15.1 Pie charts of taxonomic distribution of troglobionts in Mammoth Cave and the cave Vjetrenica in Bosnia and Herzegovina. Data for Vjetrenica from Sket (2003) and Ozimec and Lučić (2009)

discovered, and new habitats are being explored, or are yet to be explored, such as the epikarst. What is certain is that Mammoth Cave is globally important as a center of subterranean biodiversity.

References

Barr TC (1967a) Ecological studies in the Mammoth Cave system of Kentucky I: the biota. Inter J Speleol 3:147–204

Barr TC (1967b) Cave Carabidae (Coleoptera) of Mammoth Cave. Psyche 73:284–287, 74:24–26

Culver DC, Pipan T (2013) Subterranean ecosystems. In: Levin SA (ed) Encyclopedia of biodiversity, 2nd edn. Elsevier, Amsterdam

Culver DC, Sket B (2000) Hotspots of subterranean biodiversity in caves and wells. J Cave Karst Stu 62:11–17

Culver DC, Deharveng L, Bedos A, Lewis JJ, Madden M, Reddell JR, Sket B, Trontelj P, White D (2006) The mid-latitude biodiversity ridge in terrestrial cave fauna. Ecography 29:120–128

Culver DC, Pipan T, Schneider K (2009) Vicariance, dispersal, and scale in the subterranean aquatic fauna of karst regions. Freshwater Bio 54:919–929

Culver DC, Trontelj P, Zagmajster M, Pipan T (2012) Paving the way for standardized and comparable subterranean biodiversity studies. Subterranean Bio 10:43–50

Darwin C (1859) On the origin of species by means of natural selection, or the preservation of favoured races in the struggle for life, 1st edn. John Murray, London

Fišer C, Blejec A, Trontelj P (2012) Niche-based mechanisms operating within extreme habitats: A case study of subterranean amphipod communities. Bio Lett 8:578–581

Lavoie KH, Helf KL, Poulson TL (2007) The biology and ecology of North American cave crickets. J Cave and Karst Stu 69:114–134

Leithauser AT, Holsinger JR (1985) Ecological analysis of the Kentucky cave shrimp, *Palaemonias ganteri* Hay, at Mammoth Cave National Park (Phase V). Unpublished final report submitted to US Department of Interior, National Park Service Contract Number CX-5000-1-1037

Olson R (2013) Potential effects of hydrogen sulfide and hydrocarbon seeps on Mammoth Cave ecosystems. In: Mammoth Cave National Park's 10th research symposium, pp 25–30, 14–15 Feb 2013

Ozimec R, Lučić I (2009) The Vjetrenica cave (Bosnia and Herzegovina)—one of the world's most prominent biodiversity hotspots for subterranean fauna. Subterranean Bio 7:17–24

Packard AS (1888) The cave fauna of North America, with remarks on the anatomy of brain and the origin of the blind species. Memoirs Natl Acad Sci 4:1–156

Pipan T (2005) Epikarst—a promising habitat. Založba ZRC, Ljubljana

Pipan T, Culver DC (2005) Estimating biodiversity in the epikarstic zone of a West Virginia cave. J Cave Karst Stu 67:103–109

Poulson TL (1992) The Mammoth Cave ecosystem. In: Camacho Ana (ed) The natural history of biospeleology. Museo Nancional de Ciencias Naturales, Madrid, pp 569–612

Romero A (2009) Cave biology. Life in darkness. Cambridge University Press, Cambridge

Shaw TR (1999) Proteus for sale and for science in the 19th century. Acta Carsologica 28:119–129

Silliman B (1851) On the Mammoth Cave of Kentucky. Amer J Sci 10:332–339

Sket B (2003) Živótinski svijet Vjetrenice. In: Lučić I (ed) Vjetrenica: Pogled u dušu Zemlje. Zagreb, Croatia, pp 147–248

Sket B, Paragamian K, Trontelj P (2004) A census of the obligate subterranean fauna of the Balkan Peninsula. In: Griffiths HW, Kryštufek B, Reed JM (eds) Balkan Biodiversity. Pattern and process in the European hotspot. 309–22. Kluwer Academic Publishers, Dordrecht, pp 309–322

Zacharda M, Fong DW, Hobbs HH III, Piva E, Slay ME, Taylor SE (2010) A review of the genus Traegaardhia (Acari, Prostigmata, Rhagidiidae) with a description of new species and a key to species. Zootaxa 2474:1–64

Zagmajster M, Culver DC, Sket B (2008) Species richness patterns of obligate subterranean beetles in a global biodiversity hotspot—effect of scale and sampling intensity. Divers Distrib 14:95–105

Zagmajster M, Culver DC, Christman MC, Sket B (2010) Evaluating the sampling bias in pattern of subterranean species richness—combining approaches. Biodivers Conserv 19:3035–3048

Mammoth Cave Microbiology

16

Kathleen H. Lavoie

Abstract

Microorganisms are a diverse group of Bacteria, Archaea, and Eukaryotes with their very small size in common. Microbes make up the majority of organisms in numbers, biomass, and metabolic diversity and are critical component of the biosphere through geochemical cycling. Caves are models for the study of astrobiology: life on other planets. This chapter reviews intraterrestrial (inside Earth) microbes in Mammoth Cave. Despite the great size and complexity of Mammoth Cave, few microbial studies have been carried out. Great changes in methods from culture-dependent to molecular genomic studies have provided new information. Geomicrobiology is at the intersection of microbial activities and geologic processes, including sulfur-based ecosytems, formation of carbonate speleothems, saltpeter mining, and manganese oxide deposits. Microorganisms also include infectious agents like tuberculosis, and parasites of humans and cave crickets, and the devastating invasive fungus that causes white-nose syndrome in bats. Microbial nature preserves could protect communities of native cave microbes adapted to low-nutrient conditions. There are many ecological and evolutionary questions to be studied along with basic research and inventory of microorganisms in Mammoth Cave.

16.1 Introduction

Microorganisms are the only group of organisms defined by their very small size. Nearly all cells (including our own) are microscopic, but microbes live mostly as single cells; only multicellular plants, animals, and some fungi are not microorganisms (although parasites are studied in microbiology). Microbes include all prokaryotic and many eukaryotic cells. Viruses are not alive because they are not cells, but are microorganisms.

Microbes are of central importance to the biosphere and to biogeochemical cycling. They maintain the atmosphere by cycling carbon, nitrogen, and oxygen. At least half of the oxygen in the atmosphere is from phototrophic microorganisms (algae and cyanobacteria) in oceans. Microbes extend our knowledge of the strategies and limits of life. With the discovery of hundreds of new planets, it is very possible that life is abundant in the universe, microbial, uses sulfur for energy, and is located below the surface but dependent on liquid water (Domagal-Goldman and Wright 2016). Caves provide model systems for what extraterrestrial life might be. We can monitor environmental change, water pollution, the quality of an environment, and the recovery of a system to stress by studying microbes. Microbes play a major role in conservation and restoration biology, and microbial communities provide important models for understanding principles of ecology and evolution.

Because we usually cannot see microorganisms directly, they are often "out of sight, out of mind." How can such small creatures change anything? What they lack in size, they more than make up for in numbers, biomass, and metabolic diversity. There is probably more microbial biomass below the surface of the Earth than all the biomass above ground. Edwards et al. (2012) describe microbes below the surface as "intraterrestrial," life inside Earth. Caves provide access for study of shallow and deep subsurface environments.

K.H. Lavoie (✉)
Department of Biology, SUNY Plattsburgh, Plattsburgh, NY, USA

H.H. Hobbs III et al. (eds.), *Mammoth Cave*, Cave and Karst Systems of the World, DOI 10.1007/978-3-319-53718-4_16

16.2 Intraterrestrials: Microbes in Caves

Microbial distribution is ubiquitous. They can be found in every environment, but whether they are active or not depends on factors including nutrient availability, temperature, pH, and presence of other organisms. Where are microbes in caves? The cave shows evidence of microbial activity: algae and cyanobacterial growth around entrances or artificial lights, the earthy smell, white filamentous fungi growing on scat, white microbial colonies on a wall (Fig. 16.1), a white "marshmallow" with legs—a cricket killed by a parasitic fungus, powdery soil mined for saltpeter, white filaments in a stream that smells of rotten eggs, some speleothems (formations), the limestone itself and the dissolution of passages.

What should we look for to find microbes in caves? Visual evidence of microbes and microbial activities in caves include dots, which are colonies of microbes on rock surfaces (Fig. 16.1); ferromanganese deposits, seen as discoloration of rock surfaces; precipitation of banded minerals; structural changes like a coating or crust; and biofilms, communities of microbes seen as slippery rocks or white filaments in streams with inputs of sulfur in caves (Barton 2006). Despite its large size and the growth and activity of microbes throughout Mammoth Cave, relatively few microbiological studies have been carried out, offering great potential for future research (Lavoie 2015).

16.3 It's a Small World: Methods

The first review of the microbiology of underground environments was published by Caumartin in 1963. A lot has changed in the methods used for the study of microorganisms since then leading to important insights into the ecology and distribution of microbes in caves. Their small size is of critical importance in understanding microbes. The very great surface area-to-volume ratio of microbes allows for rapid diffusion of materials in and out of cells and for rapid metabolism and cell division when food is available. It is even hard to tell when a bacterial cell is dead; microbes often exist in a dormant state with little or no metabolic activity, but those same microbes can rapidly become active if environmental conditions change.

The study of microbes has always been complicated by their small size and low morphological diversity. We cannot use conventional observations that we use for cave crickets or cavefish, yet we need to know which microbes are there, and how many, their activity and interactions. Using a microscope to look for microbes in the environment is

Fig. 16.1 A female *Hadenoecus subterraneus* cave cricket with white microbial colonies on the wall behind her in the New Discovery Entrance to Mammoth Cave

difficult. Even in a nutrient-rich agricultural soil, you would only find isolated areas with a few cells, and your chances of seeing microbes in nutrient-poor cave soils would be a thousand times lower.

16.3.1 Microscopy

Microscopes gave us our first sight of the microbial world, both traditional light microscopy with staining, like the Gram stain, and electron microscopy which allows us to examine objects at extreme close up (Fig. 16.2). Bacteria do not vary much in what they look like, so microscopy is a useful tool, though limited. We can extend the usefulness of microscopy by using fluorescent-labeled antibodies that bind to only specific bacteria, allowing for quantification.

16.3.2 Cultures

The use of traditional culture media in Petri dishes with incubation at cave ambient temperature is still critical. The use of low-nutrient media for cave microbes or media designed to grow specific types of bacteria, like sulfide oxidizers, has increased the success of cultures. The great majority of environmental microbes still cannot be grown in culture, but biochemical testing for identification can only be done on pure cultures. Presence alone does not guarantee activity, and it does not tell us whether we had one cell to start with or a million.

16.3.3 Molecular Techniques

Non-culture techniques became common in the 1990s and have revolutionized microbiology. The general idea is that a specific protein or nucleic acid is selected, its component sequence determined, and the sequences are compared among organisms. Closely related organisms have similar molecular patterns. These molecular techniques describe the molecule being compared with the suffix "-omics" (e.g., genomics, proteomics, transcriptomics). Genomic studies are showing unexpected diversity and many new and unique cave microorganisms with unknown abilities (Barton 2006). Molecular and cultural methods both have value, showing different aspects of microbial communities.

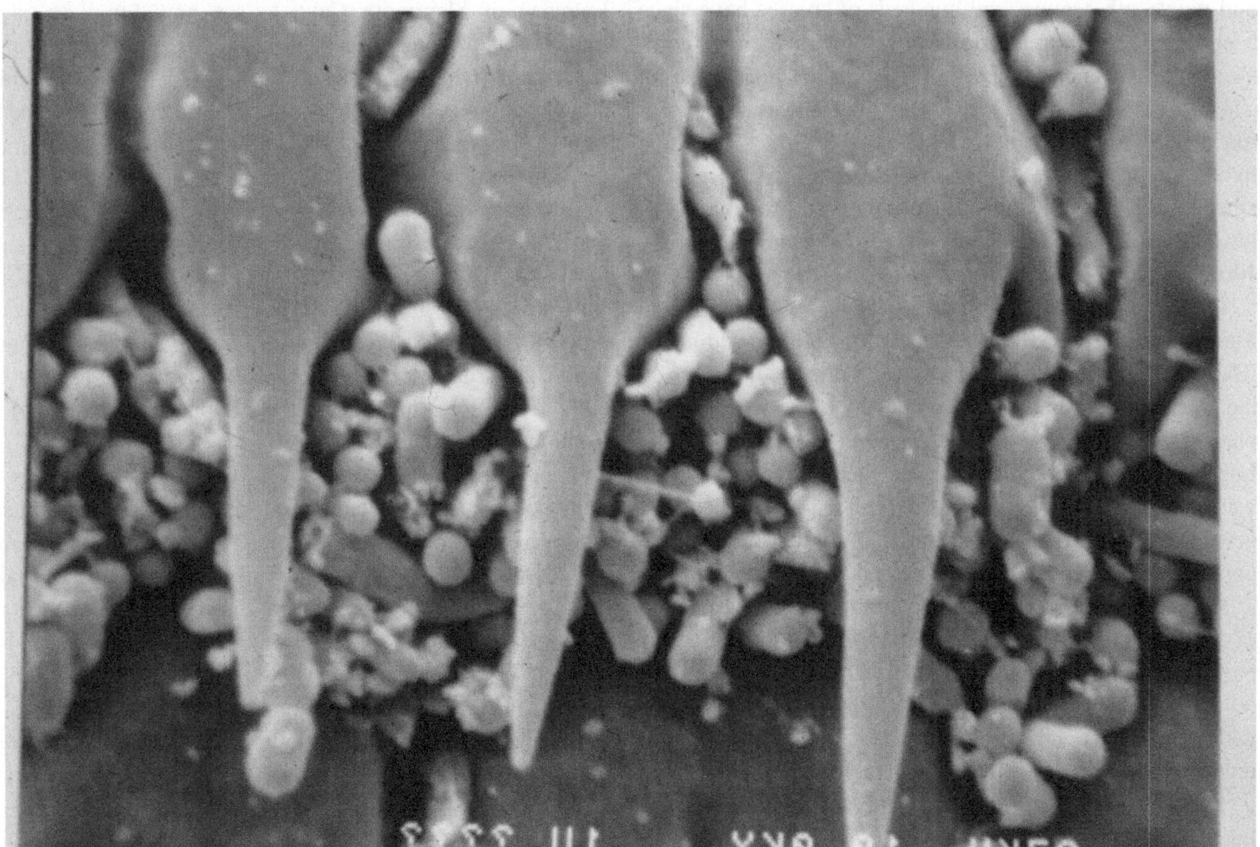

Fig. 16.2 Scanning electron micrograph of chitinous "teeth" with bacterial cells from the crop of a cave cricket

16.4 A Survey of Intraterrestrial Cave Microbes

You are probably familiar with classification of organisms at the level of kingdom, but above kingdoms are domains: Bacteria, Archaea, and Eukarya. Bacteria are prokaryotes with DNA free in the cell and are the most diverse domain. Recent discoveries of many species known only from their DNA, including from caves, have nearly doubled their known diversity (Hug et al. 2015). Archaea are also prokaryotes, but very different from the Bacteria. Eukarya include all organisms with genetic material inside a nuclear membrane, and the familiar kingdoms of animals, plants, fungi, and protists. All life comes from a common ancestor to the bacteria and a branch to a shared lineage for the Archaea and Eukarya (Hug et al. 2015) (To interpret phylogenetic trees, see A Field Guide to the Tree of Life: http://evolution.berkeley.edu/evolibrary/news/160505_treeoflife). Let us review the different types of microorganisms and some of what we know about them in Mammoth Cave.

16.4.1 Protists and Algae

The first paper on cave protists (protozoa) appeared in the mid-nineteenth century, and hundreds of species have since been identified from aquatic cave habitats and moist environments like guano, algae, soils, and parasites (Gittleson and Hoover 1969). Most species are the same as those from forest litter, but some may be truly cave-adapted, both free living and parasites of troglobionts. In Mammoth Cave, amebas were the most commonly observed protzoans, followed by ciliates, and then flagellates (Gittleson and Hoover 1969).

Protists are important in several aquatic environments in Mammoth Cave, usually located on or in bottom sediments (Barr and Kuehne 1971). Water with a direct connection to the surface, such as Echo River, shows higher densities of plankton and some seasonal changes compared to Crystal Lake, an isolated body of water perched above the current water table. Thompson and Olson (1988) found at least 13 genera of protozoa across eight orders from the stream in the upper room of Parker Cave on the sinkhole plain.

Algae are largely phototrophic and not of importance in caves except around entrances and artificial light sources as part of lamp flora (Smith and Olson 2007). Barr and Kuehne (1971) found increased algae in Mammoth Cave associated with heavy rains and spring snow melt, from algae washed into the cave and growth using organic chemicals.

Show caves are in a constant battle to remove algae and lint left by visitors without damaging underlying formations (Saiz-Jimenez 2012). Bright light results in more graffiti from visitors, and the heat dries out surfaces and decreases relative humidity, which may be lethal to cave-adapted organisms. Brightly lit areas in the Frozen Niagara entrance had no cave animals, but a switch to LED lighting made the areas habitable again.

LED lighting was tested by Olson (2006) to see whether it would reduce the growth of phototrophs in Frozen Niagara, the most heavily visited and the best-lit entrance to Mammoth Cave. He photographed a test area, cleaned it with bleach, and set up both gold-phosphor fluorescent lighting and yellow LED lamps, using existing white light as a control. After two and a half years, the traditional white lighting showed the heaviest growth of lamp flora, fluorescent lighting supported limited growth, and no growth using yellow LED lighting.

16.4.2 Fungi

Fungi growing in caves are identical to surface forms (Vanderwolf et al. 2013). They grow from spores in the cave or brought in by flooding, air, or animals. Fungi are important decomposers and recyclers. Simple filamentous forms appear first (Fig. 16.3a), and larger, more complex mushrooms appear last (Fig. 16.3b).

I did experiments in Little Beauty Cave and the Austin Entrance of Mammoth Cave to see how fungi on cave (wood) rat droppings changed with time, scat shape, and interactions with insects (Lavoie 1982). I mixed the same amount of ground up cave rat scat with water and reshaped it like the original rat fecal pellets, a single scat resembling raccoon, and spread a thin layer directly on the cave mud to simulate cricket guano. All groups of fungi were similar in the timing of appearance, but the thin layer proved difficult for the fungus to concentrate enough nutrients for mushroom formation. The numbers of beetle and fly larvae were reduced on the thin layer because the larvae had no refuge inside the scats from predators like staphylinid beetles. If early fungi get a head start, their hyphae can block colonization by invertebrates.

Food spoilage by microbial growth is a way microbes can monopolize a food resource and keep it away from much larger consumers. Microbes can produce dangerous compounds during growth, such as mycotoxins. Most animals reject moldy food if they have a choice. The abundance of cedar in cave rat (wood rat) middens, and nests may be brought in by the wood rats as a way to decrease mold growth on stored materials (Fig. 16.4).

Today the best-known fungus in caves is *Pseudogymnoascus destructans* (formerly *Geomyces destructans*) that causes white-nose syndrome (WNS) in bats, killing them by the millions across the USA and Canada. The fungus is cold-adapted and infects skin of hibernating bats. The infection is irritating and causes the bats to wake up from

Fig. 16.3 **a** Fluffy white *Mucor* fungus growing on a rat latrine in Little Beauty Cave, MCNP. Different ages of droppings have different fungi (photograph by Scott Spicer). **b** Growth of tiny white mushrooms on a highly-leached acorn in the New Discovery Entrance (photograph by Scott Spicer)

Fig. 16.4 A pack rat (*Neotoma* sp.) in her nest in White Cave showing cedar and greens (photograph by Rick Olson)

hibernation, use up their limited fat reserves faster, and leave their hibernacula early in search of food before the flying insects return. Most infected bats die of starvation.

P. desrtuctans is an introduced species from Europe where it does not cause the high mortality seen with North American bats (Puechmaille et al. 2011). Apparently the fungus coevolved with European bats over thousands of years, but it is an invasive species in North American bats that have not developed any resistance. Since its discovery in a cave in New York in 2006, WNS continues to spread across the USA and Canada, making it to MCNP in 2012–13, where it was found in a northern long-eared bat, *Myotis sepertroinalis* from Long Cave, the largest hibernacula in MCNP. Toomey et al. (2013) reviewed actions taken at MCNP starting in 2009 before white nose was detected. Continued surveillance and monitoring of hibernacula and summer bat roosts is done to document population changes. Visitor education provided an opportunity to increase understanding of bats and the value of bats in ecosystems via public announcements, pre-tour briefings by guides, and posters.

WNS has drastically changed the way we cave. Because humans may spread the fungus, great care is taken to decontaminate shoes, clothing, and equipment between caves. Caves on Federal Lands are closed or have greatly reduced access, except for a few caves open to visitors, like Mammoth Cave. WNS continues to spread and is following the major flight routes of infected bats, although it made a big jump to Washington state in 2016. For the latest on WNS, see Chap. 17.

We are not likely to find fungi unique to caves because of their high energy demands, but Vanderwolf et al. (2013) says that low nutrients, stablility, and low temperatures favor fungi adapted to cave conditions, and some may be true troglobionts. We are still looking.

16.4.3 Archaea and Bacteria

Archaea and Bacteria are both domains of prokaryotic cells, with DNA free in the cell; however, they are not closely related. Metabolically, they show huge diversity and can utilize any chemical reaction that potentially has energy. They convert energy from forms that are unusable to higher organisms and produce microbial biomass that can be eaten by animals up the food chain (For more information on food chains and pyramids, see Chaps. 13 and 14). You are familiar with bacteria as pathogens and many species that ferment foods, but the majority is beneficial to the environment and us. The focus of microbial study in caves is on the diversity of bacteria and their contributions to the ecosystem.

Archaea are often found in extreme environments, like thermal springs and salt marshes, but are widely distributed. Methanogens are Archaea that produce flammable methane gas in marshes and in the guts of mammals. Very little work has been done on Archaea in caves, but Jarrell et al. (2011) speculate that Archaea are adapted to chronic energy stress, which might be a factor in differentiating the ecology of Bacteria and Archaea. Archaea may be important in nutrient-limited cave ecosystems by contributing to nutrient cycling, through sulfur oxidization, methane production, and nitrogen fixation and cycling. Archaea compete successfully in all mainstream environments and are dominant in soils low in nitrogen with low nitrification rates. Archaea in caves need more research.

Bacteria that commonly grow in caves are members of the *Actinobacteria*, a group of filamentous bacteria that produce exospores. *Actinobacteria* make up 10–33% of total soil microbes (Janssen 2006) and are widespread in caves. Metabolically, their main role in nature is in decomposition of organic matter.

References to cave wall slime, wall fungus, and lava wall slime all refer to the often-dense growth of *Actinobacteria* and associated microbes in many caves. Individual colonies have a branched appearance (Fig. 16.5) and are often white to yellow in color, but there are also tan, red, pink, blue, and pumpkin orange colonies. Actinobacterial colonies are often hydrophobic, with water beading up on the surface; the water reflects cavers' lights, described as "cave silver." The typical earthy smell associated with caves is a chemical called geosmin produced by *Actinobacteria. Actinobateria* can influence formation development by repelling water causing pitting or irregular surfaces around colonies, and by production of corrosive compounds that alter calcite deposits. Many *Actinobacteria* fix atmospheric nitrogen, particularly in extreme environments, but the role of *Actinobacteria* in nitrogen fixation in caves has not been studied.

Actinobacteria are well known for their production of secondary metabolites including antibiotics such as streptomycin and tetracycline. The majority of our antibiotics (75%) come from *Actinobacteria*. Antibiotic production by microbes in nature may give them a competitive edge over other microbes at high enough concentrations, or the chemicals may have other functions like signal molecules or for predation. Frisch et al. (2003) isolated bacteria from Mammoth Cave that produced potential drugs that blocked cancer, tuberculosis, and angiogenesis, but it takes many years before such discoveries are brought to actual treatment.

Fig. 16.5 Close-up of isolated actinobacterial colonies showing branching. Water beads up on the colonies at the bottom of the picture reflecting caver's lights ("cave silver") (photograph by Thomas Lavoie Photography)

The biomass and activity of microbes in limestone caves in MCNP were studied by Feldhake (1986). He measured microbial metabolic rates in 12 sites in four caves, with comparisons to overlying forest soils. Except for a site rich in cricket guano, Feldhake found that organic matter content, microbial activity, and biomass were much lower in the cave than in forest soil. Autotrophic activity was very low at two of twelve sites and absent at the remaining ten.

An exception to studies that show low numbers and activity of microbes in caves is one that compared the microbial activity, density, and diversity of two aquatic sediment sites in Mammoth Cave (Rusterholtz and Mallory 1994). This study was one of the first to compare high- and low-nutrient culture media. They had high numbers of bacteria and detected active metabolism in 53–58% of the population, despite very low total organic carbon. The diversity of populations was extremely high, with 42% of the isolated species similar to surface microorganisms with no dominant species and the remainder unidentified. These studies should be repeated with today's genetic techniques.

Organic chemical utilization by microbes from water samples collected at different levels in the Styx River drainage in historic Mammoth Cave was studied by Byl et al. (2013). They detected distinct community patterns with highest activity from upper level passages that were comparable to results from a surface stream. Communities from lower levels were slower and used fewer varieties of starting chemicals, but after five days the communities adapted to use almost all of the tested chemicals. The distinct community patterns they observed may vary by season or rainfall.

We are really just beginning the study of bacteria in caves, aided by advances in technology. Despite its large size and complexity, relatively few studies of Bacteria and none of Archaea have been done in Mammoth Cave. One interesting question is the origin of purple wall stains in Mammoth Cave at Mariannes Pass and major areas of purple associated with faults in Long Cave (Olson and Toomey 2016). We do not know yet if the purple deposits are microbial, mineral, or some of both; but we will know soon.

16.5 Genomic Studies

Fowler et al. (2009)[1] suspended biobeads, an inert support surface, in cave streams and pools within MCNP to grow biofilm communities. After 1 year, samples were returned to the laboratory where DNA was extracted and used to produce clone libraries to identify the types of bacteria present in each community. Phyla and class of *Proteobacteia* are compared for two samples (Fig. 16.6); Owl Cave, which has inputs of organics and possibly toxic chemicals, located in Cedar Sink near Park City, and Eyeless Fish Trail (EFT), a pristine, low-nutrient stream accessed by the Austin Entrance on Flint Ridge. The total DNA extracted from each sample is very different: 3463 ng/g from Owl Cave, with 34 clones, and only 476 ng/g from EFT, with 38 clones, supporting lower nutrient input into EFT. In both samples *Proteobacteria* dominate, with 58% from Owl and 79% of clones from EFT, but with different classes. The proportion of unclassified bacteria is 21% in Owl and 6% in EFT. The dominant bacteria from clone libraries of soils are *Proteobacteria* (39%), *Acidobacteria* (20%), and *Actinobacteria* (13%) with all other groups making up less than 10% each (Janssen 2006). The distribution in Owl Cave closely resembles soils, suggesting increased surface input compared to EFT.

The *Proteobacteria* are a large group of Gram-negative bacteria. Both sites (Fig. 16.6) are dominated by *Alphaproteobacteria*, which are a diverse group including chemohetero- and chemoautotrophs. *Betaproteobacteria* are mostly chemoheterotrophs, but include some that fix nitrogen. *Deltaproteobacteria* are found only in Owl Cave and include sulfate reducing bacteria, including anaerobes. The *Gammaproteobacteria*, particularly dominant in Owl Cave, include many familiar Gram-negative bacteria. The distribution is consistent with a polluted environment in Owl Cave, although no indicators of fecal pollution (coliforms) were identified. The different proportion of unknown bacteria shows higher diversity in Owl Cave probably due to more surface inputs. *Nitrospira* are only found in the pristine EFT and are involved in nitrogen cycling. *Planctomycetes*, nearly double in EFT, are an unusual group of bacteria that have stalks for attachment to surfaces, and some of them also oxidize nitrate.

Taking a closer look at the same data at a finer-scale classification, to the level of genus and species, results in the phylogenetic tree shown in Fig. 16.7 for EFT (Fowler 2009, see footnote 1). The higher-level groupings to the right are what we saw in the pie charts. Clones from EFT are labeled MACA-EFT# and are grouped with their closest relatives in GenBank. The *Alphaproteobacteria* include many relatives that are stalked for attachment, like the *Planctomycetes*. Many *Beta-* use one carbon compounds (e.g., methane, methanol). The few *Gamma-* are mostly novel, or related to sulfur cycling bacteria. (For more detail on bacteria of interest, consult https://microbewiki.kenyon.edu).

There are many opportunities to apply genomic and other molecular techniques to increase our knowledge of

[1]An unpublished poster titled "Concentration and Diversity of Bacteria in Clastic Sediments and Limestone Biofilms of Mammoth Cave, Kentucky" by Rick Fowler, Rick Olson, Hazel Barton, and Shivendra Sahi, a progress report to the National Cave and Karst Research Institute in Carlsbad, NM. The poster is stored in the Mammoth Cave National Park Curatorial Facility, accession number 818.

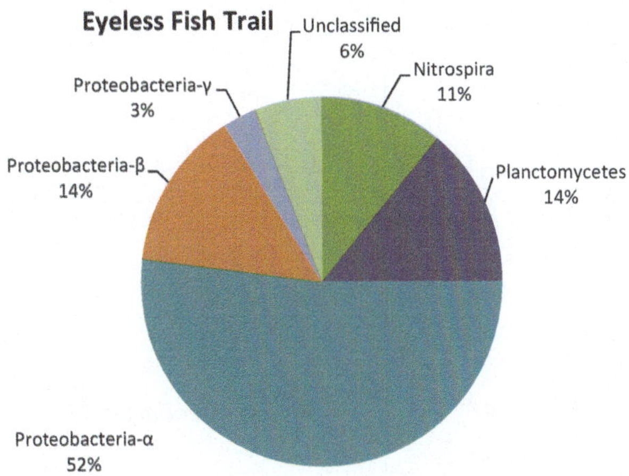

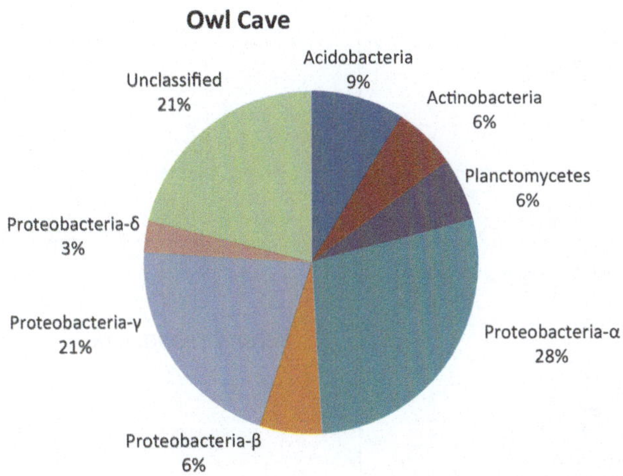

Fig. 16.6 Pie charts of bacterial phyla and Proteobacteria classes of clones from Eyeless Fish Trail and Owl Cave (Fowler et al. 2009, see footnote 1)

bacterial diversity in Mammoth Cave and for the study of Archaea.

16.6 Geomicrobiology

Geomicrobiology is a relatively new field that studies the intersection of microbial activities and geological processes. Microbes are important agents either actively or passively in chemical reactions that influence geological formations on scales from localized to landscape. Biogeochemical cycling of nutrients including carbon, phosphorous, sulfur, and nitrogen are important ecological roles of microbes. Many chemical reactions are both biotic and abiotic, but microbes are probably responsible for all or most low-energy reactions. Development of new tools and techniques in both biology and geology are contributing to a better understanding of the relative contributions of both fields.

Geomicrobiological processes are at work in caves (reviewed in Barton and Northup 2007; Engel 2010; Lavoie et al. 2010) in formation of some speleothems, mineral deposits, biokarst, and the formation of karst caves including Mammoth Cave by dissolution of carbonate rock by acidic water. Rainwater is acidic (pH 5.6), and additional acid comes from microbial activities as water moves through soil overlying limestone. Sulfuric acid speleogenesis is a chemolithotrophic microbial process for forming caves from production of sulfuric acid. The first conference in 1994 on the geomicrobiology of caves was sponsored by the Karst Waters Institute (Sasowsky and Palmer 1994).

16.6.1 Sulfur-Based Ecosystems

One of the earliest studies of microbes in caves using molecular techniques was done by Angert et al. (1998) in nearby Parker Cave, southwest of MCNP on the Sinkhole Plain that drains through Mammoth Cave (Meiman and Palmer 2009). Sulphur River is one of five parallel stream passages in Parker Cave and has a strong odor of hydrogen sulfide from the Phantom Waterfall, a soft pile of white bacterial filaments, sulfate, and elemental sulfur about 2.5 m high and 1 m wide. The likely source of sulfur is underlying oil field brines. The floor in the upper room of the cave has a highly acidic pH of 0.13 and is coated with elemental sulfur, and the ceiling has an acidic layer of microbial biofilm. The terrestrial community in the area is more diverse than other areas in the cave due to the greater microbial food base. A microscopic study of white filaments from the Phantom Waterfall by Thompson and Olson (1988) showed known sulfur-oxidizing bacteria, *Beggiatoa* and *Thiothrix*, with elemental sulfur granules. Sulfuric acid is produced as a waste product. Thompson and Olson speculate that this community is based on bacteria using energy from sulfur completely isolated from the usual indirect photosynthetic input of energy.

Comparing the sequence of a bacterial gene from the microbial filaments with known species, Angert et al. (1998) showed that the Parker Cave community had the greatest similarity to sulfur oxidizing bacteria from deep-sea hydrothermal vents and other sulfur-based environments. Others are related to species that fix CO_2 as a source of carbon. They speculate on possible impacts of growth of these microorganisms on dissolution and precipitation of minerals in caves.

Sulfur inputs are uncommon in Mammoth Cave. Hydrocarbons and hydrogen sulfide seeps near Mariannes Pass in Historic Mammoth were investigated by Olson (2013). A sulfur spring in this area was described by Bullitt in 1845. The seep is deeply weathered into the limestone and smells of hydrocarbons. White microbial biofilms in the seep

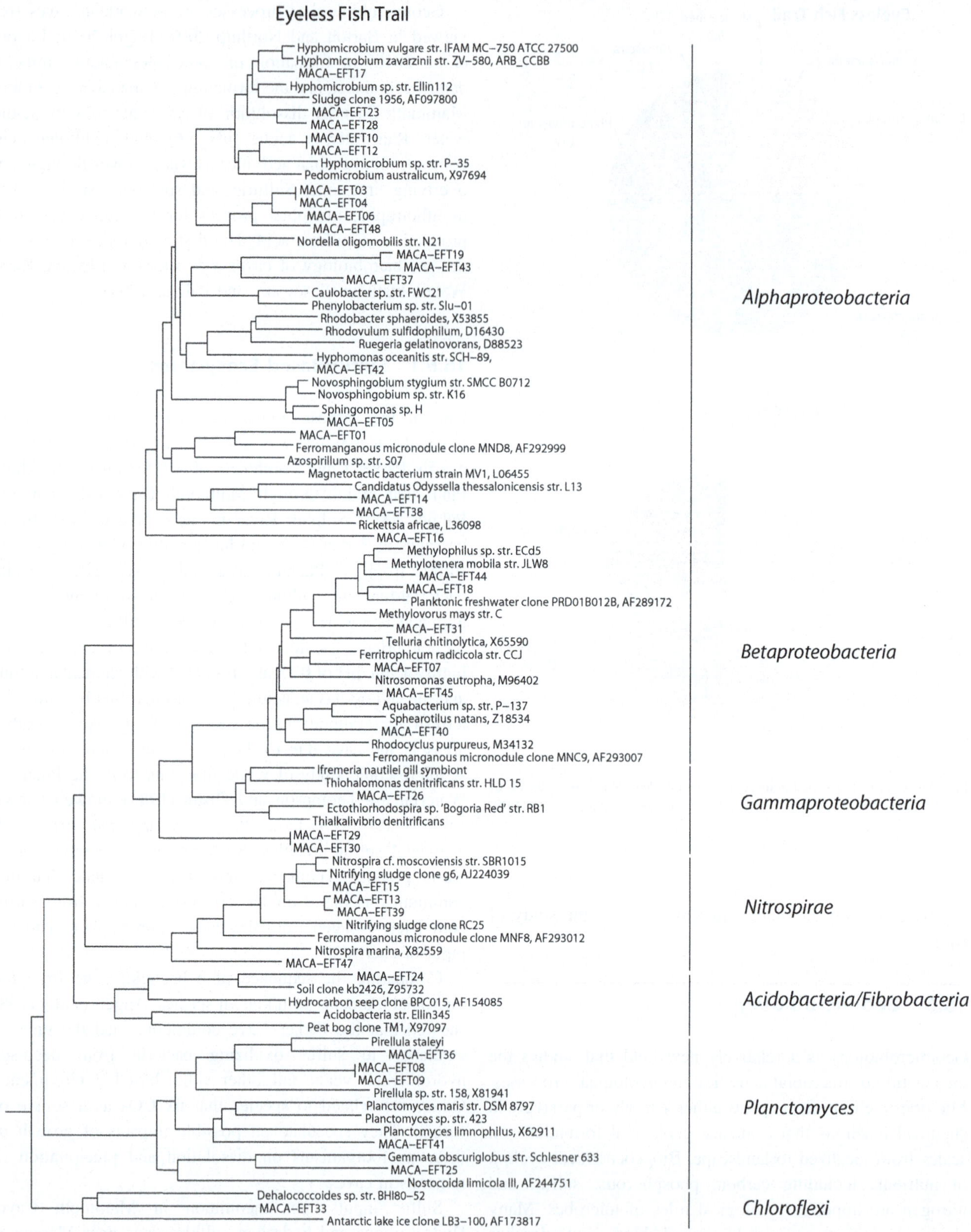

Fig. 16.7 A phylogenetic tree from Eyeless Fish Trail showing isolated clones (MACA-EFT#) and their nearest relatives (Fowler et al. 2009)

support thousands of springtails, with beetles and crickets nearby. Given the ubiquity of H_2S rich water and hydrocarbons under the entire south central Kentucky karst, sulfur inputs in Mammoth Cave need further investigation.

16.6.2 Carbonate Speleothem Formations

Most speleothems in caves are secondary calcium carbonate deposits ($CaCO_3$). A wide range of microbes and microbial processes (Barton and Northup 2007; Engel 2010; Lavoie et al. 2010) can produce extracellular polymeric substances (slime) and precipitate carbonate. Studies of microbial involvement have been carried out on stalactites, stalagmites, helectites, moonmilk, and other speleothems. Bacteria are important nucleation sites for calcite crystal growth that is influenced by the type of bacteria and abiotic factors like nutrients, temperature, and salinity. Biotic mechanisms include corrosion from release of organic acids that alter the crystal structure of the bedrock or formation, or precipitate minerals. Biotic and abiotic mechanisms can operate at the same time. I know of no studies of microbial involvement in speleothem development in Mammoth Cave.

16.6.3 Saltpeter

The best-known example of geomicrobiology in caves is saltpeter, or niter—KNO_3. Historically, caves were mined for saltpeter throughout the American Southeast to produce gunpowder for personal and strategic use during the War of 1812 when US harbors were blockaded by the British (Duncan 1997). Gunpowder is about 75% saltpeter with varying amounts of charcoal and sulfur (see Chap. 3 for more History)

The microbiology of nitre formation in cave soils is a two-step process known as nitrification that converts ammonium ion (NH_4^+) to nitrite (NO_2^-) and then bacteria add oxygen to the nitrite to produce nitrate (NO_3^-). Nitrate is also made by bacteria from insect parts in bat guano. Nitrate can be used by many organisms, or it can be converted to nitrogen gas (N_2) by denitrification. These processes are common worldwide in soils with a good source of organic compounds. Typically, there is a mix of nitrates in cave sediments, mostly with calcium and manganese. The conversion to saltpeter requires the addition of wood ash as a source of potassium and then heating of the leached solution to crystallize out the nitre (Hubbard 2005).

The biggest question in our understanding of saltpetre production at Mammoth Cave is the original source of the nitrogen. Many suggestions have been made, but the historic source was probably the large population of bats in Mammoth Cave. In shallow caves, nitrogen was probably leached from rich surface layers of organic matter and leaves that seeps down into the cave. Cave soils can regenerate niter if allowed to rest undisturbed in contact with the walls and floor of the cave, but Olson and Krapac (2001) investigated niter regeneration in Mammoth Cave for over 6 years and found regeneration rates so slow that groundwater percolation could not account for the original high niter concentrations.

One of the earliest studies of microbiology in Mammoth Cave was by Fliermans and Schmidt (1977), using species-specific fluorescent antibodies to study the presence, distribution, and population densities of *Nitrobacter*, a chemoautotrophic nitrifier. They found high population densities of *Nitrobacter* in Mammoth Cave soils and suggested that it may be responsible for the enrichment in nitrates seen in productive saltpeter soils. Leaching the soil to remove niter is the first step in production of saltpeter, and *Nitrobacter* populations did not change, while total bacteria decreased by 57%. They concluded that the original high niter content was due to bats. Repeating this study with new genetic techniques would be interesting.

16.6.4 Manganese Oxides: Ferromanganese Deposits

Black coatings exposed to flowing water in limestone caves may be poorly crystallized deposits of manganese oxide produced by microbial action (Fig. 16.8). Microbes oxidize soluble Mn^{2+} to trivalent or tetravalent manganese. White et al. (2009) did a systematic study of manganese oxide deposits from caves in central and eastern USA, including Mammoth Cave. They found that all samples contained both manganese and iron, but in different ratios; samples from Mammoth Cave had about four times more manganese than iron. They also reported enrichment of the black deposits in transition metals (copper, cobalt, nickel, vanadium, and zinc) at the fractional percent level, which are a million times greater than concentrations in the surrounding rock and water.

16.6.5 Infections and Parasites

While most microorganisms are neutral or beneficial to us, some microorganisms cause disease. Tuberculosis (TB) is a serious bacterial lung infection that even now infects (active and inactive) one third of the human population of the world. Historically known as consumption or the white plague, in the 1900s an estimated 110,000 Americans died each year from TB, second only to pneumonia and influenza (CDC 1999). Today's leading killers, heart disease and cancer were fourth and eighth, respectively. The treatments for TB in

Fig. 16.8 Black ferromanganese deposits alternating with white calcite on flowstone in White Cave (photograph by Scott Spicer)

1900 included good food, lots of fresh air, and inactivity, which led to the establishment of Sanatoriums (Sucre, n.d.). It is small wonder people were willing to live in the TB Huts deep in Mammoth Cave in hope of a cure. In 1839, Mammoth Cave was purchased by Louisville physician Dr. John Croghan (NPS History 2015). In 1841, he allowed 16 TB patients to move into wooden and stone huts in the Star Chamber beyond Giants Coffin. Cool conditions required open fires for warmth and light, resulting in soot deposits still evident today. Bushes were brought in to cheer up the patients and tours passed by the huts regularly. The deaths of some patients and the worsening conditions of others ended the experiment. Dr. Croghan died of TB in 1849 (see Chap. 3 for more History)

The diet and parasite load of ancient humans can be determined by an examination of their paleofeces. *Giardia,* a protozoan found in polluted waters that can cause diarrhea 2–4 weeks after drinking, has been reported from numerous caves and springs around the world. Human paleofeces from Salts Cave in MCNP show infestation with *Giardia*. Three paleofeces samples dated to 2420 ± BP had *Giardia* cysts (Ruppert 1994). One of eight paleofeces samples from Salts Cave in MCNP showed eggs of *Ascaris* (Fry 1974), a nematode worm that is 15–35 cm long, which is still the most common human nematode infection worldwide.

Crickets can be parasitized by horsehair worms. The infection begins when a cricket drinks from pools contaminated with worm eggs. The juvenile worm leaves the digestive tract and enters the body cavity of the cricket. The worm grows to adult size and bursts through the side of the cricket to drop into water pools under the roost where they mate and lay eggs to complete their life cycle. We Studier et al. (1991) found a horsehair worm infection rate of 9.6% among *Ceuthophilus stygius* camel crickets and 0.5% in *Hadenoecus subterraneus* cave crickets within MCNP. The difference is because *Ceuthophilus* must drink water and *Hadenoecus* gets most of their water from their food. Infected female *Ceuthophilus* had a reduction in eggs.

The cricket "marshmallow" in Fig. 16.9 is covered with a parasitic fungus (*Beauveria spp*). It is a parasite of many different insect host species that is used in insect pest management (Goettel et al. 2005). The fungal hyphae release enzymes that attack and dissolve the insect cuticle, allowing it to penetrate and grow into the insect body. Once inside the insect, it produces a toxin called beauvericin that weakens the host's immune system, and grows until it fills the entire

Fig. 16.9 A cave cricket "marshmallow" in the Frozen Niagara Entrance that has been killed by the growth of a parasitic fungus (photograph by Elizabeth Lavoie)

body cavity. When conditions are favorable (RH > 92%), the fungus will grow out through the softer parts of the insect's body, producing the characteristic "white bloom" appearance. These external hyphae produce spores that are released into the environment to infect the next insect on contact, completing the cycle.

16.6.6 Cave Cricket Microbes

Cave crickets (*Hadenoecus subterraneus*) are like little cave cows. Organisms that consume plant detritus, decaying fruit, and herbivore dung ingest a variety of microorganisms along with their food. If ingested, microbes survive and grow in the digestive tract or excrete enzymes that remain active in the gut, and then they can extend the digestive and metabolic capabilities of the organism. Cows do not actually digest their food; microorganisms in their rumen digest the food and make chemicals that feed the cow.

A similar situation is found with some orthopteran insects including crickets, grasshoppers, and cockroaches. The crop of cave crickets is a very thin-walled structure that lies between the esophagus and hindgut. The inner wall of the crop contains chitinous "teeth" that aid in mixing and movement of food through the digestive system (Fig. 16.2). Crickets can eat up to three times their body weight in food to the point of physical distortion. They waddle back to the cave and hang out, digesting their food over the next several weeks before leaving the safety of the cave to forage again (Studier et al. 1986).

Studier and Lavoie (1990) found that cave crickets rapidly lost weight in water-saturated air only 2 °C above cave ambient temperatures and die in a few hours if held above room temperature (23 °C), possibly due to loss of control over growth of crop microbes. Many bacteria and yeast make gases or toxic metabolites, like ethanol, at elevated temperatures. Some of these crickets, as well as an occasional field-collected specimen, had crops visibly distended with gas, occasionally to the point of rupture. Crop enzyme activity was optimum at 23 °C, above cave temperature (15 °C). When cave crickets were fed diets rich in either carbohydrates or protein and compared to the natural diet, enzymes responded rapidly to the different diets, as expected if microbes were producing the digestive enzymes (White 1989).

Whatever the reasons for the extreme thermal sensitivity of *H. subterraneus*, even a modest increase in cave ambient conditions could have profound negative effects on cave crickets. An increase of 2–6 °C over the next 50 years from climate changes would greatly increase metabolic demands and evaporative water loss of cave crickets, thereby forcing more frequent foraging and exposure to surface conditions and predators. Crickets could be extirpated, with loss of the major source of fixed carbon energy inputs into caves in central Kentucky. Poulson (1991) agrees and speculates that the physiological tolerance data are consistent with the narrow latitudinal distribution of *Hadenoecus* cave crickets between the Ohio River and northern Tennessee. He concludes that community change in caves may be a sensitive indicator of global climate change.

16.7 Human Impacts and Microbial Conservation

Microbes are clearly impacted by human activity. It is important to understand microbial colonization patterns, dispersal mechanisms, and potential effects on human health when studying microorganisms in caves (Saiz-Jimenez 2012). Human impacts are particularly evident in remote areas and with archeological materials.

Evidence of microbes associated with humans was done by comparing areas of Mammoth Cave and Carlsbad Cavern that had high versus low impacts from visitation (Lavoie and Northup 2006). We used swabbing and cultures to look for human-associated microbes (*E. coli* and *Staphylococcus aureus*) and bacteria that could be tracked in from the surface (high-temperature *Bacillus*). We found some trends, complicated because we do not know how long these microbes actually survive in the cave environment, but humans directly alter communities of native microbes in caves.

Shapiro and Pringle (2010) investigated human impacts on fungal diversity in caves including Dogwood Cave and Diamond Caverns that are hydrologically connected to Mammoth Cave. They sampled soils with a range of human impact, including two sites that may never have had human contact. They did not isolate any fungi from the area that had never been visited, as predicted by Caumartin (1963), who thought fungi would not naturally be found in caves without inputs from humans or animals, air, or water. In these caves, fungal diversity rises with moderate levels of disturbance and peaks in minimally disturbed sites. They concluded that impacts of human disturbance are highly localized.

Boston et al. (2006) have offered suggestions on preserving native cave microbes while removing or reducing contaminants, including wooden structures, and the difficulty of telling if materials are natural or anthropogenic (Fig. 16.10). Consideration of microbes should be a factor in choosing what cleaning methods and techniques are used to reduce collateral damage both to the microbes and speleothems. Reducing addition of organic carbon is also important in the low-nutrient cave environment to prevent overgrowth of non-native microbes. Humans continually shed hair, skin cells, and microbes, along with lint from our clothing and crumbs of food, which are great food resources

Fig. 16.10 Fungi colonizing old wood (note sneakers for scale). The white fungi are growing out from the old wood in search of new food. Is the wood brought in naturally or by humans? What invertebrates might live there? (photograph by Scott Spicer)

for non-native microbes. Microbial nature preserves could be established to protect native microbes adapted to low-nutrient conditions. We need to practice clean caving in every cave we visit or study, made all the more important by white-nose syndrome.

16.8 Conclusions

Microorganisms are critically important components of every ecosystem, including caves. Because of the extremely small size of microorganisms, we still have much to learn about their many activities. New techniques have resulted in changes in our understanding of microbial ecology and diversity, and provide many opportunities for future study.

Caves do not exist in isolation from the surface. Caves, speleothems, archeological resources, organisms, and microorganisms can all be harmed by direct visitation and any surface activity that alters quality and quantity of inputs of water, nutrients, and air exchange (Jones et al. 2003). Caves are conduits into the subterranean world, and pollution can impact water quality and cave life. In order to protect caves and what lives in them, we need to understand regional hydrology and what is happening throughout the entire drainage basin. Our knowledge of cave microbes is limited. Despite its great size and complexity, very few microbiological studies have been done in Mammoth Cave. There are many interesting ecological and evolutionary questions along with basic research and inventory to be done on intraterrestrial microbes. Microorganisms and their habitats need conservation along with larger organisms.

Acknowledgements I thank Tom Poulson, Diana Northup, Penny Boston, Louise Hose, and the late Gene Studier for many discussions that have provided insights into the contributions of microbes in caves. I thank Cody Leclair, Doug Soroka, and Lynn Kleina for comments on a draft of this chapter and the Editors for improvements.

References

Angert ER, Northup DE, Reysenbach A-L et al (1998) Molecular phylogenetic analysis of a bacterial community in Sulphur River, Parker Cave, Kentucky. Amer Mineral 83:11–12, Part 2:1583–1592

Barr TC, Kuehne RA (1971) Ecological studies in the Mammoth Cave system of Kentucky. II. The ecosystem. Annal de Spéologie 26 (1):47–96

Barton HA (2006) Introduction to cave microbiology: a review for the non-specialist. J Cave Karst Stu 68(2):43–56

Barton HA, Northup DE (2007) Geomicrobiology in cave environments: Past, current, and future perspectives. J Cave Karst Stu 69:163–178

Boston PJ, Northup DE, Lavoie KH (2006) Protecting microbial habitats: preserving the unseen. In: Hildreth-Werker V, Werker JC (eds) Cave conservation and restoration, Nat Speleol Soc, AL, pp 66–83

Bullitt AC (1845) Rambles in the Mammoth Cave during the year 1844 by a visitor. Reprint, 2nd edn. Cave Books 1985

Byl P, Trimboli SR, Toomey R et al (2013) Antibiotic resistance and substrate utilization by bacteria affiliated with cave streams at different levels of Mammoth Cave. In: Trimboli S (ed) Mammoth Cave National Park's 10th research symposium: Celebrating the diversity of research in the Mammoth Cave region, p 33–39. http://digitalcommons.wku.edu/mc_reserch_symp/10th_Research_Symposium_2013/

Caumartin V (1963) Review of the microbiology of underground environments. Nat Speleol Soc Bull 25:1–14

CDC (1999) Achievements in public health, 1900–1999: control of infectious diseases. Morbid Mort Weekly Rep 48(29):621–629 (CDC.gov)

Domagal-Goldman SD, Wright KE (eds) (2016) The astrobiology primer v2.0. Astrobio 1698:561–653 doi:10.1089/ast.2015.1460

Duncan MS (1997) Examining early nineteenth century saltpeter caves: an archaeological perspective. J Cave Karst Stud 59(2):91–94

Edwards KJ, Becker K, Colwell F (2012) The deep, hot energy biosphere: intraterrestrial life on earth. Ann Rev Earth Planet Sci 40:551–568

Engel AS (2010) Microbial diversity of cave ecosytems (Chap. 10). In: Loy A, Mandl M, Barton L (eds) Geomicrobiology: molecular and environmental perspectives. Springer, Heidelberg, pp 219–238. doi:10.1007/978-90-481-9204-5_10

Feldhake J (1986) Microbial activity and biomass of limestone caves. MS thesis, University of Cincinnati, 204 pp

Fliermans CB, Schmidt EL (1977) *Nitrobacter* in Mammoth Cave. Int J Speleol 9:1–19

Frisch RR, Schaffer P, Haak M et al (2003) Identification of three bacteria from Mammoth Cave, Mammoth Cave National Park, Kentucky, USA that are potential anti-angiogenesis, anti-tuberculosis, or anti-microbial producing organisms. Presentation 103rd general meeting of the American Society for Microbiology, Washington DC

Fry GF (1974) Ovum and parasite examination of Salts Cave human paleofeces. In: Watson PJ (ed) Archaeology of the Mammoth Cave area. Academic Press, New York

Gittleson SM, Hoover RL (1969) Cavernicolous protozoa: Review of the literature and new studies in Mammoth Cave, Kentucky. Annal de Spéléol 24(4):737–776

Goettel MS, Eilenberg J, Glare TR (2005) Entomopathogenic fungi and their role in regulation of insect populations. In: Gilbert LI, Iatrou K, Gill S (eds). Comp Mol Insect Sci 6:361–406, Elsevier, Amsterdam

Hubbard Jr DA (2005) Saltpetre mining. In: Culver DC, White WB (eds) Encyclopedia of caves. Elsevier, New York, pp 492–495

Hug LA, Baker BJ, Anantharaman K, et al (2015) A new view of the tree of life. Nat Microbiol 1:16048. doi:10.1038/nmicrobiol.2016.48

Janssen PH (2006) Identifying the dominant soil bacterial taxa in libraries of 16S rRNA and 16S rRNA genes. Appl Envir Microbiol 72(1):1719–1728. doi:10.1128/AEM.72.3.1719-1728

Jarrell KF, Walters AD, Bochiwal C et al (2011) Major players on the microbial stage: why archaea are important. Microbiology 157:919–936

Jones WK, Hobbs III HH, Wicks CM et al (2003) Recommendations and guidelines for managing caves on protected lands. Special Publication 8, Karst Waters Institute

Lavoie KH (1982) The influence of biotic and abiotic factors on the successional decomposition of cave rat dung. Dissertation, University of Illinois at Chicago

Lavoie KH (2015) "A grand, gloomy, and peculiar place": microbiology in the Mammoth Cave region. In: Wagner R, Engel AS (ed) Life in extreme environments: microbial life of cave systems. DeGruyter, Berlin, pp 47–78

Lavoie KH, Northup DE (2006) Bacteria as indictors of human impact in caves. In: Proceedings of the 2005 national cave and karst management symposium. 30th anniversary issue, pp 119–124

Lavoie KH, Northup DE, Barton HA (2010) Microbial-mineral interactions: geomicrobiology in caves: a review. In: Jain SK, Kahn AA, Rai MK (eds) Geomicrobiology. Science Publisher, USA, pp 1–46

Meiman J, Palmer A (2009) The Pennyroyal Plateau, Kentucky. In: Palmer AN, Palmer MV (eds) Caves and karst of the USA. Nat Speleological Soc, Huntsville, AL pp 116–118

Olson R (2006) Control of lamp flora in developed caves. In: Hildreth-Werker V, Werker JC (eds) Cave conservation and restoration. Nat Speleological Soc, pp 343–348

Olson R (2013) Potential effects of hydrogen sulfide and hydrocarbon seeps on Mammoth Cave ecosystems. In: Trimboli S (ed) Mammoth Cave National Park's 10th research symposium: celebrating the diversity of research in the Mammoth Cave region pp 25–31

Olson R, Krapac IG (2001) Origin and regeneration of nitrates in Mammoth Cave sediment. J Cave Karst Stud 63(3):114

Olson R, Toomey RS (2016) The effect of faulting on past and present hydrogeology in Long Cave, Mammoth Cave National Park. In: Trimboli SR, Dodd LE, Young D (eds) Proceedings for celebrating the diversity of research in the Mammoth Cave region: 11th research symposium at Mammoth Cave National Park (April 18, 2016). http://digitalcommons.wku.edu/mc_reserch_symp/11th_Research_Symposium_2016/Complete_Proceedings/

Poulson TL (1991) Effects of recent and future climate change on terrestrial cave communities at Mammoth Cave. Cave Res Found Ann Rep 47–48

Puechmaille SJ, Wibbelt G, Korn V et al (2011) Pan-European distribution of white-nose syndrome fungus (*Geomyces destructans*) not associated with mass mortality. PLoS One 2011 6(4): e19167

Ruppert LK (1994) Evidence for the endoparasite *Giardia lamblia* in human paleofeces from Salts Cave, Mammoth Cave National Park, Kentucky. Thesis (MA), Western Michigan University

Rusterholtz M, Mallory LM (1994) Density, activity, and diversity of bacteria indigenous to a karstic aquifer. Microb Ecol 28:79–99

Sasowsky ID, Palmer MV (eds) (1994) Breakthroughs in Karst geomicrobiology and redox chemistry: Karst Waters Institute special publications 1. Symposium abstracts and field guide. Leesburg WV

Saiz-Jimenez C (2012) A review: microbiological and environmental issues in show caves. World J Microbio Biotech 28:2453–2464

Shapiro J, Pringle A (2010) Anthropogenic influences on the diversity of fungi isolated from caves in Kentucky and Tennessee. Am Midl Nat 163:76–86

Smith T, Olson R (2007) A taxonomic survey of Lamp Flora (Algae and Cyanobacteria) in electrically lit passages within Mammoth Cave National Park, KY. Int J Speleol 36(2):105–114

Studier EH, Lavoie KH (1990) Biology of cave crickets, *Hadenoecus subterraneus*, and camel crickets, *Ceuthophilus stygius* (Insecta: Orthoptera): metabolism and water economies related to size and temperature. Comp Biochem Physiol Part A: Physiol 95:157–161

Studier EH, Lavoie KH, Chandler CM (1991) Biology of cave crickets, *Hadenoecus subterraneus*, and camel crickets, *Ceuthophilus stygius* (Rhaphidaphoriae); Parasitism by hairworms. J Helmint Soc Wash 58(2):248–250

Studier EH, Lavoie KH, Wares WD II et al (1986) Bioenergetics of the cave cricket, *Hadenoecus subterraneus*. Comp Biochem Physiol 84A:431–436

Sucre R (nd) The Great White Plague: the culture of death and the tuberculosis sanatorium. Accessed 6 Dec 2013 at http://www.faculty.virginia.edu/blueridgesanatorium/death.htm

Thompson DB, Olson R (1988) A preliminary survey of the protozoa and bacteria from Sulphur River in Parker's Cave, KY USA. Bull Nat Speleolo Soc 50:42–46

Toomey RS, Thomas SS, Gillespie J et al (2013) White-nose syndrome at Mammoth Cave National Park: actions before and after its detection. In: Trimboli S (ed) Mammoth Cave National Park's 10th research symposium: celebrating the diversity of research in the Mammoth Cave region, pp 206–207

Vanderwolf KJ, Malloch D, McAlpine DF et al (2013) A world review of fungi, yeasts, and slime molds in caves. Int J Speleol 42:77–96

White CR (1989) Digestive enzymes of the cave cricket, *Hadenoecus subterraneus*. In: Lindsey K (ed) 1988. Cave Res Found Ann Rep 61–63

White WB, Vito C, Scheetz BE (2009) The mineralogy and trace element chemistry of black manganese oxide deposits from caves. J Cave Karst Stud 71(2):136–143

Bats of Mammoth Cave

17

Steven C. Thomas and Rickard S. Toomey III

Abstract

Bats that inhabit caves have been misunderstood and maligned for centuries. Yet they seem mysteriously to draw researchers and visitors alike to their underground dwellings for a glimpse into their world. Indeed some species of bats spend more than 80% of their life in caves. Still they must leave the cave to obtain food when it is accessible. The food for these nocturnal true-flying mammals in the Mammoth Cave area is insects, exclusively. The consumption of large quantities of night-flying insects is one of several benefits bats provide to surface ecosystems and to humans. In cave ecosystems, insects consumed by bats on the surface provide nutrients (energy) to these generally nutrient-poor environments primarily in the form of guano (Culver and Pipan 2009). Despite the critical role bats play in ecosystems, many species face a variety of threats to their survival and some have experienced marked population declines. In fact, Mammoth Cave was formerly one of the largest bat hibernacula in the world. Indiana bats (*Myotis sodalis*), and to a lesser degree gray bats (*M. grisescens*), were prominent species in Mammoth Cave 150 years ago, but today they are listed as federally endangered. Relatively few bats use Mammoth Cave today and with the latest threat—a new fungal disease of bats called "white-nose syndrome"—this number may be greatly reduced in the near future. Nevertheless, measures to protect and restore bats and bat habitat have been implemented in Mammoth Cave, and opportunities to take additional conservation actions continue to arise.

17.1 Bat Biology

Bats are the only true-flying mammals. Skin, connected to their arms, sides, legs, tail, and extra-long fingers, enables bats to maneuver in flight. Although bats have good vision, they rely on their amazing echolocation (sonar) ability to find their way and their insect prey in the dark. Bats produce ultrasonic pulses of sound that echo back to their ears from objects ahead and are analyzed almost instantaneously to allow them to "see" in total darkness. Normally, bats eat more than half their body weight in insects every night for about six months of the year. During the rest of the year, when insect food is not available, bats in the Mammoth Cave region must either hibernate or migrate to warmer areas to survive. Bats that hibernate in caves must live on stored fat until spring. To conserve energy, the bat's vital functions—heart rate, breathing rate, metabolic rate, and immune system—slow or shut down. Bats arouse occasionally during the hibernation period to drink, eat, defecate, or even mate. Bats typically hibernate in sections of caves where the air temperature ranges from 2 to 14 °C (36–57 °F). Some species, such as gray bats and Rafinesque's big-eared bats (*Corynorhinus rafinesquii*), use caves year-round, although different caves are used during the winter and summer. Warm areas of certain caves occupied by some species of bats during the summer serve as either maternity (nursery) roosts or as bachelor (day) roosts. Cave-dwelling bats in the Mammoth Cave region typically give birth to one baby (pup), sometimes twins, in May or June. Mating usually

S.C. Thomas (✉) · R.S. Toomey III
Cave Research Foundation, Cave City, KY, USA
e-mail: swrbthomas@juno.com

R.S. Toomey III
e-mail: myotis.toomey@gmail.com

H.H. Hobbs III et al. (eds.), *Mammoth Cave*, Cave and Karst Systems of the World,
DOI 10.1007/978-3-319-53718-4_17

occurs in the fall; ovulation and pregnancy occur in the spring. Pups normally fly at about three weeks old and are nursed for about one month. Remarkably, some cave-roosting bats can live for over 30 years.

Bats can contract and transmit rabies, but they are no more likely to carry rabies than many other mammals. Only a fraction of a percent of wild bats have rabies; nevertheless, bats should not be handled. Bats are occasionally associated with a fungal disease called histoplasmosis. This respiratory disease is caused by breathing in spores found in connection with bird or bat droppings. The majority of human cases are asymptomatic or involve flu-like symptoms. White-nose syndrome, an exotic fungal disease of hibernating bats, has been devastating bat populations in the eastern North America since it was first detected in the winter of 2006–2007 (see Chap. 16). This disease was named for the white fuzzy growth on the nose, ears, and wings often seen on affected bats. Bats with the disease may display unusual behavior such as flying outside during the day in near-freezing weather. This quickly uses up their fat reserves at a time when insects are not available for food. Human health implications are not known; however, there is no information indicating that people or other animals have been affected after exposure to the cold-loving fungus. Bats are the primary predator of night-flying insects—many of which are crop and forest pests. A continued loss of large numbers of insectivorous bats to threats such as white-nose syndrome may dramatically affect the surface and subsurface ecosystems in the Mammoth Cave region.

17.2 Methods for Studying Bats

Due to their generally secretive nature and nocturnal habits, studying bats can be a particular challenge. The study question(s) to be addressed will usually guide the approach and methods used. Some studies use bat sign as evidence of the animal's presence to provide clues on past bat use (e.g., guano deposits, roost staining, culled insect parts, bones, desiccated remains). Other common ways bats can be studied without catching them include (1) acoustic surveys (recording high-frequency bat calls using one or more bat detector that is either stationary or mobile); (2) winter colony (hibernacula) surveys using direct visual counts of hibernating bats or digital photography; and (3) summer colony surveys (external) using direct counts with night vision goggles with infrared (IR) light sources, near IR video cameras with IR light sources, or thermal IR video cameras, and done when bats emerge at dusk.

Capturing bats for investigation can be accomplished by hand-capturing, hand netting (insect net), mist netting (fine-meshed black nylon nets stretched between two poles), or harp trapping (square frame with two or more sets of vertical monofilament lines strung like a harp with a canvas bag suspended below to collect and hold the bats). Valuable information (e.g., species, sex, age, reproductive condition, body mass, health) can be obtained from captured bats; researchers can collect blood, tissue, hair, or fungal swab samples as well. Having bats in hand allows scientists to mark them to gain information such as population size estimates, survivorship, home range estimates, migration activities, roost fidelity, foraging behavior, resource selection, social relationships, and bat echolocation reference calls. Frequently used methods for marking bats include forearm bands, radio telemetry tags, and microchip (PIT) tags. Understanding the relationship between roosting requirements of cave-dwelling bats and cave microclimate is important for effective bat/cave management and conservation (see Chap. 10). Therefore, researchers use instruments to collect manually or automatically information on cave microclimate parameters such as air temperature, rock temperature, relative humidity, and airflow (change in wind velocity and direction). Radio telemetry tags with temperature sensors can be glued to the back of a bat to transmit skin temperature measurements regularly to a nearby antenna and datalogger.

17.3 Role of Bats in Mammoth Cave Ecology

Most cave-dwelling bats are generally considered trogloxenes, as they spend part of their life in caves but cannot complete their life history underground. Like cave crickets and woodrats, cave-dwelling bats are often critical components of the cave terrestrial ecosystem, and to a lesser extent, they can contribute food energy to the cave aquatic ecosystem. Generally, caves are energy-poor, and organisms like bats that obtain their food outside of a cave will regularly bring nutrients (energy) into a cave in the form of guano, urine, and occasional dead bats or carcasses. These, deposited at roost sites and along flyways, are used as sources of food by cave organisms that cannot leave the cave to find food (called obligate species). In a few instances in the Mammoth Cave area, bat guano may constitute the primary nutrient source for the system, providing the foundation for cave communities, including microbes, fungi, and invertebrates (see Chaps. 13 and 16). In most other caves, bat guano may be a minor but still important energy input. Given the energy contributions of bats to cave ecosystems, bat protection efforts propagate along the food web when biological communities are sustained through natural guano input. Because the immense aggregations of bats at prehistoric and historic roosts are gone, the amount of nutrient input into Mammoth Cave by bats is now greatly diminished. In addition to very small amounts of scattered guano along flyways in the areas of Mammoth Cave still used by bats, a small guano pile is located at the back of Dixon Cave directly below an area on the ceiling used by active bats year-round. Thus, nutrient input into Mammoth Cave by bats has changed as a result of the major declines in bat numbers over the past 150–200 years.

17.4 History of Bat Use of Mammoth Cave

17.4.1 Prehistoric Bat Use

The prehistoric use of Mammoth Cave by bats as revealed by analyses of the paleontological record is addressed in Chap. 11.

17.4.2 Historic Bat Use

Until sometime in the last 100–200 years, the largest contemporary bat roost at Mammoth Cave National Park (as opposed to fossil roosts discussed in the Paleontology chapter) was located in the area around the Historic Entrance of Mammoth Cave. This roost included winter hibernation areas for large numbers of several species of bats. In addition, some bats also used the area as a summer roost. Information on this roost comes from three lines of evidence. Historic accounts provide much basic information on where and when bats were roosting. They also provide some comments on the numbers of bats. These historic accounts are supported by traces of past bat use such as guano and staining left on walls and ceilings by roosting bats. The third line of evidence is actual remains of bats (mainly bones, but occasionally mummified bats) that are found in the area. These remains provide important information on which species of bats were using which areas. Colburn (2005, 2006) and Toomey et al. (2001) provide additional details on the bat use of the areas discussed in this chapter. Figure 17.1 summarizes the bat use of different parts of this area. It also shows the extent of mapped staining on ceilings and walls from bats and preserved bat guano.

The number of bats that used this area was truly immense. Silliman (1851) estimated that millions of colonial bats roosted within a few kilometers of the entrance. Using the extent of bat stain, Tuttle (1997) estimated that the area sheltered 9–13 million bats at its highest use. He concluded that Mammoth Cave's Historic Entrance area housed one of the largest hibernating colonies of bats yet identified, and historic and paleontological analyses strongly support his conclusion.

Over the last 200 years, most of the bats that formerly roosted in the Historic Entrance area have abandoned those roosts. Human use of the area and physical changes to cave conditions caused by that use have rendered most of this section of the cave unsuitable for bat roosts. Direct disturbance and microclimate changes caused by entrance and passage modifications account for most of the habitat loss in this area. Today, the area is mainly used for roosting by tri-colored bats (*Perimyotis subflavus*) in all seasons and a few big brown bats (*Eptesicus fuscus*) (in the winter). Many other species (including most that formerly roosted in the area) visit the area and use the Historic Entrance as a swarming site.

Because the types of evidence and nature of the bat use vary by specific place in the Historic Entrance area, we will discuss the evidence and use by location. For each roosting site, we will discuss some of the historic accounts that provide information on roosts, summarize the evidence from staining and guano, and provide information on what bones tell us about who was using the location.

17.4.3 Houchins Narrows

The primary use for Historic Entrance and Houchins Narrows was a flyway to enter and leave the cave. We are aware of little or no discussion of bats in historic accounts of this area. It is unlikely that most colonial bat species (such as Indiana, little brown (*Myotis lucifugus*), gray, or Rafinesque's big-eared bats) utilized this area for a winter roost; the same would be true for tri-colored bats. Cold incoming air during the winter would make this area too cold and subject to freezing. Bone deposits preserved along the side of the passage indicate that all of the above species used the area. The colonial species would have used it as a flyway. Tri-colored bats also may have used it in the spring and fall as a staging area as they do today. Remains of big brown bats and eastern small-footed bats (*Myotis leibii*) are found also in Houchins Narrows. Often these two species will have winter roosts in cold, exposed, windy areas, so they did use the area probably as a hibernation site, likely in small numbers. Additionally, bone deposits in Houchins Narrows are also instrumental in reconstructing the pre-development fill levels of the passage.

17.4.4 Audubon and Little Bat Avenues

In some cases the names of places can provide important clues about historic bat use. This is the case for the Big Bat Room (now Audubon Avenue) and Little Bat Avenue. Lee (1835) designated the area now known as Audubon Avenue as the "Big Bat Room" on his map of the cave. He also indicated "Clusters of Bats" on the running profile through

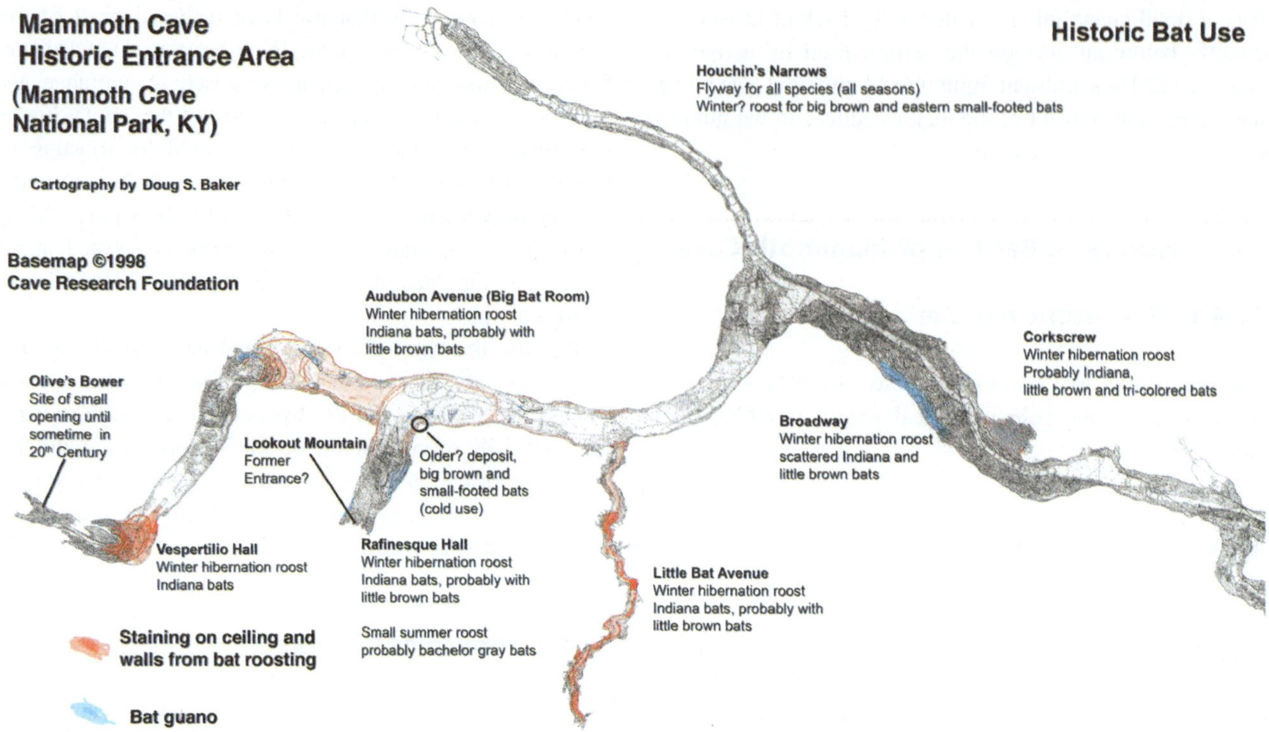

Fig. 17.1 Map of the Historic Entrance area of Mammoth Cave showing evidence of historic bat usage. *Orange shading* shows areas with staining on the ceilings indicating past bat roosts (*darker shading* indicates more intense or longer bat use). *Blue shading* shows areas where past bat guano is located. Base map derived from Baker (1998)

this area. He also labeled the Little Bat Room (now Avenue) and noted the presence of bats in it.

Staining and guano indicate significant roosts in Audubon Avenue. This is especially the case for the area between Rafinesque Hall and the west end of the passage (Fig. 17.1). Staining occurs on many of the small ledges along the ceiling, indicating colonial bat use. Bone indicates that many species used Audubon Avenue. These include Indiana, little brown, gray, eastern small-footed, tri-colored, and big brown bats. Winter colonial roosting by Indiana, probably little brown, and possibly gray bats occurred along the passage. The season that eastern small-footed and big brown bats used the passage is not clear. A photograph in Bailey et al. (1933, p. 459) shows linear aggregations of bats clinging to wall steps in Mammoth Cave. Although the caption labels the location as Little Bat Avenue, it actually shows an area on the wall of Audubon Avenue. This photograph demonstrates that at least some colonial bat use of this area occurred until at least the 1930s; however, he does not indicate which species was hibernating at that time.

Historic writings also support the presence of large numbers of bats in Little Bat Avenue. In 1845, Bullitt (p. 18) noted the importance of this location:

> The Little Bat Room Cave – a branch of Audubon Avenue, – is…but little more than a quarter of a mile in length, and is remarkable for its pit of two-hundred and eighty feet in depth; and as being the hibernal resort of bats. Tens of thousands of them are seen hanging from the walls, in apparently a torpid state, during the winter, but no sooner does the spring open, than they disappear.

Fifty years later, Hovey (1896: 76) reaffirmed hibernation of colonial bats in Little Bat Avenue:

> a branch from it [Audubon Avenue], running to Crevice Pit, was called Little Bat Room – a title that clings to it yet. Here myriads of bats take up their winter quarters, congregating for the purpose from all the region around. Deposits of bat-guano abound, and this is supposed to be connected with the quantities of nitrous earth, which is richest here.

In 1897, Rhoads (pp. 59–60) described little brown bats and Indiana bats hibernating in Little Bat Avenue:

> In a low, wide passageway (Little Bat Avenue), about one-fourth of a mile from the entrance to the cave, I found a cluster of little brown bats, which hung like a swarm of bees from a hollow space in the ceiling, just above the level of my head as I stood on the floor. The circular space covered by them was about 18 inches in diameter, and from this were suspended, head downward, nearly 150 bats in a compact, conical mass, several layers deep.

Hovey (1912: 25) also noted the hibernating bat colonies in Little Bat Avenue "named for the myriad of bats which in winter may be found here." The present day Little Bat

Avenue shelters almost no hibernating bats, and the passage floor and ledges have been swept clean by nearly 200 years of mining and tourist activities. Ceiling stain is extensive and among the heaviest encountered in all of Mammoth Cave's main level, attesting to the avenue's heavy and repeated utilization as a hibernaculum for colonial bats (i.e., little brown and/or Indiana bats). Despite Hovey's (1896: 76) claim that "deposits of bat-guano abound," only small amounts of pelletized guano can be observed there today. Bones of several bat species have been identified, including Indiana, little brown, big brown bats, and possibly eastern small-footed and gray bats.

The fossils, traces, and historic accounts for this area support the conclusion that parts of Little Bat Avenue were used as a hibernaculum until at least the turn of the twentieth century. Large numbers of bats used this hibernaculum probably over a long period of time. The Indiana bat appears to have been the dominant bat in the hibernaculum. Both little brown and gray bats also may have hibernated in the area. This hibernaculum probably represents the second most intensive use found in the Historic Entrance area. Big brown and eastern small-footed bats may have roosted in the area as well; however, it is not clear what season their use represents.

17.4.5 Rafinesque Hall

Rafinesque Hall has a variety of evidence of bat use. Ceiling stain ranges from light to moderately heavy. Bat guano lacking insect remains is preserved on large breakdown blocks and between rocks; this type of guano, which represents winter metabolism of stored fat demonstrates winter roosting in the area. In addition to bat guano, extensive raccoon scat is also found in this area. This raccoon guano has large numbers of bat bones in it and represents raccoons coming into a winter roost and plucking hibernating bats off the walls. Overall, the evidence indicates a fairly important colonial hibernation site. Bones indicate that Indiana bats (and possibly also little brown bats) hibernated here.

A thick guano deposit containing insect remains indicates a summer bat roost as well (on the east side of the passage). This guano is in a small area and is compacted into a mass about one foot thick. It was produced by a small, summer colonial roost. The most likely candidates for a summer roost would be a bachelor colony of either gray or Rafinesque's big-eared bats. We have not found material to indicate either of the species directly, but the roost type is more consistent with gray bats.

The area also has bones of both big brown and eastern small-footed bats. However, it is not clear what season these bats represent or whether these bones might be an older deposit from when there was an opening at Lookout Mountain.

17.4.6 Vespertilio Hall

Vespertilio Hall at one time probably housed one of the largest aggregations of bats in the Historic Entrance area. Evidence for this includes historic accounts, historic collections of live bats, and paleontological evidence. The name Vespertilio Hall refers to the large number of bats that were found in the area. *Vespertilio* is the scientific name that Linnaeus gave to many types of bats including several that are now in the genus *Myotis*.

Hovey (1912, pp. 26–27) and Hovey and Call (1897) describe the bats in this location at the end of the nineteenth century:

> Soon after leaving the Mushroom Beds the avenue again widens somewhat, though the ceiling is mainly low. But in the central portions the ancient waters had sculptured out an inverted kettle in the midst of a somewhat pronounced hall, and this is the rendezvous of myriads of bats. From the name of the genus which is so abundantly here represented we have given the locality the appellation of Vespertilio Hall. Thousands of bats, in winter season, suspended in great clumps, may here be seen. A single catch one night gave Doctor Call six hundred and seventy individuals, most of which went to the United States National Museum.

The specimens referred to in the above quote include a series of Indiana bats now housed at the Smithsonian Institution, National Museum of Natural History.

The ceiling has very well developed and extensive staining which is related to the presence of large numbers of bats over a long time. Raccoon scat, bat remains, and guano also indicate the presence of a roost. All evidence indicates a large, colonial hibernaculum of at least Indiana bats possibly with little brown bats.

The Vespertilio Hall roost is in an upper-level, dead-end passage, and recent temperature data indicate that the area maintains an almost invariant temperature of 11.5 °C (52.7 °F) (see Chap. 10 for more details). This is clearly too warm for an Indiana bat hibernation site. Such hibernacula typically range between 3 and 7 °C (37–45 °F) (US Fish and Wildlife Service 2007). Because of its location in a dead-end passage, the temperature of the Vespertilio Hall roost is virtually insensitive to changes in cave airflow outside of the passage arm itself. For some reason, and at some point in time after 1896, when Dr. Call collected 670 bats in Vespertilio Hall, this part of the cave changed from a variable temperature zone to an almost constant temperature zone. The most likely explanation is the existence of a former opening to the surface in the vicinity. Such an opening would have allowed cold air to be drawn in via the chimney-effect during winter. Popcorn and speleothems showing directional air movement also suggest a prior opening in the area of Olive's Bower.

17.4.7 Broadway

Lee (1835) notes clusters of bats in Broadway shortly beyond the salt peter vats. This is consistent with the presence of bat guano deposits preserved along the south side of the passage (Fig. 17.1). The area was certainly utilized by all of the species that inhabited the Historic Entrance area. Most of the bats probably used it as a flyway. The guano indicates a winter roost on the south side of the passage between the Rotunda and Corkscrew. A lack of identifiable bone does not allow identification of which species was hibernating in the area. Lee's mention of clusters of bats suggests that little brown, Indiana, or gray bats would be likely candidates.

17.4.8 Corkscrew

According to Randolph (1924, p. 92), nineteenth century guide William Garvin found the Corkscrew when "In his frequent passings in and out of the Cave, he occasionally observed bats flying with reckless speed and suddenly disappearing in a small aperture, far above the Kentucky Cliffs...." Bat stain and guano in the Corkscrew indicate that the area was a significant winter hibernation roost. Bones found in the area indicate that big brown, red (*Lasiurus borealis*), gray, eastern small-footed, little brown or Indiana, and tri-colored bats all used the area, although many of them may not have been associated with the winter roost. The winter roost was probably used by tri-colored bats, as well as by colonial bats such as Indiana, little brown, or gray bats.

17.4.9 Dixon Cave

Dixon Cave is geologically a continuation of Mammoth Cave that was sealed off by the collapse that formed the Historic Entrance. It is one of the most important modern bat roosts on the park, but evidence of bat use of the site extends to before we have counts and direct observations. Extensive saltpeter deposits that were mined from Dixon Cave at the turn of the nineteenth century provide important evidence that bats utilized the cave over a very long time (probably thousands of years). However, what we are able to say about pre-twentieth century bat use of the cave is much more limited than what we can say about use of the Historic Entrance area. There are two factors that contribute to this limitation. The primary reason is that saltpeter mining in Dixon Cave was much more intensive than in the Historic Entrance area; that is, it disrupted the floor much more completely. The digging means that many fewer bat bones can be found in the cave that are not apparently very recent. In addition, the sheer size of Dixon Cave makes it difficult to characterize ceiling staining accurately.

Hovey (1893) visited Dixon Cave in March with Ed Bishop. He observed that "... at the time of our visit it was also appropriated by myriads of hibernating bats, clinging in great clusters like swarms of bees." Staining visible in the cave, suggests that colonial bats have used the cave over long periods (although as noted above, mapping the stain is difficult). Analysis of the bones from Dixon Cave indicates that numerous species of bats used the cave. The most ubiquitous species identified in bones from the cave are (in descending order) Indiana, big brown, eastern small-footed, tri-colored, and little brown bats. Gray, Rafinesque's big-eared, and possibly southeastern (*Myotis austroriparius*) bats were also present but in much smaller numbers. The bones represent bats present in the cave over at least the last thousand years.

17.5 Present Bat Use

Bat use of Mammoth Cave during the twentieth and twenty-first centuries is a drop in the bucket when compared to historic and prehistoric times—both in terms of numbers of individuals and species. Long gone are bats roosting in the millions or even tens of thousands. What remains today of the immense colonies of bats in the area around the Historic Entrance are scattered tri-colored bats (in all seasons), a few big brown bats near Houchins Narrows during the winter, and several other species (including most that formerly roosted in the area) that visit the area and use the Historic Entrance as a fall swarming site (in the few hundreds). Dixon Cave is used during the winter by slightly more than a thousand bats from five species, during the summer by less than a hundred bats from three species, and during the late summer/fall swarming period by roughly a thousand bats from eight species. The area around the Colossal Cave Entrance is used by less than five hundred bats from four species during the winter, by a few hundred bats from six species during the summer, but the entrance does not appear to be a fall swarming site. For each of the three aforementioned areas of Mammoth Cave presently used by bats, we will discuss some of the accounts that provide information on the type and extent of bat use.

17.5.1 Historic Entrance Area

No aggregations of *Myotis* have been reported in the area around the Historic Entrance of Mammoth Cave since the Bailey et al. (1933) study in 1929 and 1930. In their discussion on Indiana bats, Bailey et al. (1933) reported that "Very few bats were seen or collected in Mammoth Cave," and on their discussion of tri-colored bats they stated "In Mammoth Cave there are now very few bats even in winter." In August 1929, L. Giovannoli collected an eastern small-footed bat and two tri-colored bats in the cave (Bailey et al. 1933). By the time assistant naturalist C.W. Hibbard conducted his assessments in 1934 and 1935, all he observed were a few stray bats in flight during the active months and an occasional tri-colored bat hibernating in obscure places during the winter. Until February 2015, no coordinated systematic effort had been undertaken to assess bat use within this area since the 1930s. Daily year-round visits by tour guides failed to produce evidence beyond scattered tri-colored bats and a lone little brown bat or two. Systematic long-term bat population monitoring in this area of Mammoth Cave by the National Park Service's Inventory and Monitoring Program began in winter 2015 and will be repeated every-other-winter. This initial count resulted in 122 tri-colored bats, 27 Indiana bats, 10 little brown bats, four big brown bats, and one federally threatened northern long-eared bat (*Myotis septentrionalis*). On March 13, 2015 the authors observed the first recent winter record of an eastern small-footed bat roosting just inside the entrance gate.

Bat use of the Historic Entrance itself has been assessed by the use of mist nets or harp traps in August and September (from 1997 to 2016), and by exit counts using night vision goggles/video cameras with external infrared lights in summer 2010 through 2016. Species captured in the Historic Entrance include tri-colored, little brown, northern long-eared, eastern small-footed, big brown, Indiana, and red bats. Nocturnal exit counts have ranged from 5 to 50 bats in June and from 70 to 127 bats in August. Given the close proximity to Dixon Cave, it is assumed that bats regularly move between the two entrances at least during fall swarming.

17.5.2 Dixon Cave

More bat research and survey effort occurred at Dixon Cave during the twentieth century than at any other cave in Mammoth Cave National Park. This happened mainly because the cave was typically occupied by more bats of more species than the other major bat caves in the park; it housed the park's largest Indiana bat colony, and it was the park's most significant fall swarming site. Today bat research and survey work in the park's three primary bat caves is spread more evenly among Dixon, Long, and Colossal caves.

Between April and November 1929, Bailey et al. (1933) and L. Giovannoli made repeated visits to Dixon Cave and collected specimens of gray, little brown, and Indiana bats. Bailey et al. (1933) described the following observations of little brown and Indiana bats in Dixon Cave:

> *Myotis lucifugus* found in "small numbers" in late April and in mid-November in "great abundance" with an estimated 1,000 individuals hanging mainly on the ceiling and some hanging on the walls and singly.
>
> A "few hundreds in all" *Myotis sodalis* were seen in late April, but on November 15 "at least 1,000" were on the ceiling.

C.W. Hibbard visited Dixon Cave in July and August 1934 and observed six to eight scattered tri-colored bats and roughly 300–350 Indiana bats. In late October 1934, he observed "a large number of *Myotis sodalis*" plus three little brown bats, "a number" of tri-colored bats, and "a few" big brown bats. In early November 1934, he saw "hundreds" of Indiana bats and "a few" little brown bats. Hibbard also saw one eastern small-footed bat and "a number" of tri-colored and big brown bats. In the late 1940s, H.B. Hitchcock banded little brown, tri-colored, and Indiana bats in Dixon and Long caves.

From the late 1950s through the late 1960s, J.S. Hall and W.H. Davis conducted studies of the bats in Dixon Cave and the Mammoth Cave area. Their work included extensive bat banding efforts to examine movements within and between caves and summer sites. Hall and Davis found that bats banded in the winter occasionally moved among Dixon Cave and other caves in the Mammoth Cave area (e.g., Coach, Colossal, Long caves) during subsequent winters. On February 3, 1957, Hall recovered a female Indiana bat in Coach Cave that had been banded by H.B. Hitchcock on December 31, 1947, at Dixon Cave. During 17 days between July 30 and October 11, 1963, Davis (1963, 1964) and others captured and banded over 12,000 bats at Dixon Cave. Species captured inside the cave included little brown, Indiana, gray, and tri-colored bats. Additional species captured outside the cave included eastern small-footed, northern long-eared, big brown, red, and evening (*Nycticeius humeralis*) bats. Interestingly, some of the bats banded at Dixon Cave that fall, quickly returned north to their summer colony sites. One Indiana bat flew more than 300 miles (483 km) to a barn in St. Josephs County, Michigan in less than 10 days.

Hall conducted bat counts within Dixon Cave during all four seasons in 1957 and found that Indiana bats were present in larger numbers during all four months (February, April, June, and November) than any other bat species. In June and October 1959, Hall collected two male southeastern bats from active Indiana bat clusters. Hall conducted annual winter bat counts in Dixon Cave from 1957 to 1960, and in 1962, 1967, and 1968. During most of that time estimated Indiana bat numbers in Dixon Cave remained between 2500 and 3000 then increased to 5000 in 1967 and 1968. Hall reported that only a few other species (little brown, big brown, tri-colored bats) were seen in small numbers.

S. Keefer estimated the winter populations of Indiana bats in Dixon Cave at 4000 (1969), at 8000 (1970), and at 4000–5000 (1971). No numbers were provided for other species in his reports.

In January 1975, S.R. Humphrey censused Dixon Cave and counted 3600 Indiana bats, 250 tri-colored bats, and 10 big brown bats. He surmised that the increase in numbers of Indiana bats in protected caves in Mammoth Cave National Park during the late 1960s to early 1970s was due to movement of Indiana bats from nearby Coach Cave where the population had declined from 100,000 to 4500 after an observation platform and building were built around the upper entrance in the early 1960s. The possibility of bats moving from Coach Cave to caves on the park had previously been established through bat banding studies by Hall and Davis.

However, estimates of the numbers of Indiana bats hibernating in Dixon Cave did not increase dramatically until the 1980s. J.B. Cope and others conducted a bat survey in Dixon Cave in early March 1978 and reported 3800 Indiana bats, 141 tri-colored bats, and one big brown bat.

Winter Indiana bat counts in Dixon Cave of 30,900 in 1982 (by R.R. Currie and J.R. MacGregor), and of 30,000 in 1983 and 26,850 in 1985 (both by R.L. Clawson) may have been overestimated because of the method used. An ocular estimate was made of the total surface area in square feet covered by the clusters of bats, that estimate was multiplied by 300 bats per square foot to obtain the estimate of the number of bats. Thus, winter bat counts after 1985 may not be directly comparable to the three previous counts. Beginning with the winter count of 16,550 Indiana bats in 1987, and every-other-year thereafter through 2003, Clawson directly counted(not estimated)the Dixon Cave bats using binoculars or a spotting scope and additional lights. Winter bat counts since 2005 have been conducted by biologists with the Kentucky Department of Fish and Wildlife Resources or the National Park Service. Starting in 2009, bats have been counted every-other-year using two methods (1) direct counting with binoculars or spotting scope and spot lights (as had been done since 1987), and (2) digital flash photography using a DSLR camera with a 400 mm telephoto lens. Indiana bats typically roost 50–70 feet (15–21 m) above the floor in this cave.

Since peaking in the early to mid-1980s, the numbers of hibernating Indiana bats in Dixon Cave have declined steadily—reaching levels observed in the late 1950s by 2007 (Fig. 17.2). In February 2015, the author (Thomas), and others, counted only 923 Indiana bats. The cause of this decline is unknown.

Since the mid-1980s, winter little brown bat numbers in Dixon Cave have fluctuated between zero and 160, gray bat numbers have ranged from 142 to 975, tri-colored bat numbers have vacillated between 3 and 494, big brown bat numbers have ranged from zero to 17, and northern long-eared bats have ranged from zero to one (Fig. 17.3).

Gray bat use of Dixon Cave is interesting. Active season use was documented in 1929 by a male collected by L. Giovannoli in early July (Bailey et al. 1933), by one individual in early June 1957 by Hall, and in 1963 when Davis (1963) and others captured 125 gray bats inside the cave in the daytime between late July and mid-September. During summer fieldwork for a paleontological inventory of Dixon Cave in 2000 through 2002, Colburn (2005) reported seeing a cluster of 50–100 bats at the back of the cave, and most appeared to be gray bats. The existence of a guano pile —and a site directly above it used by active Indiana bats (apparently year-round)—near the end of Dixon Cave was documented by Hall during the late 1950s and by R.R. Currie and J.R. MacGregor in February 1982. Perhaps the species composition of this "active site" changes among *Myotis* species both within and among years. Winter bat surveys since the late 1980s have found gray bats primarily using the back third or so of the cave for hibernation.

This summary demonstrates that Dixon Cave has been an important habitat for several species of bats year-round. It is currently listed in the highest priority category for recovery of Indiana bats (US Fish and Wildlife Service 2007). The continual decline in winter numbers of this species at Dixon Cave is a concern and an opportunity.

During the mid-1930s and mid-1960s, proposals to open Dixon Cave for public tours, or to open the other end of the cave, were thwarted through efforts by individuals such as C.W. Hibbard and W.H. Davis. Until a bat-friendly steel gate was installed at the entrance to Dixon Cave in May–June 1995, most unauthorized visitors were kept out for decades by a chain link fence around the top of the sink above the entrance. The fence was removed in the early 2000s.

Bat use of the entrance to Dixon Cave has recently been assessed by the use of mist nets or harp traps in July 2004 [M.W. Gumbert, the author (Thomas), and J.D. Kiser] which yielded 129 individuals of eight species, and in August and September bat workshops in 1997, 2004, 2006, 2014, and 2015. Nocturnal bat exit counts using night vision goggles/video cameras with external infrared lights have been conducted at the entrance in summer and fall 2010 through 2016. Exit counts have ranged from 28 to 340 bats in June/July, from 145 to 648 bats in August/September, and from 30 to 129 bats in October.

17.5.3 Colossal Cave

The entrance to Colossal Cave is an artificial opening created in the late 1890s for commercial access. A paleontological inventory of Colossal Cave in 2000 and 2001 by Colburn (2005) did not find any evidence of significant

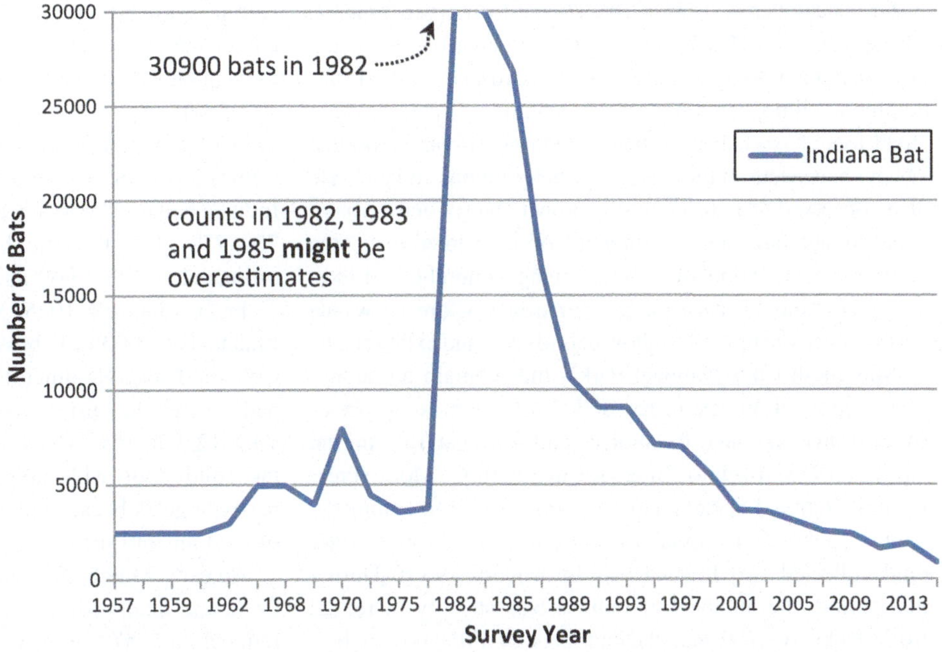

Fig. 17.2 Number of hibernating Indiana bats (*Myotis sodalis*) counted at Dixon Cave in Mammoth Cave National Park, Kentucky from 1957 through 2015

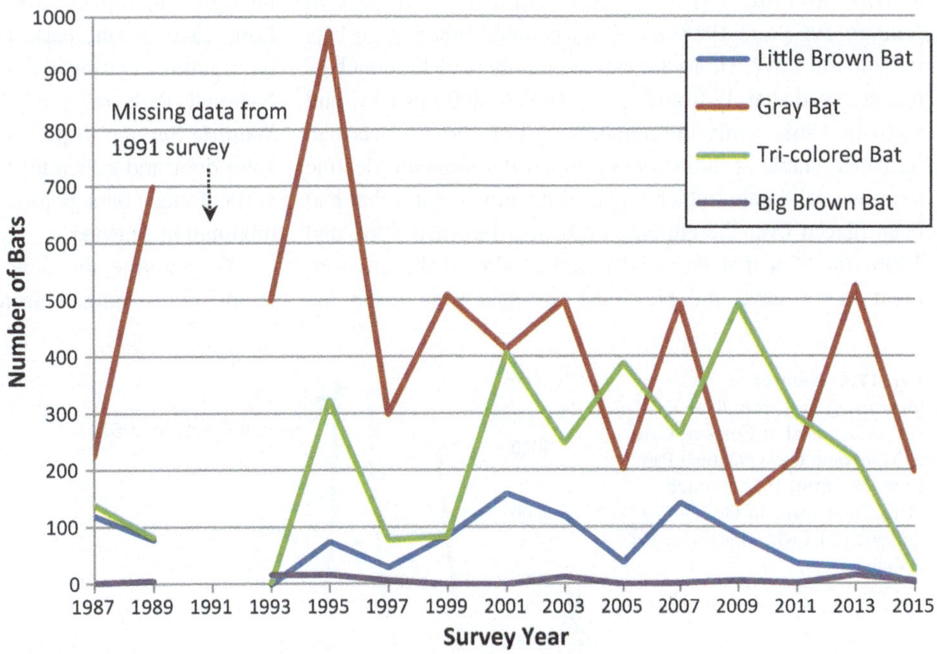

Fig. 17.3 Number of hibernating bats from four species counted at Dixon Cave in Mammoth Cave National Park, Kentucky from 1987 through 2015. *Note*: *gaps* in the *lines* indicate that data from the 1991 bat survey are missing

prehistoric bat use. The only evidence of bats using Colossal Cave before cave tours ended by the 1920s (that we are aware of) comes from the collection of two male tri-colored bats by W.W. Yothers in late August 1912. Nevertheless, Colossal Cave, now connected to Mammoth Cave, is considered one of the three primary bat caves in Mammoth Cave National Park.

In September 1929, Bailey et al. (1933) collected over one dozen (mostly male) little brown bats in the cave, and in November 1929, he collected a few additional little brown bats, a few Indiana bats, and a tri-colored bat. He remarked that "small numbers" of little brown bats were observed in the cave in late September, while in November 1929, he counted about 5000 bats in the cave, "probably half of which were of this species," he stated. He reported seeing "large numbers" of Indiana bats, 200 tri-colored bats, and even "a few" eastern small-footed bats (of which one specimen was collected) in addition to the roughly 2500 little brown bats he estimated seeing among the 5000 total.

On December 6, 1934, C.W. Hibbard observed four bat species in Colossal Cave and considered Indiana bats abundant (102 of which were collected by F. Laird), tri-colored bats common, little brown bats uncommon, and eastern small-footed bats to be rare. In mid-January, Hibbard reported, "Many bats occur in this cave, especially myotis sodalis." He also observed "a few" little brown bats, one eastern small-footed bat, and "a number" of tri-colored bats. On August 2, 1935, Hibbard reported seeing "a number" of Indiana and tri-colored bats in the cave three days before a new gate was installed on the cave to allow bats to enter and exit the cave.

Mammoth Cave National Park's files contain an anonymous report of an "estimated 6000" bats from four species (little brown, Indiana, tri-colored, and gray bats) during an April 9, 1953, trip into Colossal Cave with C. Mohr (President of National Speleological Society). J.S. Hall conducted winter counts in Colossal Cave from 1957 through 1962 (and collected two Indiana bats in January 1959). During that period, the numbers of Indiana bats steadily increased from 1000 to 6700 and the numbers of little brown bats increased from 2500 in 1957 to 4000 in 1959, then declined to 2000 in 1962 (Fig. 17.4). Hall returned to the park in January 1967 and 1968 and again counted hibernating bats in Colossal Cave. He found that the numbers of Indiana bats had declined sharply from 6700 in 1962 to 2000 in 1967 and 1310 in 1968, while the numbers of little brown bats had remained stable at 2000. He attributed the dramatic decline to a new solid gate (which replaced the grated gate) that had been placed over the entrance sometime between 1962 and 1966. Hall felt that the solid gate had altered the environment in the cave and prevented the bats from using the entrance. In response to Hall's recommendation, slots were cut in the solid gate in either 1967 or 1968 to allow the movement of bats and winter air into the cave.

Despite the solid door containing slits, Indiana and little brown bat numbers continued to decline from pre-solid door (1962) levels. S. Keefer estimated the winter populations of Indiana bats in Colossal Cave at 600 (1969 and 1970) and at 700–750 (1971). Little brown bat winter numbers were estimated at 400 (1969 and 1970) and at 450–500 (1971).

In late January 1975, S.R. Humphrey counted only 14 Indiana bats, 600 little brown bats, and one tri-colored bat in Colossal Cave. He noted that rock temperatures at the roost had increased from 4–6 °C (39–43 °F) in 1960 to 10.7–12.3 °C (51.3–54.1 °F). Humphrey recommended that the solid door with holes cut in should be replaced by bar-type gates because the door was still restricting the flow of cold air into the cave.

An early March 1978 count of Colossal Cave bats by J.B. Cope and others resulted in continued low numbers of 105 Indiana bats, 570 little brown bats, 27 tri-colored bats, and one big brown bat. Cope echoed the call of previous bat biologists to replace the "gate-wall" at Colossal, Bat, and Long caves on the park. He recommended using prison cell type gating. During the summer of 1980, Mammoth Cave National Park received approval from the US Fish and Wildlife Service to proceed with its proposal to replace the solid door and concrete/stone wall with a gate that contains vertical steel bars before the beginning of the 1980–1981 hibernation season.

We presume the new gate was installed because cave temperatures ranged from 6.8 to 7.8 °C (44.2–46.0 °F)

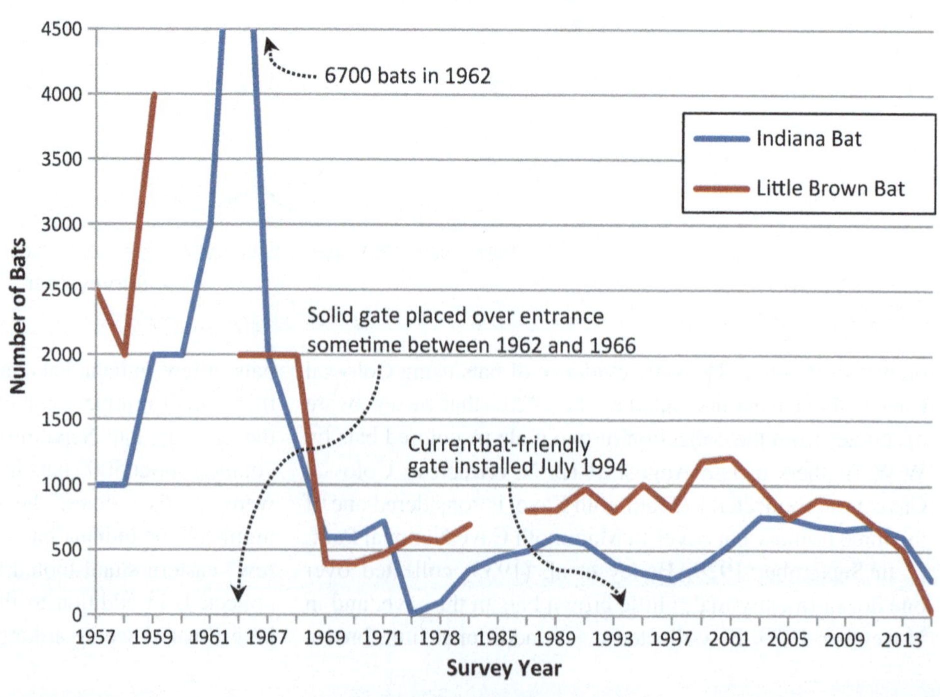

Fig. 17.4 Number of hibernating bats from three species counted at Colossal Cave in Mammoth Cave National Park, Kentucky from 1957 through 2015. *Note*: *gaps* in the *red line* indicate years when bats were not counted

during the bat count in mid-February 1982 by R.R. Currie and J.R. MacGregor. There was no mention of the solid door in their report. Both Indiana bat (349) and little brown bat (702) numbers had increased since 1978. Tri-colored and big brown bats also were observed, but no numbers were reported.

The numbers of hibernating Indiana bats counted by biologists with the Kentucky Department of Fish and Wildlife Resources and the US Fish and Wildlife Service in Colossal Cave during the every-other-year censuses in the 1980s and 1990s remained fairly stable (between 284 and 610), while little brown bat numbers increased gradually to 1192 by 1999 (Fig. 17.4). The current bat-friendly angle iron steel gate was installed in July 1994 (Fig. 17.5). However, numbers for both Indiana and little brown bats have yet to return to early 1960s levels—this is especially true for Indiana bats. In the early 2000s, Indiana bat numbers in the cave remained fairly stable, ranging from 544 to 760, while little brown bat numbers steadily declined from 1219 in 2001 to 515 in 2013 (Fig. 17.4). White-nose syndrome was confirmed in this entrance to Mammoth Cave in February 2013. In January 2015, the authors discovered that little brown bat numbers had fallen to only 38—four of which exhibited signs of the deadly fungal disease. Indiana bat numbers had also declined in January 2015 to 279. Bat counts during the twenty-first century at Colossal Cave were accomplished by biologists with the Kentucky Department of Fish and Wildlife Resources or the National Park Service and volunteers.

Limited bat banding has been carried out in Colossal Cave. In late November 1957, Hall banded 181 (93 males, 88 females) Indiana bats in order to study movements within and among caves. Some of these bats were recovered in Long Cave on the park in late January 1968. Also, two Indiana bats banded in late December 1960 were recovered in Colossal Cave in late January 1962 and mid-January 1967. In late January 1968, seven banded bats were recovered in Colossal Cave that had previously been banded in the

Fig. 17.5 Current bat-friendly angle iron steel gate installed in July 1994 at the entrance to Colossal Cave in Mammoth Cave National Park, Kentucky. NPS Photo

cave five to eight years previously. A female Indiana bat banded near Norvell, Michigan (Jackson County) on May 27, 2004, was observed hibernating in Colossal Cave (357 miles, 575 km away) during winter bat counts in 2005, 2007, 2011, and 2013 (Rockey et al. 2013).

Bat use of the entrance to Colossal Cave has been recently assessed by mist nets or harp traps in late July 2005 by M.W. Gumbert and the author (Thomas) which yielded 147 individuals of seven species and by exit counts using night vision goggles/video cameras with external infrared lights in summer and fall 2010 through 2016. Nocturnal exit counts have ranged from 7 to 306 bats in July, from 92 to 588 bats in August/September, and from 2 to 7 bats in October.

Additional bat research projects carried out at Colossal Cave include (1) pre-hibernation weight study on Indiana bats in late November 2008 by B. Slack, M. Armstrong, and others and (2) spring/fall harp trapping (with banding) to evaluate pre- and post-hibernation body condition and mass of cave-hibernating bats from spring 2011 to fall 2015 (Lacki et al. 2015).

17.5.4 Other Bat Caves in the Mammoth Cave Area

In addition to the important bat use areas of Mammoth Cave described above, there are several nearby caves, not part of Mammoth Cave, that deserve mention due to their use by bats. Significant bat caves that lie within Mammoth Cave National Park include Long, Bat, Wilson, Lee, and Hickory Flat caves. Two very important bat caves, Coach Cave and James Cave, are located on private property just south of the park. Long Cave is noteworthy for its historic and prehistoric bat use and its recent substantial year-round use by bats. Long Cave may have been occupied in the winter by as many as 50,000 Indiana bats in December 1947, as reported by H.B. Hitchcock. Today that number is closer to 1000. However, gray bats began colonizing Long Cave during the winter of 1997, and today approximately 25,000 gray bats hibernate in the cave (authors' estimate). Based on nocturnal bat exit counts and harp trapping/mist netting, over 500 bats from seven species use the cave during the summer. Bat Cave has substantial bat bone deposits from multiple species that indicate significant use of the cave over approximately the past 11,000 years (Colburn et al. 2015; Jansky 2013). In December 1959, Hall observed about 1,000 little brown bats along with a few Indiana, big brown, tri-colored, and southeastern bats. Since the early 1970s, Bat Cave has been the winter home for 200–400 bats of four primary species, including Indiana bats. A handful of male gray bats and a few scattered tri-colored bats use the cave during the summer. Wilson Cave has been used by bats for a long time but only in moderate numbers. In November 1957, Hall observed over 500 Indiana bats and about a dozen tri-colored bats. In February 1999, the author (Thomas) counted 207 little brown, 65 Indiana, 33 tri-colored, five big brown, and one gray bat. In this century, hibernating little brown bat numbers have ranged from zero to 123, Indiana bats from three to 140, tri-colored bats from five to 68, and big brown bats from one to seven. Lee Cave is used during the summer by a colony of about 75 male, or non-reproductive female, gray bats. In 2011, during the first known winter bat count, the authors found 164 little brown bats, 66 Indiana bats, 49 tri-colored bats, and three gray bats hibernating in the cave. Lee Cave also contains evidence of prehistoric bat use.

Hickory Flat Cave is worth mentioning because it contains the largest known natural Rafinesque's big-eared bat hibernaculum anywhere. Over 1570 bats of this species have been documented in the cave during the winter (authors; on January 30, 2014). Rafinesque's big-eared bats are known to hibernate in smaller numbers, from 70 to 270, in four other caves in Mammoth Cave National Park. The park has at least eight caves that house summer nursery colonies of Rafinesque's big-eared bats that range in size from roughly 25 to 200 bats.

Together Coach Cave and James Cave are used as winter roosts by approximately 400,000 endangered gray bats. Coach Cave was occupied in the winter once by an estimated 100,000 Indiana bats (January 1960), but they have now been replaced by gray bats. Both caves are inhabited in summer by several thousand male and non-reproductive female gray bats.

17.6 Conservation of Bats in Mammoth Cave

The number of bats in Mammoth Cave is a miniscule fraction of what it was in prehistoric and historic times. Unfortunately, the downward trend continues today. The various reasons for this decline range from human use and resulting physical changes to an exotic fungal disease. Since Mammoth Cave bats may travel more than 350 miles (563 km) away, they can face more disparate threats such as habitat loss, pesticides, mercury bioaccumulation, and wind turbines. Early bat conservation efforts (mid-1930s and mid-1960s) focused on limiting direct human disturbance at Dixon Cave and replacing solid gates at the entrance to Colossal Cave. In the early 1990s, new bat gates of heavy angle iron horizontal bars were installed at Historic, Dixon, and Colossal entrances. These gates were designed not to impede bats or airflow, while limiting human access; they have largely worked as designed. Despite these new bat-friendly gates, overall winter bat numbers have mysteriously continued to decline in Mammoth Cave.

At the Historic Entrance where the new bat gate (installed in July 1990) replaced sheet metal panels mounted on a masonry and steel gate, too much cold winter air was allowed into the cave—because the entrance passage (Houchins Narrows) had been greatly enlarged during the nineteenth and early twentieth centuries. The additional cold dry winter air threatened sensitive cultural artifacts, cave terrestrial organisms, and greatly increased the rate of rockfall in the Historic Entrance area. Thus, in 1999, multiple plexiglas panels were mounted as baffles on the gate to restore natural microclimate conditions experimentally during winter months for species known to have occupied historical hibernacula (see Chap. 10). Ecological restoration of the Historic Entrance area of Mammoth Cave toward facilitating the return of bats is an ongoing effort.

The discovery of white-nose syndrome in Mammoth Cave bats in 2013 has presented new bat conservation challenges and opportunities. The rapid and widespread mortality associated with white-nose syndrome is unprecedented in hibernating bats. A loss of large numbers of bats to this disease may substantially affect the surface and subsurface ecosystems in the Mammoth Cave area. In response, conservation efforts at the cave have included disease surveillance, research, bat population and cave microclimate monitoring, warning signs outside entrances, restricting human access to colonial bat roosts, screening/intervening with visitors prior to cave tours, post-tour cleaning and/or treatment of footwear (to reduce the risk of human-assisted transmission of the fungus that causes the disease), assuring cooperators/researchers observe decontamination requirements, and equipment dedication. The arrival of this devastating disease has created unique opportunities for bat conservation and white-nose syndrome related outreach and education.

Additional recent measures taken to conserve bats in Mammoth Cave include placing interpretive displays outside the entrances, restricting human access to the entrance areas during bat swarming in the fall, reduction of lighting outside the Historic Entrance, and public bat education efforts/events. Conserving bats in Mammoth Cave has been and will continue to be a cooperative effort among a variety of federal and state agencies, universities, non-governmental organizations, and individual partners.

Although Mammoth Cave National Park is still in the midst of the white-nose syndrome outbreak, and we are not able to obtain final numbers in terms of bats lost, some general comments can be made about what is currently known. Limited bat counts and other work on the park suggest significant (>50%) decreases in at least four species (northern long-eared, little brown, Indiana, and tri-colored bats). On the other hand, data suggest that both gray and Rafinesque's big-eared bat numbers are stable or increasing in spite of the disease being present in their winter roosts.

17.7 Conclusions

Bats have been part of Mammoth Cave's fauna since prehistoric times. In the past, colonies of perhaps 9–13 million bats played a critical role in the food web of the cave by contributing energy (nutrients) to the ecosystem. However, due to a "mammoth" decline in numbers and distribution of bats over the past 150–200 years, that contribution is now vastly diminished. Direct disturbance and/or microclimate changes caused by entrance/passage modifications account for most of the decline in Mammoth Cave.

Since 2013, observed winter declines of up to 93% (e.g., little brown bats at Colossal Cave) were likely an effect of the devastating disease white-nose syndrome. As we write, two species of bats that have used Mammoth Cave for millennia are endangered, one is listed as threatened, and as a result of white-nose syndrome, more species are likely to be listed in the near future. Naturally, low reproductive rates combined with the high mortality observed in populations with the disease will likely prevent affected bat populations from recovering quickly. White-nose syndrome is a threat that will make bat conservation ever more difficult. But with this latest challenge comes opportunities…chances for protecting and researching bats that survive the disease, for reaching out to present and future bat/cave conservationists, for educating those who are drawn to Mammoth Cave and its wonders, so that these amazing and resilient creatures may continue to dwell in the world's longest cave.

References

Bailey V, Bailey FM, Giovannoli L (1933) Cave life of Kentucky. Am Midl Nat 14(5):385–635

Baker DS (1998) Mammoth Cave, historic entrance sheet, map (draft), scale 1:600. Cave Research Foundation

Bullitt AC (1845) Rambles in the Mammoth Cave during the Year 1844. Morton and Griswold, Louisville

Colburn ML (2005) Paleontological inventory project: vertebrate remains found in select passages and caves at Mammoth Cave National Park, Kentucky. Final report submitted to the National Park Service, July 2005. Illinois State Museum Landscape History Program, Technical Report No. 2005-1199-007. On file at Mammoth Cave National Park

Colburn ML (2006) Vertebrate remains from four localities at Mammoth Cave National Park, Kentucky. Masters thesis, University of Illinois at Springfield

Colburn ML, Toomey RS, Widga C, Olson R (2015) Holocene paleontology of Bat Cave, Edmonson County, Kentucky. J Cave Karst Stud 77(2):91–98

Culver DC, Pipan T (2009) The biology of caves and other subterranean habitats. Oxford University Press, New York

Davis WH (1963) Operation Chiroptera, 1963. Bat Res News 4(4): 30–31

Davis WH (1964) Fall swarming of bats at Dixon Cave, Kentucky. Bul Nat Speleol Soc 26(4):82–83

Hovey HC (1893) Mammoth Cave in March. Science 21(531):189–190

Hovey HC (1896) Celebrated American caverns especially Mammoth, Wyandot, and Luray. Cincinnati: Robert Clarke & Company. Reprinted in 1976 by Johnson Reprint Corporation, New York

Hovey HC (1912) Mammoth Cave of Kentucky with an account of Colossal Cavern. John P. Morton & Company, Louisville

Hovey HC, Call RE (1897) The Mammoth Cave of Kentucky, an illustrated manual. John P. Morton and Company, Louisville

Jansky KJ (2013) Identifying Myotis species using geometric morphometrics and its implications for the fossil record and conservation. Electronic theses and dissertations. Paper 1145. http://dc.etsu.edu/etd/1145

Lacki MJ, Dodd LE, Toomey RS, Thomas ST, Couch ZL, Nichols BS (2015) Temporal changes in body mass and body condition of cave-hibernating bats during staging and swarming. J Fish Wildl Manage 6(2):360–370; e1944-687X. doi:10.3996/042015-JFWM-033

Lee EF (1835) Note on the Mammoth Cave, to accompany a map. Kentucky library research http://digitalcommons.wku.edu/dlsc_kl_non_mat/12

Randolph HF (1924) Mammoth Cave and the cave region of Kentucky illustrated. The Standard Printing Company, Inc

Rhoads SN (1897) Some notes on the mammals of Mammoth Cave, Kentucky. J Cincinnati Soc Nat Hist 19(2):53–61

Rockey CD, Stumpf JP, Kurta A (2013) Additional winter recoveries of Indiana bats (*Myotis sodalis*) banded during summer in Michigan. Notes Northeastern Naturalist 20(3):N8–N13

Silliman B Jr (1851) On the Mammoth Cave of Kentucky. Amer J Sci Arts, Second Series 11(33):332–339

Toomey RS III, Colburn ML, Olson RA (2001) Paleontological evaluation of use of caves: a tool for restoration of roosts. In: Kurta A, Kennedy J (eds) The Indiana Bat: biology and management of an endangered species. Austin, Bat Conservation International, pp 79–85

Tuttle MD (1997) A mammoth discovery. Bats Magazine 15(4):3–5

US Fish and Wildlife Service (2007) Indiana bat (*Myotis sodalis*) draft recovery plan: First revision. US Fish and Wildlife Service, Fort Snelling Minnesota

Environmental Issues Relevant to the Mammoth Cave Area

18

Rickard A. Olson

Abstract

Environmental problems affecting the Mammoth Cave area range in scale from local to global. In park caves, issues include corroding copper wire and aluminum light fixtures, lamp flora from cave lights, lint, dust, graffiti, excessive bolt holes drilled in cave walls for infrastructure, fumes from gasoline lanterns, creosote treated wood in cave streams, and management of cave atmospheric conditions. Surface issues affecting park caves include runoff from parking lots and highways, sewage treatment, sinkhole dumps, prescribed fire, hydrocarbon extraction, pipelines, and pollution plus impoundment effects on the Green River. On a global scale, there are exotic species brought in by commerce or war, air pollution, and the rising concentration of carbon dioxide in the atmosphere from combustion of fossil fuels. This along with other greenhouse gasses is causing warming on a global scale and probably severe weather extremes.

18.1 Introduction

Ecological impacts affecting Mammoth Cave and other caves in the regional karst landscape come from myriad sources, near and far. Most environmental issues affecting Mammoth Cave result from activities the park has little or no control over, and some are directly linked to park operations. Designated as a World Heritage Site (Bishop 1982), Mammoth Cave serves as the core of an International Biosphere Reserve (US MAB 1995). These designations by the United Nations Education and Scientific Organization indicate that Mammoth Cave and the regional karst landscape are exemplary on a global scale and have goals to work toward ecologic and economic sustainability. However, to avoid any misunderstanding, these designations do not affect sovereignty of the USA over the park and environs.

18.2 Climate Change, Fossil Fuel Combustion, and Invasive Exotic Species

Climate change, and especially associated weather extremes, could have serious direct and indirect effects on cave ecosystems. For instance, severe drought impairs the productivity of plants, and there is no water to transport what food there might be into cave streams. To the other extreme, major floods can cause direct damage to cave life and wash dissolved organic matter out of cave passages.

Air pollution from national and even global sources can impact vegetation that bats, wood rats, and cave crickets rely upon directly or indirectly for food supply. Such a negative turn in vegetation status could then reduce the supply of guano from cave crickets, woodrats, and bats that many cave adapted organisms rely upon for their food supply. Ozone causes foliar damage to vegetation (Jernigan and Carson 2013), and mercury from coal combustion has many potential negative effects. For example, tissue samples from bats at Mammoth Cave National Park (MCNP) have mercury concentrations of up to 1–2 mg/l (Van der Heiden and Webb 2013). Acid deposition linked to coal combustion causes direct damage to vegetation, leaches nutrient cations

R.A. Olson (✉)
Cave Research Foundation, Cave City, KY, USA
e-mail: rickardolson16@gmail.com

H.H. Hobbs III et al. (eds.), *Mammoth Cave*, Cave and Karst Systems of the World,
DOI 10.1007/978-3-319-53718-4_18

such as calcium and magnesium from soil, and mobilizes aluminum, which is toxic to plants and aquatic life such as fish (Olson 2001). All component ecosystems within the karst landscape, including those in caves, are largely dependent on the photosynthetic productivity of plants.

Other sources of impacts to vegetation include 224 species of exotic plants, many of which are invasive such as sericea lespedeza (*Lespedeza cuneata*), garlic mustard (*Alliaria petiolata*), crown vetch (*Securigera varia*), and spotted knapweed (*Centaurea stoebe*), any of which can crowd out native plants important directly or indirectly to cave species. Exotic diseases such as chestnut blight (*Cryphonectria parasitica*) and butternut canker (*Sirococcus clavigignenti-juglandacearum*) have already seriously changed park forests. Ecosystems within Mammoth Cave have mostly been spared impacts from invasive exotic species; however, there is an Asian millipede (*Oxidus gracilis*) that has become so numerous at times in the Frozen Niagara section that park staff has gone in with vacuum cleaners to reduce the ick factor for visitors. Because they were feeding upon lamp flora near lights, and native cave life tends not to feed on algae near lights, impact on native populations of cave life was not apparent. MCNP has an Invasive Species Early Detection Web site available at http://science.nature.nps.gov/im/units/cupn/monitor/invasiveplants.cfm. A fungus called *Pseudogymnoascus destructans* arrived at MCNP in 2012. It causes White Nose Syndrome, which has killed millions of cave-dwelling bats in the USA (See Chaps. 16 and 17 for details).

18.3 River Pollution and Effects of Impoundments

On a regional and local scale, the Green River brings pollutants from upstream into the park. Not only do these contaminants affect riverine and riparian life, but also back-flooding via springs carries these pollutants deep into caves, affecting both aquatic and terrestrial cave ecosystems. For example, disposal of oilfield brine into Green River from the Greenville, Kentucky area in 1960 caused serious pollution events (Cushman et al. 1965). These events may have contributed to the near extinction of the Kentucky cave shrimp (*Palaemonias ganteri*), and its subsequent listing as endangered.

Impoundments on Green River used formerly for navigation or currently for flood control have had major impacts to river and cave aquatic ecosystems. For instance, Lock and Dam #6 on Green River (Fig. 18.1) adjoining MCNP degraded habitat for nine aquatic species listed as endangered, including the Kentucky cave shrimp, and eight species of freshwater mussels, and most recently a fish called the diamond darter (*Crystallaria cincotta*) (Olson 2005). This dam, built in 1905, breached on November 25, 2016. The sudden draining of the pool behind the dam caused some slumping of saturated banks along the Nolin and Green Rivers, and thousands of mussels were temporarily stranded above water level before being moved back into the water. Overall, though, this is a positive development, and negotiations by the park Superintendent and Chief of Science and Resources Management with high-level government officials for dam removal were already well advanced. Green River Reservoir dam upstream of the park can alter water levels, flow velocity, and flow direction in base-level cave streams. The Nature Conservancy has worked with the US Army Corps of Engineers to implement a release pattern that mimics natural flood pulses (Postel and Richter 2003). Unnatural sediment deposition may result in the decline in both the diversity and abundance of species in the cave aquatic ecosystem due to burial of vital rock-gravel habitats and preventing transport of organic matter from headwater inputs (Poulson 1992).

18.4 Creosote Treated Wood in River Styx and Echo River

In the 1930s, the Civilian Conservation Corps built an impressive elevated boardwalk from River Styx to Echo River. This 1100-foot (335 m)-long structure facilitated tours to Cleaveland Avenue and beyond by eliminating the need for boats in River Styx. Unfortunately, to prevent fungal decay, this structure was made of wood treated with creosote, which is toxic to living things in general, including cave life. Decades later, the boat tours were discontinued but the boardwalk remained. Over a period of nine years, volunteers with the National Speleological Society (NSS) safely removed the boardwalk before it became too rotted for safe travel (Carson and Vanhoozer 2006). The pilings were extracted from the mud by NSS volunteers using John Vargo's puller built for this difficult work. The timbers and planking were carried to Vanderbilt Hall where they were cut into pieces and bagged (Fig. 18.2). Up to 900 bags on a given day were carried to Mammoth Dome, passed up the tower by this army of caver volunteers, carried to the entrance, passed up the stairs, and loaded onto a truck for proper disposal in a landfill. Over the past quarter century, NSS caver volunteers have removed countless tons of material abandoned in the cave, cleaned lamp flora from lit passage walls, swept up lint, and helped in any way asked.

Fig. 18.1 Lock and Dam#6 just downstream of the park. This facility was decommissioned in 1951 and breached on November 25, 2016. Restoration of free flow will enhance habitat for many endangered species and provide an economic boon to Edmonson County via paddle sports, which have become very popular in the free flowing section of Green River in the park. NPS photograph courtesy of Brice Leech

18.5 Fire

Fire suppression or excessive fire can cause negative impacts to both water budget and foraging habitat for species that commute between cave and surface ecosystems (Olson 2002). There are significant plant communities in the Mammoth Cave region that are wholly or in part dependent upon fire. Most of the Sinkhole Plain south of the park was grassland locally referred to as Barrens prior to settlement, and without fire, forests of various types will take over depending upon the physical habitat type. Oak-hickory forest/woodland is largely fire dependent and will be taken over by beech-maple forests if fire is absent. Woodrats feed heavily upon acorns, so this is one example of a connection between a fire-dependent vegetation community and a cave-dwelling mammal. Cedar-oak glades on southwest facing limestone slopes are a special karst-linked community type that is mostly dependent upon the dry conditions created by karst drainage and a sunny aspect, with only a limited role for fire because eastern red cedar is not fire tolerant. The risk associated with excessive fire is significant; in 2010 prescribed fire in the park killed over 4000 acres (1620 ha) of native Virginia pines, which are not fire adapted (Olson et al. 2013). These areas will have to be treated for invasive exotic plants and managed intensively for decades to insure a return toward desired future conditions. The impact on cave ecosystems is not known, but the scorched areas are underlain by many cave passages (Fig. 18.3).

18.6 Drilling for Hydrocarbons and Potential Impacts from Pipelines

Drilling for oil and gas in the vicinity of the park has taken place over many decades (Fig. 18.4). Development of gas fields has been prominent in the past decade but has not yet included fracking. This pressure fracturing technique requires a great deal of water, which could locally lower the water table and affect water supply into karst aquifers in the park. Fracking also requires the use of toxic chemicals, so the potential exists for contaminants to be introduced into park cave streams in the process of injection and recovery of these chemicals. In 1994, an oil well blew out just outside

Fig. 18.2 NSS volunteers remove the last creosote treated piling from River Styx in 2005. From *left* to *right* are Tim Alkema, John Mason, Steven Peterson, John Vargo, Larry Matiz, Ruth Vargo, Pam Saberton, Ken DeJonge, Kevin Betz, and Kim Nelson. John Vargo invented the puller used to extract boardwalk pilings driven deep into sediment

the park boundary, and crude oil flowed hundreds of feet down Dry Branch into the park. Most of the oil was captured with absorbent materials before going underground. Recently, it has been proposed that a 70-year-old gas pipeline passing very close to Mammoth Cave's drainage basin boundaries be repurposed for carrying natural gas liquids. The potential impact of liquid spills is far greater than with a gaseous product, so great concern has been expressed by park managers (Bruggers 2015).

18.7 Parking Lot and Highway Runoff

In developed areas, including within the park, runoff laden with oil and heavy metals poses a significant threat. Barr (1976) documented such contaminants entering Mammoth Cave's Historic Entrance from the Visitor Center parking lot. In response, the park installed parking lot filters in 2001, and these are moderately effective at reducing copper and zinc. Removal of diesel fuel from parking lot runoff has been very effective (McMillan et al. 2013). Major highway and rail transportation corridors are sources of chronic contamination and catastrophic spills. A major die-off of aquatic cave life resulted from a hydrocarbon spill along I-65 (Brucker 1979). There have been many smaller spills over the years, and after much discussion the Kentucky Transportation Cabinet agreed to build runoff filtration and spill retention structures at every sinkhole or stream crossing on I-65 as it is widened (Olson 2013).

18.8 Dump Cleanups

Beginning in 1996, I organized a dump cleanup program called "Don't Mess With Mammoth Days" (Olson 2003). Years ago, there was no trash pickup outside of cities, and so people used sinkholes and ravines as convenient places to dump their trash. After rural trash pickup became available, then it made sense to start the cleanup process. Over the past two decades, many dumps have been cleaned up in Barren, Edmonson, and Hart Counties, both inside and beyond park boundaries. For over 10 years, Ms Peggy Nims, Environmental Education Coordinator for the American Cave Conservation Association (ACCA) in Horse Cave, KY, has

Fig. 18.3 Successional forest on Flint and Joppa Ridges was killed by prescribed fire in 2010, and it will be decades before these areas can support surface or cave wildlife as they once did. These fires left the park with 4000 acres of standing dead trees (see Fig. 12.5)

played a leading role in this dump cleanup program. As a result of her efforts, especially through organizing Alternative Spring Break groups from major universities, the ACCA received KY EXCEL's 2015 Community Conservation Award.

18.9 Sewage Treatment

Sewer line breaks have caused serious pollution in the park. Today, periodic sampling for coliform bacteria in cave waters near tour trails where problems have occurred in the past helps insure that visitors and wildlife are not exposed to potentially dangerous bacterial contaminants. In the 1970s, a sewage lagoon at the old Job Corps site on Flint Ridge periodically overflowed during rainy weather and entered Eyeless Fish Trail in Unknown Cave. This was resolved by moving the Job Corps site to a less vulnerable site and ultimately converting the sewage lagoon to a prairie plantation. In similar fashion, the park's antiquated sewage treatment facility located on the Green River floodplain was dismantled after completion of a regional sewage treatment system that includes Park City, Cave City, and Horse Cave along with the park. This was a major partnership between MCNP, the Environmental Protection Agency, and local municipalities. The record since then has been generally better with a few exceptions. In August of 2004, a major sewage spill was detected in Logsdon River within the Roppel section of Mammoth Cave (Schumann 2004). Caveland Environmental Authority (CEA), which manages the entire regional sewage system, was working on sewer lines just north of Cave City at the time, at the headwaters of Logsdon River. However, when asked about a spill, they denied any loss of sewage. More recently, on May 15, 2014, a spill occurred in the park while CEA was working on the lines. This spill contaminated Mammoth Dome and the Historic Entrance waterfall plus other sites in the cave. Then, on May 27, 2014, an even larger spill occurred while CEA

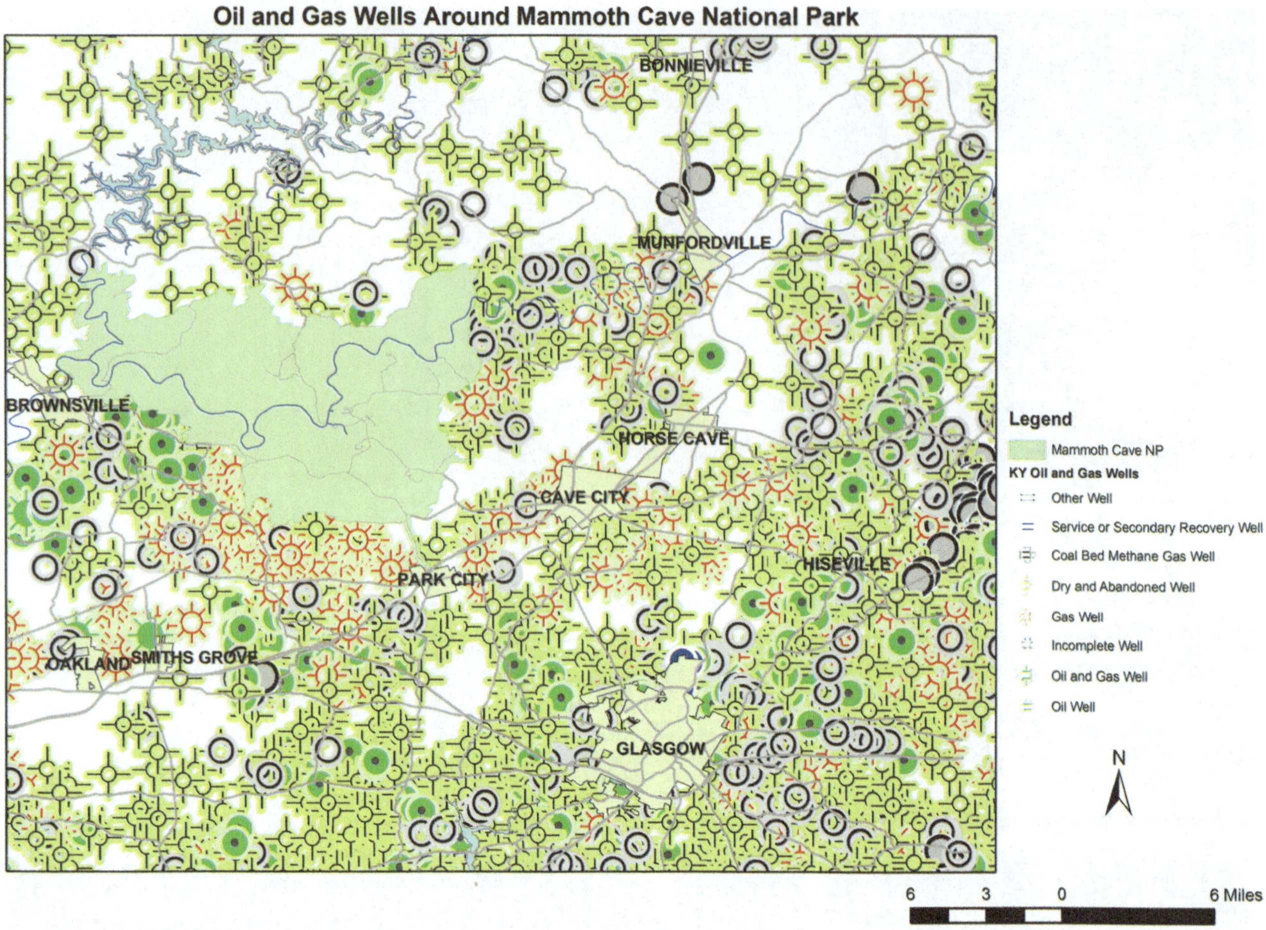

Fig. 18.4 Thousands of oil and gas wells have been drilled in the vicinity of Mammoth Cave National Park. Environmental laws are far more strict than in the past, but the potential still exists for contamination of karst aquifers that flow into the park. NPS map courtesy of Rick Toomey

was again working on the sewer lines near the Carmichael intersection in the park. Thousands of gallons of sewage entered Mammoth Cave at a waterfall know as the Cataracts, which ultimately drains into Roaring River. This underground stream is the designated critical habitat for the endangered Kentucky cave shrimp. All of these sewage spills demonstrate that even with state of the art infrastructure, human error can still result in environmental damage.

18.10 Cave Entrance Modification and Artificial Entrance Development

Over time, there have been many artificial entrances created as well as changes made to natural cave entrances in park caves for a variety of reasons. The entrances have been classified using the best available information (Olson et al. 1997). Changes to natural entrances can have profound effects on cave resources. For example, an old gate in the Historic Entrance to Mammoth Cave was replaced in 1989 to restore bat habitat. However, the entrance passage had been enlarged, so the influx of cold winter air greatly increased. This triggered the collapse of a 40-ton slab onto War of 1812 saltpeter mining artifacts in the Rotunda and heavy condensation dripping on the saltpeter works at Booths Amphitheater. In addition, mold started growing on thousands of prehistoric archaeological remains that had been stable for the past 2000–5000 years (Olson 1996). These problems were greatly mitigated by installation of Plexiglas panels on the entrance gate to approximate natural airflow rates. Data intensive cave atmospheric modeling is being pursued to simultaneously restore bat habitat and conserve cultural resources.

When the MCNP Association took control of land parcels in order to build the park, they closed many cave entrances to prevent illegal access. In a April 9, 1932, letter from Cave Manager Marty Charlet to Association Secretary George Zubrod, he explains that the Cox, Unknown, Salts, Bedquilt, Hazen, and Proctor entrances had all been physically closed. Dossey Domes Cave was scheduled to be closed up with concrete and rock the following week. This approach to cave management was not limited to these caves, and this may be how the opening to the surface at Olives Bower came to be filled in with the consequence that a major bat hibernation site was lost due to the lack of cold air entering in winter.

Artificial entrances can change the natural pattern of water and/or airflow in cave passages with serious consequences. For instance, the abandoned Visitor Center ventilation shaft has leaked water from surface facilities and utility conduits down the shaft into Houchins Narrows for decades. This water is causing War of 1812 saltpeter mining pipes to grow mold, which will eventually destroy them.

18.11 Wildlife Disturbance

Disturbance to wildlife, especially hibernating bats, can have serious effects linked to depletion of fat reserves. For this reason, bat hibernacula are off limits for human entry, except for monitoring, from September through May. Disturbance to other keystone species such as woodrats could affect not only these species, but also the communities that are supported by their guano. Cave gates have largely eliminated illegal entry as a source of wildlife disturbance in Mammoth and nearby caves (Fry 1996).

18.12 Actions with Unintended Consequences

In response to increased visitor numbers, the park began a self-guided tour along the Historic Route in 1967. Abundant graffiti on cave walls is still evident today, and mounds of sediment left from the War of 1812 saltpeter mining operation became noticeably smoothed from all the off-trail foot traffic. A memo from Superintendent John Aubuchon to the Regional Director dated December 27, 1967, stated: "The only serious fault with the semi-guided technique is that it does allow vandalism to take place. Many names and dates were scratched into the soft wall rock, and restoration is difficult. In view of the overall success of the semi-guided technique, we intend to continue it during the 1968 season. We had hoped to extend the technique to the Frozen Niagara tour, also, but our plans must be modified because of the recent decision to keep the New Entrance closed due to unsafe conditions." The self-guided tour was not popular with the guides, and according to Joe Duvall (personal communication), who has guided at Mammoth Cave since 1961, a vote was held among the guides, who were overwhelmingly against this plan. But, as is too common, the guides were not consulted. The self-guided tour finally ended in 1973. There was discussion in the 1990s by the Chief of Interpretation about the possibility of having self-guided tours in Frozen Niagara and into Cleaveland Avenue via the elevator, but fortunately this did not happen.

Historically, cave trails were covered in fine cave sediment dug from "borrow pits" at various locations along tour trails. One of these borrow pits, located in Rafinesque Hall just beyond Bunker Hill in the Historic Section, was deemed unsafe and was therefore targeted for stabilization in a project statement by the Chief of Maintenance. The project was funded at a cost of $123,200, and one day in 2004, a crew of laborers was sent to stabilize the edge of the borrow pit. In the process, a historic dry laid stone wall of the type built by saltpeter miners during the War of 1812 was knocked into the borrow pit along with a scatter of prehistoric torch fragments. In cave management, our first duty is to cause no further harm, but without involvement of a Cave Resources Specialist, historic and prehistoric artifacts were damaged.

18.13 Impacts from Electric Cave Lights Plus Corrosion of Aluminum Fixtures and Copper Wire

Heat from inefficient electric lighting can dry areas that would naturally be moist. As the temperature rises, relative humidity drops. Adapted to humidity near saturation, cave life is highly vulnerable to desiccation compared to similar surface species, and mineral deposition may also be altered. For these reasons, energy efficiency in cave lighting is very important. Fluorescent lights are very efficient, but contain highly toxic mercury, which is released when lamps are accidentally broken in the cave. Another major consideration in cave lighting design is the growth of cyanobacteria (otherwise known as blue-green algae), algae, mosses, and other plants induced by the lights in cave passages. Collectively, these photosynthetic organisms growing near lights in caves are called "lamp flora" (Mulec 2012). The great quantities of lamp flora found along tour trails are alien to cave ecosystems, and chemicals such as bleach used to control their growth constitute a major distortion to cave ecosystems (Olson 2006). Selection of light wavelengths least utilized by algae (Fig. 18.5) and to a lesser extent by cyanobacteria has been demonstrated to significantly reduce the growth of lamp flora in wet passages (Toomey et al. 2009). The park newspaper stated "LEDs (Light Emitting Diodes) will be placed in wet areas of the cave, providing sufficient light that produces less heat and prohibits the growth of algae and other plants." (Anonymous 2005), but the Chiefs of Maintenance and Interpretation approved full spectrum lights in algae-prone areas, so Mammoth Cave still has a significant lamp flora problem (Fig. 18.6).

One last issue linked to lighting in Mammoth Cave is corrosion of aluminum fixtures and copper ground wires. The constant high humidity in caves hydrates the otherwise protective oxide coating on aluminum, resulting in

Fig. 18.5 Special LED lamps can avoid wavelengths of light absorbed by chlorophyll and therefore reduce the growth of lamp flora. This design, developed by enLux Lighting, has *70 yellow*, *9 green*, and *1 blue* LED yielding a color temperature of 2050 K. This lamp grows vastly less algae than full spectrum lights, and shows the cave well

conversion to a clear to white alumina gel and the mineral gibbsite, neither of which are native to Mammoth Cave (Olson 1991). Handrails made of aluminum have largely been replaced with stainless steel, but hundreds of aluminum light fixtures were installed beginning in 2003, and many of these are currently showing major corrosion. The Chiefs of Maintenance and Interpretation were advised about the vulnerability of aluminum in the cave; the cave floor will have to be cleaned up and the fixtures replaced with more appropriate designs. Miles of 0.25 in. diameter solid copper ground wire were installed in the cave in the 1950s. Fortunately, the new cave electric system does not need these wires, but they were abandoned in place along with miles of insulated cable. Volunteers with the NSS have worked for years and will continue for many more, to carefully remove the old wiring (Vanhoozer 2010). This provides a double benefit to the park in that copper, toxic to invertebrates such as the Kentucky cave shrimp, is removed as a threat, and the park gets an infusion of cash from sale of the metal to a recycler.

18.14 Torch Throwing

In the 1800s, it was a challenge to show visitors how big Mammoth Cave was, given the dim lanterns used at the time. One solution was to tie a cotton rag into a twisted loop, soak it in kerosene, hook it onto the end of a special stick with a metal end, light it, and then throw the flaming mass up onto a ledge. Torch throwing was very popular with both visitors and the guides for about a century, but the side effects were serious. The cave smelled of kerosene, soot blackened the walls where torches landed, and cultural artifacts were set on fire inadvertently. Assistant Superintendent Vaughn Baker called for a civil

Fig. 18.6 Full spectrum lights installed as part of the new lighting system beginning in 2003 have resulted in continued significant lamp flora problems in Mammoth Cave. Brown material growing on the algae is mold. The $560 (in 2003) aluminum fixture lighting the area has almost corroded through, leaving residues of white aluminum oxide on the cave floor that will have to be cleaned up by NSS volunteers. This is a perfect example of what not to do in cave lighting

discussion among park employees about the wisdom of continuing torch throwing. Many guides felt that the tradition of torch throwing was more important than protecting the cave, but the superintendent decided to end the practice in 1991.

18.15 Fumes from Hydrocarbon Fueled Lamps

Most lights used in the cave are electric, but Coleman gas and Dietz oil lanterns are used for some tours and during power outages. Generally, impact to air quality in the cave is minor because the tour moves through passages and does not stay in any one place for long. However, fumes from hydrocarbon combustion in the cave do reach worrisome levels if many lamps are left burning in place for hours, such as during the annual Cave Sing. These fumes may affect cave life as well as pose a health hazard to people attending the performances. In addition, in 2010, the Interpretation Division at the park switched from LED lanterns to Coleman gas lanterns for the River Styx tour, which caused heavy releases of fumes in the cave. This problem was temporarily solved by seasonally installing compact fluorescent light fixtures in the River Styx passage. In 1972, a partial can of Coleman fuel exploded as a team of biologists passed by in River Styx. One member of the team had third degree burns on his hand, but it could have been much worse because his whole upper body was on fire before other team members extinguished the fire by rolling him on the floor. The passage was filled with smoke from a picnic table and plastic buckets that caught on fire. Coleman fuel is particularly volatile and explosive and should not be used in park caves.

18.16 Concluding Thoughts

There have been many challenging environmental issues in and near Mammoth Cave, and probably always will be, especially as long as the human population continues to grow and we insist on an ever-expanding economy (Olson 1995). In spite of these realities, the outlook for the Mammoth Cave region is fairly positive in the near term. The park is heading toward development of a Cave and Karst Management Plan after decades of insisting that one was not necessary. This will provide guidance for what is and is not acceptable in park caves. Oversight will be most effective if the park has a Cave Management Specialist tasked with being an advocate for the cave. Historically at the park, ideal cave management was supposed to result from consensus among various park professionals. Given the foibles of people and competitive agendas, the results have often been disappointing. For example, priority was given to human comfort over hibernating bats when the Historic Entrance was blocked in February 2004, and a large electric heater was installed in Houchins Narrows during cave trail construction. At the same time, Mammoth Dome Tower was being painted and overwhelming paint fumes filled the cave, which were noticeable at Carmichael Entrance over three cave miles away. Even so, environmental protection for the caves and karst of the Mammoth Cave area are at an all-time high. Air quality is monitored continuously and water quality is monitored periodically. Aquifer protection is a high priority in and near the park, and protective structures are being built along I-65. The CSX railroad needs similar protective structures, and these could be built along with an expansion in rail lines from one (since 1859) to, however, many we need. Rail is much more energy efficient than trucks for commerce, and efficiency is our only real hope if numbers of people and the economy to support them continue to grow.

Some endangered species, such as the Kentucky cave shrimp (*Palaemonias ganteri*), are clearly better off than when first listed, and the American bald eagle is off the endangered list. Gray bats have clearly increased in numbers, and Indiana bats had at least stopped their decline. Now, of course, White Nose Syndrome is having catastrophic effects. Population levels of endangered mussels are harder to gage, but the park has supported a mussel rearing facility in cooperation with the Kentucky Department of Fish and Wildlife Resources to boost populations in the Green River. Furthermore, the park is working with neighboring cities and the US Army Corps of Engineers to remove Lock and Dams #5 and #6 downstream of the park, the former of which backs up water into the park. L&D#6 washed out on November 25, 2016, partially restoring free flow to miles of the Green and Nolin Rivers within the park plus greatly enhancing habitat for endangered mussels and the Kentucky cave shrimp. Given what the park and its partners have done in recent decades and the plans for continued ecological restoration projects, the environmental future for Mammoth Cave looks fairly bright for a few decades.

References

Anonymous (2005) Enlightening the darkness. The official newspaper of Mammoth Cave National Park, XI:1

Barr T (1976) Ecological effects of water pollution in Mammoth Cave – final technical report to the National Park Service, Contract no. CXSOOOS0204

Bishop S (1982) Mammoth Cave National Park named to world heritage list. Cave Research Foundation newsletter, pp. 1–2

Brucker R (1979) Conservation at Mammoth Cave. Cave research foundation 1979 Annual Report, pp. 40–41

Bruggers J (2015) Mammoth Cave park sees pipeline threat. http://courier-journal.com/story/tech/science/environment/2015/08/11

Carson V, Vanhoozer R (2006) A mammoth volunteer victory. NSS News, pp 12–13

Cushman R, Krieger R, McCabe J (1965) Present and future water supply for Mammoth Cave National Park, Kentucky. U.S. Geological Survey Water Supply Paper 1475-Q, United States Government Printing Office, p. 617

Fry J (1996) Eighteen cave gates and airlocks: conclusion of a three-year project to restore cave entrance dynamics at Mammoth Cave National Park. In: Proceedings of the fifth annual Mammoth Cave National Park science conference, pp 69–83

Jernigan J, Carson B (2013) Cumberland Piedmont Network ozone and foliar injury report—Little River Canyon National Preserve, Mammoth Cave NP and Russell Cave NM: Annual report 2011. Natural resource data series. NPS/CUPN/NRDS—2013/578. National Park Service. Fort Collins, Colorado. Published Report-2204474

McMillan S, West A, Solomon D, Diehl R, Roland V, Embry I, Toomey RS (2013) Evaluation of storm water filters at Mammoth Cave National Park, Kentucky, 2011–12. In: Mammoth Cave National Park's 10th research symposium proceedings, pp 188–192

Mulec J (2012) Lampenflora. In: White W, Culver D (eds) Encyclopedia of caves, 2nd ed. Academic/Elsevier Press, Amsterdam, pp 451–456

Olson R (1991) Aluminum handrail corrosion in Mammoth Cave. In: Proceedings of Mammoth Cave National Park's second science conference, pp 47–49

Olson R (1995) Infinite growth in a finite world. In: Linn RM (ed) Sustainable society and protected areas. George Wright society, Hancock, Michigan, pp 189–194

Olson R (1996) This old cave: the ecological restoration of the historic entrance ecotone of Mammoth Cave, and mitigation of visitor impact. In: Proceedings of the fifth annual Mammoth Cave National Park science conference, pp 87–95

Olson R, Fry J, Meiman J, Ward B, Henrickson S, Bradybaugh J (1997) Cave entrance management: principles and practice at Mammoth Cave National Park. In: Proceedings of the 1997 karst and cave management symposium, 13th national cave management symposium, Bellingham, Washington, pp 146–149

Olson R (2001) Ecological effects of acid deposition and nitrogen enrichment on terrestrial and aquatic systems in Mammoth Cave National Park. Unpublished report on file at Mammoth Cave National Park, 13 p

Olson R (2002) The ecological foundation for prescribed fire in the Mammoth Cave area. In: Proceedings of the ninth Mammoth Cave science conference, National Park Service, pp 54–65

Olson R (2003) Don't mess with Mammoth days in the Pike Spring Basin. In: 16th national cave and karst management symposium proceedings, Gainesville, Florida, October 13–17, 2003, Greyhound Press, pp 101–104
Olson R (2005) The ecological effects of Lock and Dam No. 6 in Mammoth Cave National Park. In: Proceedings of the 2005 George Wright Society conference on parks, protected areas, and cultural sites. Hancock, Michigan, pp 294–299
Olson R (2006) Control of Lamp Flora in developed caves. In: Hildreth-Werker V, Werker J (eds) Restoration and conservation of caves, Nat Speleological Soc, Huntsville, AL, pp 343–348
Olson R, Scoggins L, Toomey RS, Burton J (2013) 2011 Vegetation map for Mammoth Cave National Park. In: Proceedings of Mammoth Cave National Park's tenth research symposium, National Park Service, pp 4–8
Olson R (2013) Spill retention and runoff filtration structures on Interstate 65 in the vicinity of Mammoth Cave National Park. In: Proceedings of Mammoth Cave National Park's tenth research symposium, National Park Service, pp 201–202
Postel S, Richter B (2003) Rivers for life: managing water for people and nature. Island Press, Washington-Covelo-London, p 155
Poulson T (1992) The Mammoth Cave ecosystem. In: Camancho AI (ed) The natural history of biospeleology, Monographs of the National Museum of Natural Sciences, Madrid, Spain, pp 564–611
Schumann F (2004) Cave research foundation trip report for August 21, 2004. Cave research foundation trip report database
Toomey RS, Olson R, Kovar S, Adams M, Ward R (2009) Relighting Mammoth Cave's new entrance: improving visitor experience, reducing exotic plant growth, and easing maintenance. In: White W (ed) Proceedings, 15th international congress of speleology, vol 2, part 2, Kerrville, TX, pp 1223–1228
U.S. MAB (1995) Biosphere reserves in action, case studies of the American experience. National Technical Information Service, U.S. Department of Commerce, 5285 Port Royal Road, Springfield, VA 22151, pp 51–58
Van der Heiden L, Webb C (2013) Mercury analysis in Rafinesque Big Eared bat populations. In: Proceedings of Mammoth Cave National Park's tenth research symposium, National Park Service, p. 59
Vanhoozer R (2010) Another successful year at Mammoth Cave. NSS News, March 2010, p. 19

The manufacturer's authorised representative in the EU is Springer Nature Customer Service Centre GmbH, Europaplatz 3, 69115 Heidelberg, Germany. If you have any concerns regarding our products, please contact ProductSafety@springernature.com

Printed and bound by CPI Group (UK) Ltd, Croydon, CR0 4YY
01/10/2026
02225884-0001